Authentic Arabian Bloodstock II

Judith Forbis

The Story of Ansata
and
Sharing the Dream

Authentic
Arabian Bloodstock II

This book is based on research and study by the author over a period of four decades. She has taken great care to ensure the accuracy of material presented. Nevertheless, the author assumes no responsibility or liability for errors or inaccuracies or for any damages, consequential or otherwise, as a result of reliance on the contents or for personal opinions expressed herein.

Other books by Judith Forbis:
   Hamdan Stables Studbook, Cairo, Egypt, 1969
   Hoofbeats Along The Tigris, J. A. Allen, London, 1970;
      Ansata Publications, Mena, Arkansas, 1990
   The Royal Arabians of Egypt and the Stud of Henry B. Babson,
      Judith Forbis and Walter Schimanski, Thoth Publishers, Waco, Texas, 1976
   The Classic Arabian Horse, Liveright, New York, N.Y., 1976
   Authentic Arabian Bloodstock, Ansata Publications, Mena, Arkansas, 1990
   The Abbas Pasha Manuscript, Judith Forbis and Gülsün Sherif,
      Ansata Publications, Mena, Arkansas, 1993
   Ansata Ibn Halima - The Gift, Ansata Publications, Mena, Arkansas, 1998

Copyright 2003 by Judith Forbis

All rights reserved. Except for brief quotations for book reviews, no part of this publication may be reproduced, stored in a retrieval system, or transmitted, in any form, or by any means, electronic, mechanical, photocopying, recording or otherwise, without prior written permission of the author.

ISBN 0-9625644-4-3
Design and Production: Joe Ferriss
Graphics: JF Graphics, Quincy, Michigan
Printed in U.S.A. by Jostens Printing and Publishing, Topeka, Kansas

Published by Ansata Publications, Mena, Arkansas, 71953

Front cover: The Bedouin Tent, Painting by Mary Haggard, ©2002; collection of Hassanain Al-Nakeeb
End sheets: Welcome to the Tent, Eugéne Giradet, Private Collection. Courtesy ACR Edition, Paris
Title page: The Three Graces, photo by Judith Forbis
Contents page photo: Horse in relief, Temple of Medinet Habu, Luxor

Contents

Foreword - Judith Forbis 7
Acknowledgements 8
Introduction - Joe Ferriss 9

Chapter I – Believing in the Magic of Your Dreams
The Journey of A Thousand Steps 13
The Valley of the Nile 18
Oklahoma, Here We Come 20
Deep in the Heart of Texas 23
Arkansas, the Natural State 25
Back to the Desert 27
Moving On 28

Chapter II – Conceptualizing the Dream
Vision, Design, and Purpose 33
Breeding Arabian Horses is a Creative Art 34
The Art of Breeding 36
What is Art 36
Art as Beauty – Aesthetics 36
What is the Purpose of Art 37
Of What is Art Composed 38
Evaluating Art or a Fine Horse 38
Upholding the Classic Spirit 42
Preparing the Palette 43
Compositional Considerations 45
    Bloodlines 45
    Pedigrees 45
    Strains and Families 45
    Study of Family Traits 46
    In-Breeding, Linebreeding, Outcrossing 47
    Prepotency 50
    Heredity 54
Colors and Values 55
Completing the Picture 57

Chapter III – Dreams of the Past, Realities for the Future
Portrait of a Breeder 63
Evaluating Ancestral Horses 64
R.A.S./E.A.O. Ancestral Stallions and Mares 67
Pictorial Section of Additional Ancestors 87

Chapter IV – Realizing the Dream

    The Ansata Breeding Program 97
    Key Stallions 98

<u>Dahman Shahwan Strain</u>
    Introduction 120
    The Bint Sabah Family through Bukra 121
    El Dahma Ancestral Chart through Bint Sabah 123
    The Bint Sabah Family through *Bint Bint Sabbah 181
    El Dahma Ancestral Chart through Bint Sabah/ *Bint Bint Sabbah and Layla 183
    The *Bint Bint Sabbah Family 184
    Fay-Sabbah, Fa-Habba, Habba, Fa Saana, Fa-Abba Families 184
    The Bint Sabah Family through Kamar 208
    The Farida Family 211
    El Dahma Ancestral Chart through Farida 216
    Halima Family 219
    Futna Family 225
    Helwa Family 237

<u>Bint El Bahreyn of Lady Anne Blunt</u>
    Introduction 243
    Bint El Bahreyn Ancestral Chart through Maisa and Elwya 245
    Elwya Family 246
    Maisa Family 247

<u>Saklawi Jedran Strain</u>
    Introduction 252
    Roda/Roga El Beida Family 253
    Roda /Roda El Beida Ancestral Chart through Wanisa, El Bataa and Ghazalah (Inshass) 256
    Moniet El Nefous Family 260
    El Bataa Family 267
    Ghazalah (Inshass) Family 271
    Ghazieh of Abbas Pasha I Family 273
    Ghazieh of Abbas Pasha I Ancestral Chart through Radia and Jemla 274
    Zaafarana Family 276
    Serra Family 281

<u>Hadban Enzahi Strain</u>
    Introduction 284
    Bint Samiha and Bint Rustem Families 285
    Venus/Hadba Ancestral Chart through Samiha and Bint Rustem 288
    Samiha Family 291
    Bint Rustem Family 296

<u>'Abeyyan Om Jurays Strain</u>
    Introduction 300
    El Abeyya Om Jurays – Inshass Mona Family 302
    El Shahbaa Ancestral Chart through Mahfouza and Mahdia 303
    Inshass Mona Family 304

<u>'Abeyyan (?)</u>
    Abeyyan – Karima Family 307
    Obeya/Karima Ancestral Chart through Hodhoda
    Hodhoda Family

<u>Kuhaylan Jellabi Strain (Saklawi Jedran?)</u>
    Introduction 310
    The Negma/Mahroussa/*Maaroufa Family 314
    The disputed Kuhaylan Jellabi and Saklawi Jedran Strain Yamama Ancestral Chart 316
    *Maaroufa Family Branch Chart 318
    El Maar Family 319
    Fay Roufa Family 321
    Maarou Family 322
    Aaroufa Family 323

  Kuhaylan Rodan Strain
   Introduction 326
   Rodania of Lady Anne Blunt Ancestral Chart through Malaka 328
   Bint Riyala Branch through Malaka, *Mawaheb Family 329
   *Watfa Family 330
   *Salomy Family 332

Chapter V - Perfecting the Dream
 A Tribute to Ansata Halim Shah 337
 Born to Rule 339
 Show Horse 341
 The European Connection 343
 Home Again 346
 Renaissance Stallion 353
 Horse of a Different Stature – Poem 356
 The Halim Shah Look 358

Chapter VI - Sharing the Dream
 Introduction 367
  Dahman Shahwan Strain
   Matriarchs:
    Sabah/Bukra Family: Ansata Delilah 370
    Sabah/*Bint Bint Sabbah Family: Ansata Sabiha 392
    Sabah/*Bint Bint Sabbah Family: Falima 399
    Sabah/Kamar Family: *Pharrah 411
    Farida/Halima Family: Malikah and Moheba II 435
    Bint El Bahreyn/*Bint Maisa El Saghira Family 462

  Saklawi Jedran Strain
   Matriarchs:
    Roga/Moniet El Nefous Family: Mabrouka and Mona (Mouna) 474
    Ghazieh/Radia Family: Zaafarana 489

  Hadban Enzahi Strain
   Matriarchs:
    Hadba/Bint Samiha Family 504
    Hadba/Bint Rustem Family 515

  Abeyyan Strain
   Matriarchs:
    El Shahbaa/Mona (Inshass) Family: Hanan 539
    El Shahbaa/*Magidaa Family: Bint Magidaa 558

  Kuhaylan Jellabi Strain (Saklawi Jedran?)
   Matriarchs:
    Yamama/*Maaroufa Family: Maar-Ree and RDM Maar Hala 564
    Yamama/*Maaroufa Family: Fada and Fa Moniet 574

  Kuhaylan Rodan Strain
   Matriarchs:
    Rodania/Riyala Family: Malaka 580
    Rodania/Risala Family: Bint Rissala 594

Conclusion
 Why Breed Arab Horses 601
 Ansata Statistics 2002 604
 Arabian Breeders Award 605

Index
 Benefactors 607
 Patrons 610
 Index of Photographs 614
 Index of Horse Entries Appearing in Chapter 4 - The Ansata Studbook 620
 Bibliography 623

Dedication

In memory of my mother, Audrey Scott L'Hommedieu, and my father, William L. Freni, for their love, wisdom, and understanding of a little girl who loved horses more than anything in the world.

To my aunt, Evelyn Scott Whitwell, and my late uncle, George Whitwell, who inspired my love for Egypt and who encouraged and supported my fledgling horse adventures.

In memory of Chessie Yeary Forbis and Nolan L. Forbis, Don's mother and father, for caretaking with pride our first horses.

Foreword

Row, row, row your boat,
gently down the stream.
Merrily, merrily, merrily,
Life is but a dream.

Judi Forbis with Ansata Selket

As children we sang this little song without giving it much thought. Now, having traveled many a mile and dreamed countless dreams, many of mine have come true in real life. Writing this book involved reliving those dreams, pulling together the threads of experience and weaving a colorful tapestry that others could unravel, reweave, and use to create their own dreams. Dreamweavers of the past have awakened us to the heroic journeys, the quests, the rites of passage, and the tasks and initiations along life's path. Ultimately, they have connected us more deeply to our own spirit. Thus inspired, we as Dreamwalkers have undertaken our own special quest - the search for the perfect Arabian horse.

When our Egyptian odyssey began, little written reference or photographic material was available from which to create a pattern. Over time we obtained pictures of early foundation stock, and we photographed horses who now, due to passage of time, have also become ancestral. These pictures and other information were shared in *Authentic Arabian Bloodstock*, affectionately referred to by its users as "the blue book." *Authentic Arabian Bloodstock II*, "the golden book," is to be used in conjunction with "the blue book," to chronicle the story of Ansata Arabian Stud and its worldwide influence after 45 years of dedication to the breed, and to inspire others to pursue their dreams of breeding the perfect Arabian horse.

Some of the classic lithographs and paintings of Arabian horses illustrated in this book, particularly those by deDreux, we saw in the Egyptian Agricultural Organization's administration offices overlooking the Nile when we went there in 1959 to purchase our first three "living works of art," *Ansata Ibn Halima, *Ansata Bint Mabrouka, and *Ansata Bint Zaafarana. As a result of keeping those artistic images in mind, many horses we have bred have closely resembled them. Although creating the perfect horse in reality is ever elusive, and the quest to obtain it never ending, Dreamweavers can always find perfection by returning to the place where they began - the dream world!

Judith Forbis

Acknowledgements

A very special thank you to everyone who has supported Ansata Arabian Stud from its formative years up to its 45th anniversary in 2003, and to those who have participated in making the products of Ansata Publications valued educational tools worldwide. The Ansata library is now in the hands of Sheikh Abdulaziz Bin Khaled Bin Hamad Al Thani of Al Rayyan Farm, Qatar, with the expectation that it will be shared with others in the future, as we have shared it in the past.

I am also grateful to our long-time Ansata staff, and to all who have provided photographs and information for this book, and to the following who have assisted in its preparation:

Yvette Van Natta, Ansata's breeding manager, for her years of dedication to the Ansata breeding program above and beyond the call of duty.

John Skroms, Ansata's night watchman and foaling manager, who has foaled out most of the Ansata horses born in Arkansas.

Sheila Theriot, Ansata's administrative assistant, for her help in coordinating various aspects of this book.

Jerry Sparagowski, whose photographs have captured the nobility, spirit, and classic essence of Ansata horses and portrayed the Ansata story from the formative years to the present.

Mary Haggard, for her exclusive design of *The Bedouin Tent* painting for the cover of this book.

Tamás and Tünde Rombauer, and Judith Wich, for their aid in translating the Von Szandtner herdbooks.

Dr. Carolyn Evertson, Ph.D., for her advice and assistance in proofing and reviewing this manuscript.

Joe Ferriss, for his exceptional knowledge about Egyptian bloodlines, for his never-ending patience and guidance in developing this and previous Ansata publications, and for his preparation of the ancestral charts appearing in this book.

Sharon Ferriss, for hosting me in her home as part of the family, and providing a peaceful working environment during the development of the Ferriss-Forbis book projects.

Don Forbis, my husband, for his unending patience during the time it has taken to create *Authentic Arabian Bloodstock II*.

Introduction

When the Forbises began their quest for the classic Arabian horse in Egypt almost a half century ago, the writings of Lady Anne Blunt and Carl Raswan were all that guided them. Primary records were scarce and ancestral photographs, rare lithographs, and other valuable data were found in unlikely places - scattered in attics, barns, in boxes or loose, some water-stained and deteriorating. In fact, the priceless H.H. Prince Mohammed Aly library was stored at El Zahraa in the stallion barn's musty tack room along with Nazeer's skeleton. Some of the old herd books in Arabic and English had not been seen by anyone other than the E.A.O. management, who let Judi view them, and the missing original *Inshass Herd Book*, as well as the long-lost manuscript of Abbas Pasha I, were located only through Judi's continued efforts to find them.

In the years since their first visit to El Zahraa, Judi has collected essential documents and photographs and meticulously recorded the lineage of the classic Arabian horse culminating in such foundational works as *The Classic Arabian Horse, Authentic Arabian Bloodstock I* and *The Abbas Pasha Manuscript*. These books have made the history and detailed breeding information about Arabian horses in general, and Egyptian Arabians in particular, accessible to all who love them.

Authentic Arabian Bloodstock II represents a milestone in the art of breeding Egyptian Arabian horses. This profusely illustrated book and its companion, *Authentic Arabian Bloodstock I*, published in 1990, document a true breeder's work and provide a visual record rarely existing for students, breeders, and researchers.

Chapter III of this book features rare photographs and the heretofore unpublished translation of General Von Szandtner's notes, including commentary by other authorities about ancestral Egyptian bloodstock. Many of the horses depicted appear in pedigrees throughout the world, and these photographs allow us to examine the differences or similarities in type that develop in a long term breeding program.

Chapter IV comprises the Forbises' Ansata herd book to December 31, 2002, a breeding record of more than four decades. Unlike most herd books, it provides photos of nearly every horse bred by or added to the stud. It also provides many details about individual horses along with a candid assessment of foundation stock and families as the Ansata program has unfolded.

Chapter V pays tribute to *Ansata Halim Shah, a unique stallion who in body, soul, and worldwide influence, comes the closest to symbolizing the perfect Arabian horse conceptualized by his breeder.

Chapter VI "Sharing the Dream," emphasizes once more the importance of sharing with others the fruits of our work if the classic Arabian horse is to be preserved and perpetuated.

This book represents the fulfillment of a dream and is dedicated to all who love Arabian horses. It is a story told from the heart of a true breeder and connoisseur of the arts with the hope that it will inspire others to follow their dreams and to be true to the artist in themselves.

Joe Ferriss

Caravan on the March. J.F. Herring, Sr. Courtesy Frost and Reed, Ltd.

CHAPTER I

Believing in the Magic of Your Dreams

Judi riding her first love, "Silver" at the East Islip, Long Island pony ring in 1942. "Going it alone" with Herman, the owner waiting.

CHAPTER I
Believing in the Magic of your Dreams

"We grow great by dreams. All big successes are big dreamers. They see things in the red fire of a long winter's evening, or through the mist of a rainy day. Some of us let these great dreams die, but others nourish and protect them, nurse them through the bad days 'til they bring them to the sunshine and light, which come always to those who sincerely believe that their dreams will come true."

Woodrow Wilson

THE JOURNEY OF A THOUSAND STEPS

From the time I was a child in West Islip, New York, all creatures great and small were my beloved friends. One day when I was three years old my parents took me for an afternoon drive in the Long Island countryside. As Fate would have it we passed a pony ring where an old man was leading a white pony ridden by a little girl. "Stop," I yelled. And stop they did. Soon they were lifting me atop a pony named Silver. This gentle white creature touched my soul and never again would my life be without horses. Every weekend they took me to ride him. Together Silver and I imagined our galloping free across verdant fields instead of around the confining dirt ring. Occasionally if Silver was rented, I rode fat black Minnie, the other pony, but I never left without waiting to ride Silver too. Somehow white and black set the tone for my life in which there have never been many shades of gray.

In time I grew too big to ride Silver and began riding horses at the South Bay Country Club Stables where a three-gaited gelding named Black Watch became my best friend. His night-black coat was illumined by a perfect white star on his forehead. Zsa Zsa Gabor, his owner, who was then married to Conrad Hilton, kindly agreed to let me ride him since she didn't have the time and was glad for him to be exercised. Together we made a formidable team and I won my first equitation blue ribbon on him. Other horses owned by the stable entered my life too; a chestnut Thoroughbred racing discard that carried me to wins in jumping and hunting competitions, and a small handsome part-Arabian black gelding that I trained, and whom jockey, Eddie Arcaro, eventually bought for his daughter to ride.

To earn extra money for my own riding lessons, or just time out on the trails, I raked leaves, mucked out stalls, and taught other children to ride during the sum-

Young Judi with Black Watch.

mers. I loved riding for hours through the maze of sandy bridlepaths towered over by whispering pines. A horse to call my own continually occupied my dreams, but it wasn't to be at that time because my parents could not let me have one. Perhaps had my heart's desire of owning a horse been fulfilled at that time, the incredible journey life had in store for me would not have materialized.

Although neither my father nor mother was horse oriented, they and my aunt Evelyn and uncle George Whitwell encouraged my horse activities. One birthday they gave me *The Black Stallion*, by Walter Farley, and after reading it, like thousands of other children, dreams of owning Arabian horses and racing them across the desert danced in my head. *National Velvet* and Elizabeth Taylor's love for horses further kindled my burning desire for a horse of my own. At that age, I had not heard of James Allen's truism: "The Vision that you glorify in your mind, the Ideal that you enthrone in your heart - this you will build your life by, this you will become." But it came true for me.

As compensation for not having a horse of my own, I drew pictures of them and sought out an art career. After attending Syracuse University's College of Fine Arts, I worked as a greeting card designer. My talented supervisor, a former employee of Walt Disney, patiently taught me the importance of technique, style, design and composition. Still, travel and adventure were foremost on my mind, and one day I saw an ad in the *New York Times*: " Secretaries wanted for U.S.. Government overseas jobs with the International Cooperation Administration" (formerly the Marshall Plan). The opportunity for globetrotting was at hand. I applied, was accepted, and after passing the interview and indoctrination at the Washington. D.C. headquarters, I was offered a choice of three countries. Having read the post report that Americans stationed in Ankara, Turkey, were allowed to ride the Turkish Cavalry's horses, the choice was easy. At age 23 - had boots, would travel!

From New York, our Turkish Mission group walked across a long red carpet and boarded a giant Pan American clipper. This aircraft was a far cry from the plane Jimmie Doolittle piloted when he took my great grandmother Havens, at age 74, on a flight over Long Island; it was also a faster mode of transportation than the slow ocean-going vessel that carried my great-grandfather,

Black Watch's first show with Judi up.

David Scott, to Persia as a missionary in the early 1900's.

After landing in Paris we changed planes and flew on to Rome. We spent two days sightseeing and, like most tourists, we tossed coins in the Fountain of Trevi to make a wish. Mine was about to come true. Enroute to Turkey the plane landed at Istanbul, formerly Constantinople, the city of graceful mosques and capital of the once powerful Ottoman Empire. After refueling, the plane took off again over the Bosporus and the bustling Golden Horn, soon to land on the drab Anatolian plains at Ankara just before Valentine's day 1957.

Together with my roommate, Sandy Aylesworth, we rented an apartment and trudged daily through the rutty mud roads to our Mission office building nearby. Desirous of learning Turkish, since many Turks didn't speak English, we enrolled for Turkish lessons at the U.S. military post. I soon located the riding club at the Turkish Cavalry stables, and anxious to realize my dream now that I was employed, I began the search to purchase a horse of my own. A Turkish Captain hearing of my quest offered to sell a white Arabian mare he no longer rode and took me to see her at a military outpost.

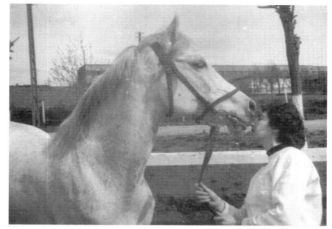
Judi with beloved Ceylan.

Beneath bruised purple skies
They brought her forth
Stamping, snorting.
Grey she was, like the day;
Proud, alert, black eyes searching,
Ears pricked to zephyrs chasing o'er the hills.

Some meetings are Divine Appointments! She was poetry in motion. She was beautiful. It was love at first sight, and now she was mine. Her name was Ceylan, *the gazelle*, and although she was prone to occasional lameness, I trained her carefully and praised her daily. We bonded and soon became part of the Ankara

Judi riding Ceylan in the civilian jumping championships of Turkey in 1959.

The Ankara Binicilik Club team with Judi in the middle on Ceylan.

Binicilik Club team. It wasn't long before our jumping team was invited to compete against the Istanbul Atli Sport Club. Traveling on board the legendary Orient Express, we disembarked amidst endless tracks at the Istanbul train station, and Turkish soldiers walked the horses to the distant stables on the opposite side of the city. Red flags bearing the symbolic Turkish crescent fluttered above the show grounds as we competed over high jumps. Our team won, and individually Ceylan and I accumulated enough points over the year to win the Civilian Jumping Championship of Turkey.

A black Ceylan also came into my life. An ex-American cavalry mount, she now belonged to a Turkish officer who offered to let me ride her. She was a stronger horse and could jump higher than Ceylan. Together we won the coveted Governor's Trophy at the jumping championships held in Ankara, and were honored to have it presented to us by the President of Turkey, Celal Bayyar.

One day while lunching at an American PX, one of my friends introduced me to a handsome young man

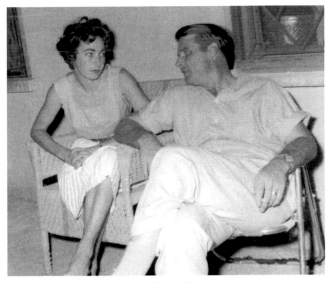

Judi and Don, new beginnings in Diyarbakir.

whose rugged good looks reminded me of Burt Lancaster. His name was Don Forbis, and he was there on assignment with an oil well servicing company. Another love at first sight! Within a few days, we were engaged to be married.

Don loved sports and was a top football player in high school before he joined the U.S. Navy to serve in World War II. After the war he studied petroleum engineering at Tulsa University on a football scholarship, and eventually, he was employed by the Halliburton Company who sent him to various countries in Europe and Arabia. Early mornings he joined me at the race track where I was legging up race horses before going to work. He too loved horses and animals, and as a boy he had ridden the family's Quarter horses in Chickasha, Oklahoma, often jockeying them in the local races. By a fateful coincidence, his first love was a white horse named Silver!

Don had been assigned to manage Halliburton's Turkish headquarters in Ankara. On March 10 and 15, 1958 we tied the civil and church marriage knots respectively, and then we moved to Southeast Turkey where he supervised the company's field operations. Our new home, a primitive unfinished brick apartment

Another love at first sight, Don Forbis.

Entrance to the fortress of Diyarbakir, one of the main gates.

building, stood just outside the ancient city of Diyarbakir, a Roman fort whose great basalt walls had been erected in A.D. 394 by Emperor Constantius. While it was hardly a honeymoon cottage, we managed to make do, lived off the local products, and began learning more about the region, its people, and their horses. The dream that began with reading *The Black Stallion* was about to become a reality.

We knew that the Turkish Jockey Club sanctioned regional races for Arabian horses, so we decided to participate. We purchased Söhret, a chestnut race mare from the Ankara track, and Mesut, a two year old locally bred chestnut colt. Don trained them and I jockeyed them - an unheard of phenomenon because women had not totally removed the veil in this pristine bastion of traditional Islamic culture, even though Ataturk had decreed it illegal to wear one. Even though I was a woman, and a foreign one at that, the officials let me ride. Good sportsmanship was not the rule, however. In the first race the male jockeys converged on Söhret, knocked the saddle under her belly and I went off! Undaunted, I tried again, this time I not only remounted, I won, and I was winning races frequently thereafter. Soon the Turks and Arabs in the region were bringing horses for Don to train and for me to jockey, and despite male protests, women were attending the races to see and cheer for the American woman they called Judi Hanim (Lady Judi) race against the men!

A woman jockey in Turkey, Judi on Söhret warming up for the race.

Stallion presentation at the E.A.O.

*Morafic in front of the E.A.O. office.

By now we were captivated by Arabian horses and took every opportunity to visit the Turkish government stud farms. We traveled among the now sedentary Arab tribes along the Turkish-Syrian border, the Jezirah area, and throughout Syria and Lebanon where we studied the history and qualities of the Arabian horse and its people. We now began to understand the "archetype" or classic essence of the Arabian horse which had made it world renowned, not only for its endurance, but for its nobility and beauty. Arabian horse books began to line our shelves. I began to study Orientalist representations of Arabian horses, and to correspond with various breeders and authorities, including Carl Raswan, who had lived in the Arab world. Raswan suggested we visit Egypt before making any decision to buy Arabian horses, and taking his advice, when it was time to leave Turkey, we traveled to Egypt and the Egyptian Agricultural Organization's El Zahraa Stud Farm in Ein Shams near Cairo.

THE VALLEY OF THE NILE

General Tibor Pettkö von Szandtner, the former stud manager of Hungary's Babolna State Stud had just retired as the Director of El Zahraa, and Dr. Mohamed Marsafi, the farm's long-standing Egyptian veterinarian, had replaced him. As we entered the palm-fringed sandy road leading to the main office, we witnessed beautiful mares and foals peacefully munching emerald green *berseem* in their sandy paddocks. We looked at each other! We were dazzled! Here were the classic type Arabian horses we had been searching for.

The silver-haired Dr. Marsafi, wearing an immaculate white suit, greeted us at the office building and invited us for tea before guiding us through the paddocks and the stallion barn. Nazeer, age 25, but still splendid in his shining white coat, came before us, posing regally beneath an arched trellis of fushia bougainvilla. There too was the elderly Sid Abouhom, emaciated, fine boned, with lively big black eyes set in a sculptured head; Antar, big, red and powerful; Gassir, fleabitten, strong of body and prideful carriage; Mashhour, long-headed but a handsome dark bay; the tall and elegant liver chestnut Alaa El Din, and then a spectacular spirited white stallion named Morafic, whose head was the most exotic we had ever seen.

We saw and admired the legendary mares of their era: Bukra, Halima, Moniet El Nefous, Mouna and Mabrouka, El Bataa, Zaafarana, Yosreia, Mohga, Farasha, Maysouna, Maisa, Kamr, Futna, *Ghazala, Helwa, Abla, Malaka, Samia, Yashmak, to name a few of the old Royal Agricultural Society's herd, along with Farouk's Inshass mares including Hafiza, Mona, Maysa, Ameena, Ghazala, Rooda, Yasmeena, and others. Later, at a nearby estate housing the remaining ex-King Farouk horses, we were shown several mares and the magnificent copper bay stallion, El Sarie, about as perfect a stallion as one could be, and Sameh, a very handsome well-balanced grey sire who was surprisingly fine-boned considering his strong body. Having traversed much of the Middle East between us, both before and after we were married, never had we imagined such a collection of high-quality Arabian horses could be found in one place.

Since the General had already culled the herd, the young horses remaining were those he planned to keep. The yearlings and two-year olds were of extremely high quality, most of them Nazeer sons and daughters, and a few by Sid Abouhom. The full sister to Morafic, a yearling named Bint Mabrouka immediately captivated me; she had the most luminous big black eyes set in a wide forehead, a long neck set on a fine-boned frame, and a prideful spirit. We bonded within moments. A two-year old full sister in blood, Bint Mouna, was near to her in quality and type, and another close relative, Maya, was also special. Bint Zaafarana and Bint Maisa El Saghira were also very appealing.

Of the colts one named Ibn Halima had a most beautiful head. Other handsome youngsters were Ibn Yosreia (Aswan), Kaisoon and Hassan, all by Nazeer, but we considered Ibn Halima to be the most balanced as well as the most classic of all. Returning to the Continental Savoy Hotel that night, we made the momentous decision to buy three yearlings to take back to America and to enjoy as riding horses.

We spent the next few days studying the mares and foals again. Don was determined if we took a filly or two we must take a colt. Bint Mabrouka (my favorite), Bint Zaafarana (his favorite) and Ibn Halima - our favorite, became our final choices. When we asked Dr. Marsafi if they were for sale, he said he would ask Dr. Afifi, the General Director, and take us to meet him at the old E.A.O. administration building downtown overlooking the Nile. Upon entering his office, its walls adorned with old DeDreux lithographs including Omar and Ciaffar, Dr. Afifi stepped forward to greet us. After

*Ansata Ibn Halima being examined by Dr. Mohamed Marsafi.

The three Egyptian yearlings that made history, *Ansata Ibn Halima, *Ansata Bint Mabrouka and *Ansata Bint Zaafarana, in 1959 at their time of selection in Egypt by Don and Judi.

tea and coffee, he said he had decided to sell us the horses we requested. The price was well beyond our reach, however, and we returned to the hotel to mull over the situation. The next day we countered with an offer. Afifi considered it, spoke privately to Marsafi, and then said they could not accept it; however, they would sell us the two fillies and Ibn Halima would be "a gift" to us. Indeed, he became a gift to the world.

Again, the hand of Fate had written. Eventually we learned from Richard Pritzlaff that he had received a letter from his friend, General von Szandtner, stating that had he still been manager of El Zahraa, he never would have sold those horses.

OKLAHOMA, HERE WE COME

We shipped the three yearlings to Oklahoma, and life was never the same again. A new adventure had begun. Douglas Marshall, an Arabian breeder from Texas was the first to visit them and remembering his days in Egypt during the war and his fondness for Egypt and its horses, he decided to replace his Arabian herd with imported Egyptian bloodlines. A new chapter in Egyptian Arabian horse history was about to be written as a result of his decision. He and his wife, Margaret, turned their Gleannloch Farms into an Egyptian Arabian showplace. Without their dedication, the Egyptian Arabian in America would never have reached the heights to which it would soon soar.

Villa Akhnaton in Egypt.

Tribes on the move in Iran.

Now that we owned horses, we needed a farm. A charming 40 acre ranch in Chickasha, Oklahoma, where Don had been raised, became available, and we were able to purchase it. After careful contemplation, we named our farm Ansata - based on the Ansata Crux, (Latin meaning looped cross) and known as the ancient Egyptian *ankh* or key of life, representing life and reproduction. We also trademarked the name with the U.S. Patent Office as an exclusive prefix used to identify Ansata-bred horses thereafter.

Don's father, Nolan Forbis, cared for our "children" until the herd grew sufficiently to warrant hiring a farm manager. Each spring, I flew home from wherever we lived at the time to manage breeding and foaling out

*Left to right our Oklahoma mare band: *Ansata Bint Mabrouka, *Ansata Bint Zaafarana, and the Babson mares Fa-Habba, Aana, Fay Sabbah and El Maar.*

the mares. In 1960 we moved to Libya, then to Iran during the Shah's reign. In 1964 we went to Greece, then back to Turkey for a short time in Diyarbakir once again; then to Columbia, South America, back to Greece, and then to Egypt after the 6-day war in 1967. It was during this time, at our home overlooking the Great Pyramids of Giza, that I began to conceptualize *The Classic Arabian Horse* and started the translation of *The Abbas Pasha Manuscript* with Gülsün Sherif. In 1970 Don and I left our beloved Egypt and went to England where I was able to research Lady Anne Blunt's diaries at the British Museum, and to befriend such benefactors of the breed as Lady Anne Lytton and Musgrave Clark.

In the meantime, from 1959 to 1970 our herd had increased from the original three imports to include their offspring as well as progeny resulting from a lease arrangement we had made with Henry B. Babson in 1959 when we had visited the Babson Farm to acquire knowledge about his imports from Egypt during the 1930's. At that time Mr. Babson agreed to lease us several Egyptian mares, hoping he would get a colt by Ibn Halima to serve as a stallion for his Egyptian herd while giving us a chance to obtain bloodlines tracing to *Maaroufa and *Bint Bint Sabbah. We chose four excellent mares, Fa-Habba, Fay Sabbah, Aana and El Maar, and with Mr. Babson's blessing, along with his

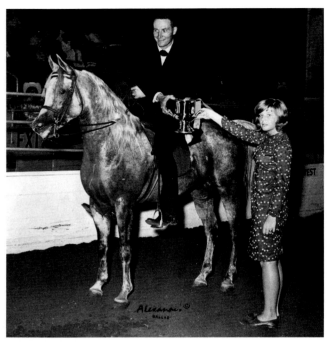

*Ansata Ibn Halima proved to be a fine campaigner in the capable hands of Tom McNair resulting in multiple National and Regional wins.

manager, Homer Watson's approval, they arrived at the farm in Chickasha in 1962 to become part of Ibn Halima's harem.

By 1964 Ibn Halima had matured into a handsome stallion and during Don's vacation, we decided to show him at the Enid, Oklahoma, show. He placed second with Don handling him, but he created a sensation among the spectators, trainers, and breeders because he was "different." Even the judge, Garth Buchanan, stopped to visit the farm a few days after the show and said she regretted not having put him first. Next we took him to Dallas where the Texas Arabian Horse Club was holding its show in conjunction with the U.S. Nationals. Ibn Halima placed fourth, but the crowd thought he should have won. After the class, Doug Marshall came back to the stalls and asked if he and Margaret could lease him to breed some of their newly imported Egyptian mares until their stallion, *Morafic, arrived from Egypt. In return Tom and Rhita McNair, their trainers, campaigned Ibn Halima at shows across America, racking up championships from East to West in halter and performance, including Regional Championships in the Park Class. With Tom at the end of the lead shank, "Ib" garnered U.S. Top Ten Stallion in 1966, 1967, and 1969 - the latter year winning with his son, Ansata Ibn Sudan, who two years later became U.S. National Champion Stallion against an entry slate of 83 champions. Over the years, many other Ansata horses, or those of Ansata bloodlines, consistently found their way into national and international winners' circles. While winning is exciting, we learned early in our career that the spectators and breeders are often the most important judges. Whether you win or lose, the show ring provides a valuable opportunity to introduce one's horses to the public.

Because of Ibn Halima's popularity, we began standing him at stud, thus turning our "riding horse venture" into the unforeseen business of running a breeding farm. Our first imports and the Babson mares began producing, and we eventually imported additional mares: *Ansata Bint Bukra, *Ansata Bint Misr (in utero), *Ansata Bint Elwya, *Ansata Adeeba, *Ansata Bint Sameh, *Ansata Bint Misuna and *Ansata Bint Nazeer. In 1967 Halliburton moved Don to Egypt, providing us an opportunity to assist others in importing, including the Marshalls, Loebers, Klines, Hecks, and Huebners, the latter for whom we selected *Deenaa and *Sakr, among others. Interest in Egyptian horses was now growing in America and around the world particularly as a result of publicity and show ring wins generated by these imports.

Realizing the value of these straight Egyptian horses, and the fact the imports of the 30's had not been preserved in a valuable breeding nucleus, I suggested to Doug Marshall that we found an association dedicated to preserving the straight Egyptian horse so the blood of the new Egyptian imports, and those still existing in the States, would not be lost in the future. He agreed and a group of like-minded breeders met at Gleannloch Farms and formally founded the organization that

Marshall aptly named, The Pyramid Society. I became its Secretary and undertook the secretarial chores from my little office in our apartment overlooking the North Sea in Great Yarmouth, England, the port where many Victorian travelers had sailed for the Orient, and where the Anna Sewell House downtown paid homage to the author of *Black Beauty*. The creation of the Pyramid Society solidified the preservation and perpetuation of this bloodstock, and the idea of breeding "straight Egyptian horses" within the guidelines of the Society's definition swept the world.

In keeping with our nomadic spirit, we decided to leave Oklahoma for greener pastures, since Don's career with Halliburton was coming to an end. However, three of our foundation mares would not accompany us. *Ansata Bint Mabrouka was laid to rest in a special gravesite near the tall cottonwoods on the Chickasha farm with *Ansata Adeeba and *Ansata Bint Sameh keeping her company nearby.

Ansata nestled in the piney woods of Lufkin, Texas.

Ansata's home on the range, the plains in Chickasha, Oklahoma.

DEEP IN THE HEART OF TEXAS

In 1973 Don retired from Halliburton, and two years later the Forbis caravan moved to the piney woods of East Texas, not far from Gleannloch and Bentwood Farms. We spent many wonderful and challenging days building the new farm and enjoying camaraderie with other breeders in the area, especially the Marshalls, who were still living in Houston even though they were busy constructing their magnificent new Gleannloch Farms in Barksdale, Texas. As the Ansata stock became recognized worldwide, we hosted seminars educating breeders about Egyptian horses and the importance of respecting animals as individuals having rights of their own. Breeders arrived from countries all over the world, and we enjoyed new international associates and clients including Philippe de Bourbon of Uruguay, Count

Federico Zichy-Thyssen of Paraguay and Argentina, Dr. Aloysio Faria of Brazil and Claudia Quentin of Argentina. At our 20th anniversary celebration, we invited Dr. Marsafi as a guest speaker from Egypt. It was a nostalgic moment as he recounted to the crowd our first meeting and "the gift" of *Ansata Ibn Halima. After the seminar, the Marshalls took Marsafi for a memorable ride in their private jet to Barksdale where he was reunited with horses who had been under his care in Egypt. It was a combined trip Marsafi would never forget.

Meanwhile the Pyramid Society's membership was increasing, and additional managerial assistance was required. The Society rented an office in downtown Lufkin and Jean Jennings was hired to handle administrative duties. I remained Secretary of the Society, and Jarrell McCracken replaced Doug Marshall as President. McCracken's syndication of *Ibn Moniet El Nefous for $1,000,000 and sending Ansata Abbas Pasha and Fa Halima to compete (and win) at the Salon du Cheval in Paris, France, created international excitement and marked the beginning of a new international era in the Egyptian horse world. Lee Cholak was hired as Executive Director and the Society's office was moved to the Kentucky Horse Park in Lexington, Kentucky. Cholak conceptualized The Egyptian Event and membership

Don Forbis presenting U.S. National Champion Ansata Ibn Sudan at the Arabian Horse Fair in Louisville, Kentucky.

in the Society as well as the number of Egyptian horses in America increased dramatically. The tax code favored investing in horses and "investor mania" gripped the horse breeding world. While it sparked a tremendous interest in Egyptian horses from a financial standpoint that resulted in the importation of many horses from Egypt, it did not always encourage the adherence to good breeding principles by the "investors" who were not necessarily horsemen.

During this time, however, many fine Ansata-bred horses came into the world at the Lufkin farm: Ansata Omar Halim, Ansata Halim Shah, Ansata Imperial, Ansata Ibn Shah, Ansata Abu Sudan, Ansata Samantha, Ansata Nile Dream, Ansata Ghazala, Ansata Aya Halima, to name only a few. Other horses including imports *Ansata Bint Zaafarana and *Ansata Bint Misuna were laid to rest there.

Left to right, Don Forbis, Margaret Marshall, Judi Forbis and Douglas Marshall during their induction into the Arabian Horse Trust Tent of Honor.

The Ansata Studfarm in Mena, Arkansas

ARKANSAS - THE NATURAL STATE

Although we enjoyed Lufkin, and it fulfilled its purpose in realizing the dream, something was lacking and we yearned for that "special place." During a visit with Walter Schimanski in Cove, Arkansas, we were led to a scenic and peaceful farm nestled at the foot of Rich Mountain in the Ouachita Valley near Mena. The 450-acre property once housed the Commonwealth College where students worked the land and in turn received their education. It was one of Nature's masterpieces - a naturalist's dream. We bought it.

Don and a group of young Amish builders undertook construction of the rustic-looking barns and houses - building them from pine lumber and native rock. In September 1981, barns completed, the Forbis caravan was once again on the move. Considering Ibn Halima's state of health that year, we were grateful he lived long enough to see the new farm he had helped to build. Soon after arrival, with his mares watching, he was laid to rest beneath a spreading oak tree at the entrance to the farm.

Arkansas became a peaceful haven for us and for many of our American and international clients. In addition to our own herd, we sold to or maintained horses for foreign clients including Dr. and Mrs. Francesco Santoro of Italy, the Stoessels of Switzerland, Dr. Aloysio Faria and Adelia Audi of Brazil, Claudia Quentin of Argentina, and Count Federico Zichy-Thyssen of Argentina, Paraguay and Uruguay, who used our farm as a base to collect, breed, and then export horses to South America. Paolo Gucci, whose farms were in U.S.A. and England, also entrusted his black Egyptian-Arabians and other mares with us to manage temporarily. As our breeding program expanded, and horses we bred won U.S. National Championships and other worldwide championships such as the Salon du Cheval, Nation's Cup, etc., the name Ansata became synonymous worldwide with Classic Type. We have always believed, "no type, no breed" and have never deviated from our goal no matter the current fad in the marketplace.

During this era our friendship with Dr. Hans Nagel of Germany took on a new relationship. He had previously leased Ansata Abbas Pasha from Bentwood Farms and used him in Germany. We imported his stallion *Jamilll, whom we showed in 1983 to U.S. Top Ten Stallion and incorporated within our breeding program. Nagel in turn leased Ansata Halim Shah, who won 1983 Reserve World Junior Champion in Paris at the Salon du Cheval and went on to make breed history in Europe as one of the most valued licensed breeding stallions of all time. We also visited the famed Babolna Stud in Hungary with Dr. Nagel who assisted us in leasing the incredibly beautiful mare, Ibn Galal I-7, "Gala" whom we showed to third place as a U.S. National Top Ten Mare. Many breeders thought she should have won the coveted National Championship and that she was one of the most beautiful mares they had ever seen. Tragically, Gala died while giving birth before leaving any foals to replace her.

Auction Sales were a part of this era, and Ansata supported several of them. At the Pyramid Society Sale I, we consigned the lovely yearling filly, Ansata Nile Glory, who was purchased for a record price of $250,000 by Leonor Romney of Somerset Farm. The sale was an outstanding success, setting an industry standard, and annual sales followed for a period of time. In February 1985, the mare, Ansata Damietta, was consigned to the Classic Arabian Horse Sale in Scottsdale, where Don Saluti paid a record $350,000 for her. When a change in the tax codes took place, however, prices fell, and auction sales were less frequent. On December 15, 1990, Ansata held a "Just Before Christmas Sale," in Oklahoma City, and on May 4, 1992, due to Mrs. Marshall's ill health, Gleannloch held a final dispersal sale at Barksdale, marking the end of an era. Many of the consignments found new owners not only in America, but in Europe and South America.

Arkansas Governor, Bill Clinton, visits with Ansata Ibn Sudan at Ansata in Mena, Arkansas.

Reflecting on earlier travels in Arabia Deserta.

Ansata Nefer Isis winning at the Qatar National Show. J. Wich photo.

BACK TO THE DESERT

One of our greatest pleasures yet to come was in assisting the Arab world throughout the 90's to recapture its Arabian horse heritage by incorporating Ansata bloodstock whose ancestors originated in Arabia. The wheel had now come full circle - back to the desert.

The sale of Ansata Haji Jamil as a birthday gift from his people to H.R.H. King Hassan II of Morocco was a milestone in exporting to the Muslim world. Further, the purchase of two mares, Ansata Majesta and Ansata Splendora, by Sheikh Abdulaziz Bin Khaled Bin Hamad Al Thani of Qatar introduced Ansata bloodlines to the Arabian Gulf. This young Sheikh dedicated himself to restoring the classic Arabian horse to his country, and he became a light unto the Arab world to do likewise. Ansata Majesta lived up to her name, becoming the first Junior National Champion Mare of Qatar. Sheikh Abdulaziz continued to import additional stock, and the Emir, H.H. Sheikh Hamad Bin Khalifa Al Thani, with assistance from Sheikh Hamad Bin Ali Al Thani, his manager, acquired Ansata Halim Shah and a select group of Ansata mares for his Al Shaqab Stud. Sheikh Abdullah Bin Nasir Al Thani followed suit, also founding his Al Naif stud on Ansata bloodstock.

Ansata-bred horses began reaching Arabia in the 90's as a result of renewed appreciation of the breed. The first horse of Ansata bloodlines, El Wajeeh, was imported to Saudi Arabia after the Gleannloch Sale by a remarkable architect and avid Arabian horse enthusiast, Abdullah Al Sweilam. The brothers H.R.H. Prince Abdullah, H.R.H. Prince Turki and H.R.H. Prince Khalid Bin Fahad Bin Mohammed Al Saud,

Representing the Arabian Horse Trust, Judi meets with H.R.H. Prince Abdullah Ibn Abdul Aziz Al Sa'ud, Crown Prince of Saudi Arabia and Saudi Ambassador to the United States, Prince Bandar Ibn Sultan Al Sa'ud in Washington, D.C.

Judi watching Ansata horses with Sheikh Abdulaziz Bin Khaled Bin Hamad Al Thani of Qatar at the Qatar International Show.

along with H.R.H. Prince Khalid S.K. Al Saud were the first to import a select group of Ansata-bred horses to their Nejd Stud located in the heart of the desert; H.R.H. Prince Khalid Bin Sultan Al Saud, Khalid Al Haddad, and Mohamed Aleysai followed suit. H.R.H. Princess Zeyn Bint Hussein and Majdi Al Saleh acquired Ansata stock for their Al Jaffar Stud in Jordan. Roushdy Abu Seda, Omar Sakr, Marty Garrett and Dr. Ali Abdel Rahim secured Ansata bloodlines for Egypt.

Another first was the importation of Ansata horses to Kuwait by Usamah Al Kazemi and Khaled Ben Shokr. As a result of Al Kazemi's visit, we were asked to assist the Arabian Horse Center of Kuwait to acquire foundation horses, mostly of Ansata bloodlines, to replace the horses they had lost during the Gulf War. The Center had been founded in 1980 by the Emir of Kuwait, H.H. Sheikh Jaber Al Ahmed Al Sabah in respect for his interest in Arabian bloodstock and to provide a focal point for this horse as opposed to the growing regard for English thoroughbreds. Fifteen Arabian horses had been imported from Germany in 1988 bringing the total to 36 horses before the Iraqi invasion of Kuwait. During the invasion most of the horses were stolen, leaving only two remaining. In 1993 five more mares were imported from Germany, and in 1998 we assisted them in acquiring twelve more. Today the Center is a focal point and public relations center for a growing group of enthusiastic breeders.

A major importation was also made by an astute young enthusiast, Mohammed J.K. Al Marzouk, who founded his Ajmal Stud with Ansata fillies and mares together with the outstanding stallion and sire, Ansata Hejazi. Other breeders in Kuwait have since developed breeding programs incorporating Ansata bloodlines including Sheikha Sarah Al Sabah, the first woman from the Arab countries to purchase directly from Ansata. The whole Arab world now had a major influx of Ansata bloodstock - either by direct importation, or through other acquisitions.

MOVING ON

Due to our continued Pyramid Society activities in Kentucky, especially when I was president, and as a result of participating in the annual Egyptian Event, we spent much time looking at horse farms in the area with the idea of purchasing one as a satellite farm. Through Mikhail Albina, a Palestinian friend who was managing Buckram Oaks Thoroughbred Stud, we learned about Lees Lake Farm located between Georgetown, Midway, Lexington and Frankfurt. Only twelve miles

Ansata in Lexington, Kentucky.

from the Kentucky Horse Park, it resembled a park with its 40 acre lake, rolling bluegrass hills and southern colonial home. A white heron greeted us at lakeside as we walked the property that would soon become ours. The farm became a special meeting place for friends during Pyramid Society activities, and provided a safe haven for our horses during the Egyptian Event. Formal and manicured, it was an absolute contrast to the wild and natural Arkansas property. Although we enjoyed owning it for five years, the Arkansas farm remained our true home, and in 2002 the Kentucky farm was sold.

From September 25-27, 1998, Ansata celebrated it's 40th anniversary with a special seminar and tribute to *Ansata Ibn Halima. Much of the Ansata story has been told in *Hoofbeats Along the Tigris*, *The Classic Arabian Horse*, *Authentic Arabian Bloodstock* and the last book, *Ansata Ibn Halima, The Gift*, a tribute to that special horse who made our dreams, and those of many others, come true. More than 20 years have gone by since he passed on to a higher realm, and as this book is written, the dreams he inspired and the Beauty he and our foundation mares bequeathed to the world stand ever before us in their descendants.

The moving Finger writes, and as this book - in celebration of the Egyptian Arabian horse - is written, it moves on. For those who dream of creating Beauty through breeding Arabian horses, may you too be inspired from the words penned by James Allen in *As A Man Thinketh*:

"In all human affairs there are *efforts*, and there are *results*, and the strength of the effort is the measure of the result.... The dreamers are the saviors of the world. As the visible word is sustained by the invisible, so men, through all their trials and sins and sordid vocations, are nourished by the beautiful visions of their solitary dreamers. Humanity cannot forget its dreamers; it cannot let their ideals fade and die; it lives in them; it knows them as the *realities* which it shall one day see and know...He who cherishes a beautiful vision, a lofty ideal in his heart, will one day realize it."

The succeeding chapters are the results of conceptualizing, realizing, nearly perfecting, and sharing the dream.

*Don and Judi Forbis receive the plaque commemorating *Ansata Ibn Halima's induction into the Arabian Horse Trust Hall of Fame. Rob Hess photo.*

Arabian horse sent by Abbas Pasha to Duke of Leuchtenberg. Albrect Adam. Courtesy Stadt Museum, München.

CHAPTER II

Conceptualizing the Dream

Ciaffar, 19th century lithograph by Alfred DeDreux. An inspiration for the Ansata program.

CHAPTER II
Conceptualizing the Dream

The greatest achievement was at first and for a time a dream.
The oak sleeps in the acorn; the bird waits in the egg;
and in the highest vision of the soul a waking angel stirs.
Dreams are the seedlings of realities....
And you, too...will realize the Vision (not the idle wish) of your heart,
be it base or beautiful, or a mixture of both, for you will always
gravitate toward that which you, secretly most love.
Into your hands will be placed the exact results of your own thoughts;
you will receive that which you earn; no more, no less.

James Allen - As A Man Thinketh

VISION, DESIGN, AND PURPOSE

Every person who seeks and hopes for the best in life must have vision, make a design, and pursue a purpose. These are requirements beyond dispute. They form the dynamics of achievement, particularly in breeding horses; they, as well as some of the factors noted next, must be considered in conceptualizing a meaningful breeding program.

Understanding: Do I understand the breed and its ideal standard, and do I know its history before I try to breed it?

Vision: What is the best type of Arabian that is desirable for me? That is, what bloodlines, family or strain of horses and what specific "look" within the breed standard, do I want to create to set my horses apart from others? Just as Rembrandt or Monet were known as artists of a distinct style, so are superb breeders noted for their distinct style of bloodstock, such as "the Babson type" or "the Ansata look" within the straight Egyptian-Arabian breed.

Purpose: Have I developed the sense of purpose and focus to enable me to obtain my goals?

Design: Am I developing plans to make my purpose effective and to make my vision come true?

Dedication: Am I prepared to dedicate my life to this profession?

Patience and Perseverance: Knowing that success succeeds and never precedes, will I have these qualities of endurance to make me a "stayer" and not give up under adverse circumstances?

Sacrifice: Am I willing to make the necessary sacrifices, knowing they will come knocking at the door along with golden opportunities?

Having determined to become a breeder, one should also understand that breeding is an art, and to become successful, it must be treated as such. Keep in mind that each breed has its own standard. That standard states those features that represent the ideal physical specimen of the breed as well as its inner qualities of spirit, courage and pride. The word that best describes the qualities of the standards is TYPE. An Arabian horse is typey or not typey according to how nearly it approaches the perfection set forth in the standard. TYPE is also an appreciation of quality in the EDUCATED EYE of the beholder. The three most important attributes to remember in the creative art of selecting and breeding Arabian horses, and in setting a goal for

"Today we speak of the art of breeding Egyptian Arabian horses as a tradition that is thousands of years old." Forbis photo, Luxor. Van Lent Jr. photo, Qatar.

one's breeding program, are TYPE, TYPE, TYPE. Breed identification (e.g., it should look like an Arabian horse, not a half-Arab, Quarter horse, Thoroughbred or other breed) is of paramount importance. TYPE, therefore, is the one word that expresses the degree of similarity to the ideal standard of perfection - the perfect horse of the Arabian breed - the horse that has captured the imagination of artists since time immemorial.

BREEDING ARABIAN HORSES IS A CREATIVE ART

When we speak of "the art of breeding," great horses, like great works of art, inspire us - but the creative process nourishes us. In our work we need to remember that great endings start with great beginnings - that no beginning can be great without individual imagination, individual creativity, and individual innovativeness. Sponsorship of art that reminds us of these things is not patronage. It is a business and human necessity.

The renderings of horses depicted on ancient Egyptian architecture are monuments to the fact that the Arabian horse is larger than life. Pharaohs prized it, and artists captured its unique type, leaving us a visual legacy of its natural beauty, grace and spirit. While temples and tombs will crumble with time, the Egyptian Arabian horse has survived through time as a living and immortal treasure.

Today we speak of the art of breeding Egyptian Arabian horses as a tradition that is thousands of years old. Then, as now, the impulse to create is an attribute of the human spirit which finds its release in this challenging endeavor and the opportunity to become a link in this long chain of history that binds breeders together over thousands of years, and that takes root in many souls who wish to be immortalized for contributing something memorable during their lifetime.

Bringing together the genetic potential from a mare and stallion that the breeder molds into a third creation of his or her ideal, transforms the breeder into a veritable sculptor of living art. Whether the resulting horse achieves critical acclaim as a work of art will be determined by whether or not the right bloodlines and families have been selected to closely approximate the breeder's ideal Arabian. A fine Arabian may well be considered a work of art. However, if the end product results from an unplanned union, the words *art* and *artist* can hardly be applied to its creation and creator. In this case, the one who is recorded as breeder of such a horse is not a true breeder, but merely the possessor of the dam at the time of her service to the stallion.

Successful breeders of the past, such as Abbas Pasha I, Ali Pasha Sherif, Lady Anne Blunt, and H.H. Prince Mohamed Aly appreciated fine art; the latter two were artists. They envisioned their ideals, understood what they wanted to accomplish, and pursued their goals regardless of the challenges and sacrifices required. Abbas Pasha sent emissaries throughout the Arab world to find the choicest Arabians among the desert tribes and bought them at any price. Ali Pasha Sherif preserved Abbas Pasha's legacy and added to it at great expense. Lady Anne Blunt and H.H. Prince Mohamed Aly, among others of the Egyptian royal family, built their foundation upon Ali Pasha Sherif's collection and in turn sponsored and donated horses to the Royal Agricultural Society, (renamed The Egyptian Agricultural Organization by Nasser). These foresighted and dedicated people individually and collectively were the bedrock upon which Egyptian breeders build today.

What the ideal a breeder formulates and seeks to produce, like that of a painter, sculptor, or dancer, is self-expression. The emphasis a breeder places upon certain traits, such as type, head, balance, legs, soundness, movement, disposition and spirit in the selection of breeding stock, declares his or her own nature. A breeder who would strive for beautiful color or incredible heads at the cost of honest structure differs from one who prefers a correctly conformed Arabian horse. This form of self-expression, this fulfillment of the creative urge, is what makes the differences in type and quality in the end result among breeders. One can own a special horse by purchase, gift, or breeding and can derive much pride, pleasure and companionship from this ownership. However, the excitement and emotional satisfaction of achieving the conceptualized horse from one's creative "vision," is not obtainable by simply possessing a horse, no matter how magnificent, that someone else has bred. That joy belongs solely to the breeder.

Arabian horses were created creatures of beauty and are Nature's testimony that Beauty is Truth, Truth is Beauty. The consistent production of them through selective breeding, employing love and the laws of heredity to perpetuate the desirable characteristics in the ancestral germ plasms and to eliminate undesirable

"A fine Arabian may well be considered a work of art."
1929 painting of Ibn Rabdan, the R.A.S. stallion described as a "world champion type". Musgrave Clark collection.

traits, is undeniable artistry. And it is with complete awareness of it as an art, and only as such, that the breeding of Arabian horses should be undertaken.

THE ART OF BREEDING

There are those who just do things, and there are those who do things with artistry. It is only through time and dedication that we can understand and practice all the arts which entitle us to be known as master horsemen or master breeders, or masters of the arts in our chosen field.

From time immemorial there has been an insistence on the interdependence of the Arts, united in universal correspondence and harmony. Color, music, poetry, dancing, sculpture and architecture, perfumes and tastes, natural and cosmic rhythms all correspond to one another like the many facets of a polished jewel.

A picturesque explanation of the profound link uniting all the Arts is noted in the *Vishnudharmottara*, in which is related the following conversation between the King and a Sage whom the King has visited to learn the craft of image-making:

King: O Sinless One! Be good enough to teach me the methods of image making.

Sage: One who does not know the laws of painting can never understand the laws of image-making.

King: Be then good enough, O Sage, to teach me the laws of painting.

Sage: But it is difficult to understand the laws of painting without any knowledge of the technique of dancing.

King: Kindly instruct me then in the art of dancing.

Sage: This is difficult to understand without a thorough knowledge of instrumental music.

King: Be then good enough, O Sage, to teach me the laws of instrumental music.

Sage: But the laws of instrumental music cannot be learned without a deep knowledge of the art of vocal music.

King: If vocal music be the source of all the arts, reveal to me then, O Sage, the laws of vocal music.

To become aware of all the arts is to understand the art of everything. This principle applies as well to the art of breeding fine horses, or training, or showing, as to any other related disciplines. Some of the following is taken from lectures presented by the late Professor Robert Dyer at Ansata breeding seminars.

WHAT IS ART?

Man is by his very nature a sentient, communicative being - a creator, an artist. When man begins to create, to record, or to communicate what he has seen or felt as he or she has experienced it, ART BEGINS. All humans create, but most seldom produce anything unique, life-enhancing or enduring. However, the need to create is inherent in human beings and our very existence is based upon it. ART, then, is life interpreted by and through the mind of the artist. He adds something to nature, and the result is neither true nor false, although the illusion may seem more real than life itself. Art is man added to nature. It is the means of communicating emotions. And indeed artists and breeders become emotional over criticism of their creations!

ART AS BEAUTY - AESTHETICS

Plotinius wrote: "What is it that attracts the eyes of those to whom a beautiful object is presented, and calls them, lures them, toward it, and fills them with joy at the sight? Undoubtedly the Principle exists; it is something perceived at the first glance, something which the soul names (as from an ancient knowledge) and, recognizing, welcomes it, enters into unison with it. Our interpretation is that the soul - by the very Truth of its nature by its affiliation to the noblest Existents in the hierarchy of Being - when it sees anything of that kin, or any trace of that kinship, thrills with an immediate delight, takes its own to itself, and thus stirs anew to the sense of its nature and of all its affinity." Could

Plotinius wrote: "What is it that attracts the eyes of those to whom a beautiful object is presented, and calls them, lures them, toward it, and fills them with joy at the sight?" Alfred DeDreux painting, photo from H.H. Mohamed Aly Collection.

there be a more perfect description of quality - or classic beauty - that makes our heart jump when a kingly stallion or exotic mare stands before us?

Dictionaries generally define the word "aesthetics" as a "sense of beauty, love of beauty, or philosophy of beauty." However, the opposite of aesthetic is *anaesthetic*, which is the diminution and/or loss of communication and excitement of ideas and emotions; and that quality of art or of a horse is determined by how well the ideas and emotions (the artist's and breeder's end result) have been communicated. Beauty must ever induce wonderment to create human emotion.

WHAT IS THE PURPOSE OF ART?

No matter whether we speak of literature, of painting, of sculpture, of music, or any of the other arts - including the art of breeding livestock, this concept of creating form out of chaos is the single common basis and foundation for all aesthetics. In art, there is something more - there is the consciousness of purpose, the consciousness of a peculiar relation of sympathy with the artist-breeder who made this thing to arouse precisely the sensations we experience. And when we come to the higher works of art, a Rembrandt or Van Gogh, or the superstar horses such as Nazeer, *Ansata Ibn Halima, *Morafic, or *Serenity Sonbolah, where sensations are so arranged that they arouse in us deep emotions (desire, love, joy, hate, envy), and the feeling of a special tie with the artist-breeder who expressed them, we feel that he or she has expressed something which was latent in us all the time, but which we never realized, and that it has revealed us to ourselves.

A work of art - or a fine horse - has value. In other words, that it is a source of satisfaction. For a work of art is not a given thing, but a thing made by man for his pleasure. It first exists in the visionary or dream world before it is made manifest in the material world. All truly great works of art were composed to communicate ideas and emotions; all differ from each other in style, degree of abstraction, perceptual and conceptual understanding of man and nature. Each artist is sharing with humanity his or her vision of the world, and way of seeing, and to do so has had to learn how to use the materials and elements of art in composition. This is likewise true of a breeder.

OF WHAT IS ART COMPOSED?

Art consists of three specific elements - *substance, form* and *technique*. These elements also apply to horses. A fine Arabian horse is also composed of ideal *type* (e.g., breed identification - one knows what kind of breed it is by looking at it), *conformation* (e.g., overall harmony and correctness of parts when standing and moving), *substance* (e.g., proper physical mass in relation to adequate bone structure) and *quality* (e.g., the difference between 24 carat gold and 10 karat gold or between a Vermeer and a modern painting of a toreador on velvet, even though it may be technically well executed).

Another important aspect of order in a work of art and a fine horse, is *unity*. Unity of some kind is necessary for our restful contemplation of the work of art as a whole, i.e., in judging the overall picture or balance of the horse, if it lacks unity one cannot contemplate it in its entirety. In a painting this unity is due to a balancing of the attractions of the eye about the central line of the picture, or in a horse, harmony from front to back - as in a two dimensional painting, and in the round - as in a three dimensional sculpture. The result of this balance of attractions is that the eye rests willingly within the bounds of the work of art, or the total form of the horse.

Design in aesthetic objects is governed by intrinsic form. In the same way we understand why in a picture a certain color demands its contrasting and complementary colors, in a poem we *know* that a word of certain length and sound texture is demanded, independent of its meaning, because of the necessities of rhyme and rhythm. In a horse, we *know* that the law of *True Beauty gives rise to form, function, and symmetry*, and that these qualities cannot be attained unless True Beauty precedes them.

Almost everyone declares that the *symmetry* of parts toward each other and toward a whole constitutes beauty recognized by the eye. It is universally accepted that in visible things, as indeed all else, the means is that only a compound can be beautiful, never anything devoid of parts, and only a whole. The several parts will have beauty, not in themselves, but only as working together to give a handsome total. *Yet, beauty in aggregate demands beauty in details; the law must run throughout.* For example, a beautiful head placed on a badly conformed body results in lack of symmetry and harmony overall and thus does not fulfill the law of True Beauty.

EVALUATING ART OR A FINE HORSE

The assumption underlying every philosophy of art is the existence of some common nature present in all the arts, despite their differences in form and content. This common nature exists in painting and sculpture, in poetry and drama, and in music and architecture (as illustrated by the story of the King and the Sage). Consistent with this idea is the understanding that the study of one art informs the study of the other arts. This is often expressed by the familiar habit of characterizing one art in terms of another art. So, how does one look at and judge a work of art?

First, look at the work and ask yourself what you see. Inventory the total, not just the parts, for all the shapes, colors, lines, textures, and spaces and notice the manner in which these elements are combined, that is, the composition. If evaluating a horse, a sculpture or a painting, stand away from it to see the overall type-style, balance-composition, substance and quality. Also assess what is missing for often it is not by the sins of commission that something is judged, but by the sins of omission.

Second, ask yourself what you know. There is a difference between seeing and knowing. You might see the form of a flower, but only special knowledge will define that particular flower as a tulip or a daffodil. Likewise, a horse may stand before you, but only special

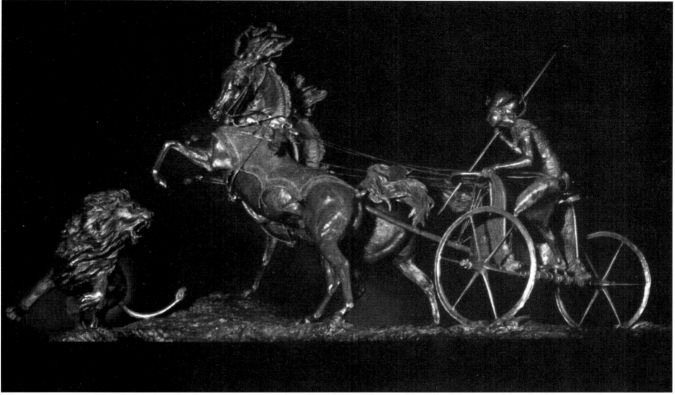

"Beauty in aggregate demands beauty in details; the law must run throughout." Tutankhamen, Pharaoh of Egypt - sculpture by Edwin Bogucki.

knowledge will define it as an Arabian, and further knowledge defines it as a good or bad representative of the breed.

Third, ask yourself what the artist-breeder was attempting to do. The artist sets the mood, describes items in certain ways through color, shape and lines to arrange these elements to create certain effects. Each art work has a composition that should function to make all parts work together in rhythm and harmony. The rules for judging any work are implicit in the work itself. Is this not true in judging the quality of a horse?

Fourth, judge how well the artist solved the problems he set before himself e.g., depicting a still life, a landscape, a storm at sea, or how the breeder has created a specific type and harmony in his horses. If you can recognize the merits of a work of art, but find it has qualities distasteful to your temperament, you are developing discriminating personal taste, e.g., a horse that has a beautiful head, but is weedy, herring-gutted, or splay-footed is faulty. If you find, also, that you like works with little or no artistic merit, but they are predominately your favorite color or subject, e.g., an ill-conformed chestnut horse that is a beautiful chestnut color, recognize that there are often factors in a work of art that you value over purely artistic values, e.g., chestnut color over correct type. In any case, you should begin to know how and why a work, such as a particular type within the standard of the breed, moves you. Aesthetic understanding and appreciation is a give-and-take process involving you and the work of art, or the horse that stands before you. In undertaking the problem of criticism, you should become a more knowledgeable individual.

A critical verdict about a breeder's horse as a "living work of art" must rest upon whether or not the breeder has employed genetic and family materials in the right manner to come close to the "ideal" by which all others are measured. While an artist uses paint to express his vision, the breeder uses genetic materials through the study of strains/families and individuals within pedigrees, i.e., the sires, dams, grandparents and great-grandparents in making breeding decisions. Judging art, as in judging horses, is subjective, and in conceptualizing Arabian horses some breeders may prefer a stronger more masculine type individual (often considered the Kuhaylan type) while others prefer a more feminine and refined individual (the Saklawi type), or a combination of both (the Dahman type) and variations thereof. However, each must conform to the classic spirit of beauty, balance, and harmony if it is to approximate the eternal, fine and noble ideal.

An inspiring example of a creative breeder is found in Lieutenant Col. Sidney G. Goldschmidt's book, *Skilled Horsemanship*, published in 1937. Goldschmidt discusses the breeding program of H.H. Prince Mohamed Aly of Egypt (breeder of *Nasr, Negma, *Aziza, *Zarife, Mahroussa, *Maaroufa, *Fadl, etc.). A descendant of Viceroy Mohamed Aly the Great, Govenor of Egypt as ruling Pasha from 1805 to 1848, and who conquered most of Arabia, the Prince had a strong sentimental interest in continuing the Arab blood which had been identified with his family so many years. Goldschmidt says, "But then the selection of sire and dam at this great stud is not on orthodox lines. The Prince has made a collection of old prints and drawings showing the traditional Arab horse, the horse of poetry and romance. These serve as his guide, and it is his aim to breed to this standard. Every sire and dam, as well as their progeny, are studied with this ideal before him, and the tests are applied with almost mathematical precision. Any that fall short in the minutest detail are ruthlessly weeded out and sold." He further notes, "Photographs of some of the horses placed alongside old prints show how surely he is approaching his goal. In reply to an unguarded question as to the utility of a stud of horses that were never ridden, we were told that the attainment of an ideal need have no utilitarian object, and just as we in Europe hang beautiful pictures on our walls to look at and to admire, so he has his beautiful horses. I must say that those we saw, especially those that had already spent their maturing months in the desert, excelled in beauty any work of art it has ever been my good fortune to see."

In his book, *Breeding of Purebred Arab Horses* published in Cairo, 1935, H.H. Prince Mohamed Aly corroborates Goldschmidt's remarks: "I have been breeding horses for thirty five years and my aim has always been to unite dignity with grace. Briefly then, my perfect horse must have all the symbols of an antique pedigree, a horse that might have stepped from some rare or ancient picture. I am an artist myself, and have an extensive collection of paintings and prints of beautiful animals, and this of course has made my choice easier." He goes on to say: "Many people have criticized me because my object in breeding was, to use an old phrase, 'Art for Art's Sake,' and not to give my horses the chance to race and prove their superiority." As Goldschmidt noted, "One should not be misled by this statement that he bred only 'ornaments', for *Nasr, his chief stallion was a successful race horse." Furthermore, the Prince believed that while cavalry officers or veterinary surgeons have technical knowledge when it comes to judging horse shows, (and here might be added trainers as well), "90% of them only understand what is useful in a horse according to their own ideas, and fail to appreciate the fancy horse, whose beauty should not prevent it from being useful as well. As a result, horses are often disqualified because their joints are too delicate, etc. I think that if, from time to time, the advice of breeders and amateurs were taken, and even that of special animal painters, who understand the value of the beautiful, there would be a distinct progress. That is to say the criticisms would be aesthetic as well as technical."

Additionally, it should be remembered that standards are based upon archetypical perfection; they are matters of body, not of performance. The so-called per-

Lithographs by C. Vernet, left; V. Adam, right; A. Dedreux, below.

"The Prince has made a collection of old prints and drawings showing the traditional Arab horse . . . These serve as his guide . . ."

41

*H.H. Prince Mohamed Aly "…My aim has always been to unite dignity with grace." Above left, *Nasr (Rabdan El Azrak x Bint Yamama) and above right, Mahroussa (Mabrouk Manial x Negma) both bred by the Prince. Carl Raswan photos.*

formance horse is NOT a type; therefore, there is no uniformity. Any horse, weedy, ill-proportioned, ugly or otherwise, that does well in performance, may be called a performance horse. However, utility never gets us anywhere in type, because it has not an aesthetic goal. Performance does not transmit itself. Each performance horse must be found anew by its own deeds. The Prince also proved this adage to himself, as he relates in his book.

UPHOLDING THE CLASSIC SPIRIT

One of the great muralists of the early 1900s was Kenyon Cox, who held firmly to what he saw as the fundamental principles of the great Renaissance and Baroque painters. Cox wrote of the Classic Spirit that "it desires that each new presentation of truth and beauty shall show us the old truth and the old beauty, seen only from a different angle and colored by a different medium. It wishes to add link by link to the chain of tradition, but it does not wish to break the chain." How true this is in relation to the art of breeding classic Arabian horses.

Cox went on to say that "the Classic Spirit is the disinterested search for perfection; it is the love of clearness and reasonableness and self-control; it is, above all, the love of permanence and of continuity. It asks of a work of art, not that it shall be novel or effective, but that it shall be fine and noble… It strives for the essential rather than the accidental, the eternal rather than the momentary – and it loves to steep itself in tradition. It would have each new work connect itself in the mind of him who sees it with all the noble and lovely works of the past, bringing them to his memory and making their beauty and charm a part of the beauty and charm of the work before him."

Artistic intention, he believed, must dominate everything, control everything, mould everything to its purpose. Its sovereignty must be absolute and complete. However, to control facts, and to bend them to one's purpose, one must know them, and know them vastly better than he who merely copies them. With a good eye and a good deal of practice an artist may copy a leg, or a hoof quite well, but to understand that leg or that hoof, so that one can distinguish the essential from the accidental, one must experience it by touching, lifting, and feeling the physical. To know how each part functions, one must ride the horse, or as the great painter Stubbs did, dissect the carcass and draw each individual layer. Whether one is an artist or a breeder, or both, one must practice and learn to control what is important to one's purpose, and to master its creation. That is a life's work and takes a lifetime of dedication.

Cox closed his chapter on the Classic Spirit by the following message, which is still apropos as we enter this new millennium, and whereby the words Egyptian Arabian horses may be substituted for the word art: "Our art is, indeed, the best alive today, but the best is but poor compared to what has been or to what should be. Let us strive to make it equal to any, and to that end let us fill ourselves with the Classic Spirit. Let us strive for perfection, recognizing that perfection is only to be attained by discipline and by self-control. Let us think not what is new, but what is good; not what is easy and attractive, but what is eternally right. Let us attach ourselves to what is noblest in the art of the past, trying to understand the reason of its nobility, and spend ourselves in generous emulation. Let us believe that passion and personality will find their way into our art, if we have them, and that it is a poor and sapless individuality that can be killed by a little hard work." Only in this spirit can a great art, or a great breeding herd, or superior individuals be created.

PREPARING THE PALETTE

Having determined to become a serious breeder, there are many tools to be used in developing the composition, but the primary color to place first on the palette, and to maintain in abundance is that universal solvent - LOVE. One of the great breeders of all time was Luther Burbank. His development of many varieties of plants, fruits, and vegetables was due to this very ingredient. When he wanted his plants to develop in some particular and peculiar way not common to their kind, he would get down on his knees and talk to them. An article in *Horizon* magazine by Manly P. Hall stated: "To Burbank, the secret of plant breeding was summarized in a single word: love. This power, greater than any other, was a subtle kind of nourishment that

"the Classic Spirit...desires that each new presentation of truth and beauty shall show us the old truth and the old beauty..."
Above left Alaa El Din at the E.A.O., above right "Arab Warriors," Eugéne Delacroix, private collection.

"The Classic Spirit…asks of a work of art…that it shall be fine and noble…it strives for the eternal rather than the momentary…"
*Ansata Ibn Halima, painting by Count Bernard de Clavière de Hust.

made everything grow better and bear fruit more abundantly. He explained to me that in all his experimentation he took plants into his confidence...asked them to help, and assured them that he held their small lives in the deepest regard and affection. He insisted that this was the secret of his green thumb...."

This is just as true when working with animals, and particularly horses. We have been given dominion - not domination - over them, and if we love and communicate with them, they respond in kind. Small breeders who love their horses may sometimes seem remarkably blessed with "beginners luck" and ongoing success, while a large breeding farm with every means available fails because selection is done by someone without vision, an "eye," or knowledge, or by committee or government administrators who see their work only as a job. The lack of love is reflected in the demeanor of the horses and in their production.

"...when working with animals...if we love and communicate with them, they respond in kind." Gigi Grasso photo, Qatar.

COMPOSITIONAL CONSIDERATIONS

Having primed the palette, how shall we begin? What techniques can be employed to create the substance of the idea? Here are some of the useful tools that were used in developing the Ansata program through certain bloodlines to achieve the Ansata "Look." (Refer also to the chapter on Planned Breeding, *Authentic Arabian Bloodstock I*, pp. 283-287.)

Bloodlines: What is a bloodline? When speaking of bloodlines, we are not referring to blood or the heart. A bloodline refers to hereditary relationship of those having all or most of the same common ancestors. In Arabia marriage is often kept within the family, for marrying outside the family dilutes the strength of the bloodline and may establish a new line. The same is true of horses or other livestock.

Pedigrees: Every living thing has a pedigree. It is the sum total of one's ancestors. In general modern usage it refers to known ancestors and particularly to both paternal and maternal ancestors of the same limited species. Learning to interpret a pedigree is a critical component in mastering the art of breeding. A pedigree is a roadmap that names bloodlines, strains, and families.

Strains and Families: The words "strain" and "family" are common to all livestock breeding, Arabian horses included. Simply defined, a strain is a family of Arabian horses that share common ancestors and physical characteristics. Historically speaking, strain names came into being among the Arabs through various and peculiar circumstances and did not necessarily relate to breeding characteristics at the time. For example, the Dahman strain took its name from an original mare who was called El Dahma, meaning very dark or black; the Saqlawi from an old "saqla" mare, meaning a kicker. The Bedouins linebred and inbred some of their horses, and eventually when the cream of the stock came to Egypt during the reign of Mohamed Ali the Great, this nucleus of horses became even more linebred and inbred by the royal families, the Blunts, and the Royal Agricultural Society/Egyptian Agricultural Organization and other private breeders who used the same foundation stock and worked within this relatively limited group. Thus certain strains, and families within the strains, became noted for particular characteristics if they were linebred to retain them. HOWEVER, JUST BECAUSE A HORSE THAT TAKES ITS STRAIN NAME STRICTLY FROM THE DAM'S TAIL FEMALE IS OF A PARTICULAR STRAIN, IT MAY OR MAY NOT REPRESENT WHAT IS CONSIDERED TYPICAL OF THAT STRAIN OR FAMILY. It is here that careful study of pedigrees, strains, and family traits is critical. New "families" are always being created within the strain to which they belong. This is illustrated below under *Prepotency*. Chapter Four in *Authentic Arabian Bloodstock I* also provides helpful information on this subject, while *The Abbas Pasha Manuscript* further illustrates the historical stories told by the Bedouins as to how strains and families got their names, why they were liked or disliked, and why they did not use stallions from certain strains.

Study of Family Traits: In almost every breed, there are favorite families prized by breeders for the persistence with which they stamp their unique characters upon their offspring. Any specific trait of an ancestor, more or less remote, whether of form, color, habits, mental traits, or predisposition to disease, may make its appearance in the offspring without having been observed in the parents. This form of heredity, known as reversion, has for a long time been recognized by breeders as "throwing back," "breeding back," "harking back," etc. The sum of the characters or physiological units that comprise a horse cannot be represented only in the external traits that are obvious to the naked eye. Many of the most important characters of the composition, in a given case, may not appear outwardly or functionally, and they can only be traced in the ancestral history and in the inherited traits of offspring. Further, it is necessary to distinguish between the more obvious and prominent characters of the individual (*dominant characters*) and the more obscure characters that can only be shown to exist by their hereditary transmission to offspring (*latent* characters). An example here would be the stallion Kazmeen. His dam had a terrible croup, as did he, and he transmitted it with great consistency as a breeding stallion. Nevertheless, he also transmitted extreme beauty in type with equal regularity; and as will be seen as we continue, one often gives up one thing to get another.

Theoretically, a defect or peculiarity may be "bred out," until it is represented mathematically by a fraction so small as to scarcely merit attention, and yet, it may again appear in a manner indicating that it has been constantly transmitted, without change, through a long series of generations. Indeed the law of heredity is not constant in its action, but limited by numerous exceptions. Therefore, when speaking of the resemblance of offspring to ancestors, the dominant characters are alone referred to; but, as these may constitute but a small proportion of the elements of the composition, a strict comparison of resemblances must include a wider range of characteristics. In this connection, the importance of a full record of the pedigrees of the breeding stock is necessary as a means of tracing the history of ancestors to determine the characters that are liable to be transmitted by atavistic descent. Analyzing the pedigrees of horses intelligently helps reveal the source of such buried genes, and one can decide whether to utilize an animal for his genetic virtues, or discard it for the faults one knows it has as part of its heritage. In other words, what is one willing to give up to get something else, for sacrifice is almost always involved: to take a step forward, one must sometimes take a step backward, since heredity often skips a generation.

The best ways to learn about ancestors is to talk with experienced and reliable breeders who may have seen them, and to obtain movies and photographs of the desired horses within pedigrees, and those related thereto, as far back as possible, attending to the tail female line, assessing the tail male line, and reading as much information as can be found about them. Place the photos from ascending to descending tail female generations under each other, and also develop a picture pedigree. Note the assets and demerits of each line. What breeds true? What does not? How have certain combinations of sires or dams reinforced the type, or altered it? Never forget, however, that photographs can lie. Study the principles of horse photography and take photographs of horses one knows. The eye can be trained when looking through the lens of a camera, because absolute focus is required on the individual at hand to get the best picture. This intense focusing enables one to judge the merits and demerits of photos, once the principles of photographing horses are understood. These principles are quite different from other kinds of photography. Knowing the physical horse, and how its appearance can be altered photographically by certain poses, is invaluable in judging other horse pictures that may be intentionally posed to cover up undesirable traits (e.g., a frontal head view which does not show the plain or ugly profile).

In-and-In-Breeding: High breeding (inbreeding and linebreeding) refers to a careful selection of breeding-stock within certain limits of a family's genetic makeup and with reference to a particular type and desired characteristics. *Inbreeding* is practiced to develop livestock that will breed true to type. It is the safest and fastest way to ensure uniformity, to perfect type, and to develop certainty in a breeding program. It is the opposite of "outbreeding" or "outcrossing." Inbreeding refers to breeding stock more closely related than the average. The most intense inbreeding combinations [true inbreeding] are: full brother to full sister, father to daughter, mother to son. Other combinations often referred to as inbreeding are grand-parent to grand-offspring, uncle to niece, aunt to nephew, first cousins to first cousins, or second cousins to second cousins. The first prerequisite for inbreeding is to start with superior animals, never with mediocre breeding stock, because good as well as bad traits will be intensified.

Linebreeding is a type of inbreeding, sometimes referred to as inbreeding at a distance. It usually refers to a series of matings back to a particular animal, or its near descendants. It is the mating of a stallion and mare that have at least one common ancestor not further back than four generations. However, a pedigree may represent linebreeding without any two ancestors being the same individual. Certain horses would be closely

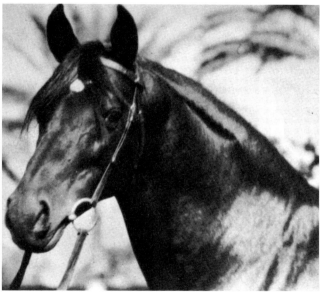

A Raswan frontal view photo of Sotamm (*Astraled x Selma) in Egypt showing pleasant expression but not describing the horse's head profile.

A side view of Sotamm when in England revealing the profile shape of his head. Photo from Crabbet Arabian Stud, Archer, Pearson and Covey.

related of the same family, for example, if all four grandparents were of the same family. This would also be close linebreeding.

With linebreeding there is less likelihood of undesirable factors coming to light, especially if a degree of outcrossing is incorporated with the linebreeding, and the matings are not arranged between very closely related stallions and mares. As in any kind of breeding, background knowledge of the stock concerned is the key to success. Without knowing the virtues and faults of the close-up ancestors of both sides of the individuals being mated, linebreeding (as well as inbreeding or outcrossing) may well have the opposite effect of that desired.

Outcrossing is the mating of unrelated pairs. Parents without a common ancestor in 5 generations or 62 ancestors, may be considered unrelated for practical breeding purposes. It is difficult to accomplish within the straight Egyptian lines because the gene pool is limited and family relationships are strong through the limited number and power of those few tail female lines. Nevertheless, some degree of it is possible through careful

IN-BREEDING:

		Nazeer
	*Ansata Ibn Halima	Halima
Ansata Ibn Sudan		Nazeer
	*Ansata Bint Mabrouka	Mabrouka
Ansata Nile Dream		*Ansata Ibn Halima
	Ansata Ibn Sudan	*Ansata Bint Mabrouka
Ansata Nile Queen		*Ansata Ibn Halima
	Falima	Fa-Habba

Father to daughter inbreeding

		Mansour
	Nazeer	Bint Samiha
*Morafic		Sid Abouhom
	Mabrouka	Moniet El Nefous
Ansata Shah Zaman		Mansour
	Nazeer	Bint Samiha
*Ansata Bint Mabrouka		Sid Abouhom
	Mabrouka	Moniet El Nefous

Full brother to full sister inbreeding

		Mansour
	Nazeer	Bint Samiha
*Morafic		Sid Abouhom
	Mabrouka	Moniet El Nefous
"Ibn Fantasy"		El Deree
	Sid Abouhom	Layla
		Shahloul
	Mabrouka	
		Moniet El Nefous	Wanisa

Son to mother inbreeding - fantasy example - this breeding was not done.

LINE-BREEDING:

		*Ansata Ibn Halima
	Ansata Halim Shah	Ansata Rosetta
Ansata Hejazi		Ansata Abu Sudan
	Ansata Sudarra	Ansata Delilah
Ansata Malik Shah		*Ansata Ibn Halima
	Ansata Halim Shah	Ansata Rosetta
Ansata Malaka		*Jamilll
	Ansata Malika	*JKB Masouda

Sire of the sire is also the sire of the dam

		Nazeer
	*Morafic	Mabrouka
*Khofo		Anter
	*Nabilahh	Farasha
Bint Magidaa		Nazeer
	Alaa El Din	Kateefa
*Magidaa		Anter
	Maysa	Mahfouza

Line breeding to two individuals

48

LINE-BREEDING (continued):

		Gamil Manial
	Mansour	
Nazeer		Nafaa El Saghira
		Kazmeyn
	Bint Samiha	
*Ansata Ibn Halima		Samiha
		Mansour
	Sheikh El Arab	
Halima		Bint Sabah
		Ibn Rabdan
	Ragia	
		Farida

Sire of the sire is grandsire of the dam

		Mansour
	Nazeer	
*Morafic		Bint Samiha
		Sid Abouhom
	Mabrouka	
*Ibn Moniet El Nefous		Moniet El Nefous
		Ibn Rabdan
	Shahloul	
		Bint Radia
Moniet El Nefous		Sheikh El Arab
	Wanisa	
		Medallela

Close linebreeding - grandson to granddam

selection. For example, a straight Babson-bred Egyptian horse could be an outcross to inbred Nazeer lines, even though there would be similar ancestors further back in their pedigrees. Another example of outcrossing would be the Inshass horses bred by Kings Fouad and Farouk. In the 50s many of these horses were an outcross for the intensely bred R.A.S./E.A.O. herd. The type difference was obvious when the Inshass herd was brought to El Zahraa and eventually melded with the R.A.S./E.A.O. stock. Although some of the Inshass horses were of the same desert strains as the R.A.S./E.A.O. horses, they were from unrelated families. Within a few generations of mating Inshass horses to the R.A.S./E.A.O. stock, the R.A.S./E.A.O. type took precedence.

Outcrossing should be made only for the specific purpose to correct certain faults that may have developed in an inbred strain. After an outcross has been accomplished and the purpose has been achieved, one should immediately breed back into the original strain or family to avoid losing what had been established previously.

Some breeders claim in-and-inbreeding produces a predisposition to disease, a delicacy of constitution, and a lack of fertility. Nevertheless, in general, breeders with outstanding reputations have aimed to establish in their bloodstock certain well-marked characteristics that adapted their horses to a particular purpose or created a specific "look" or type, and in so-doing they resorted to some form of inbreeding. For example, to express their ideal type, or standard of excellence, they found it necessary to limit their selection to breeding-stock that had the characters they wished to perpetuate.

OUTCROSSING (lesser degree):

		Mansour
	Nazeer	
*Ansata Ibn Halima		Bint Samiha
		Sheikh El Arab
	Halima	
Fa Halima		Ragia
		*Fadl
	Fabah	
		*Bint Bint Sabah
Sabrah		Fa-Serr
	Serrasab	
		Fay-Sabbah

EAO bred sire to Babson bred mare of related tail female lines

OUTCROSSING (greater degree):

		Ibn Fayda
	El Moez	
Sameh (INS)		Bint Zareefa (INS)
		El Deree
	Samira (INS)	
*Serenity Sonbolah		El Samraa
		Mansour
	Nazeer	
Bint Om El Saad		Bint Samiha
		Shahloul
	Om El Saad	
		Yashmak

Inshass bred sire to E.A.O. bred mare of unrelated female lines.

Horses are not improved by breeding except by the increased stability gained in dominant characters, and the certainty with which these characters are transmitted, because the offspring, at the time of birth, can only be possessed of the characters they have derived from their ancestors. It is generally found that in-and-in breeding has only been resorted to in the case of some favorite individual or individuals that were superior in certain respects to the average members of the herd or family which they represent, and the object has been to secure, in their offspring, a predominance of their most highly-valued characters. Examples of very close breeding (inbreeding) are illustrated by pedigrees of Ansata Nile Dream and Hafiza E.A.O./Egypt (father to daughter), Ansata Shah Zaman, Ibn Fa-Serr and Sidra (full brother to full sister), Farfoura/E.A.O. Egypt) (brother and sister in blood), while son to mother is perhaps the strongest of all inbreeding (see fantasy pedigree) and even then there are degrees of inbreeding depending on the extent of inbreeding or linebreeding within the overall pedigree.

Prepotency: In livestock breeding the parent that apparently exercises the greatest influence upon the dominant characters of the offspring is said to be prepotent. When certain desirable characters have been developed in a few individuals, they can only be ingrained into the entire herd by making them the dominant characters of the stallions that are to be used and securing in them prepotency in their transmission, thus prepotency in the transmission of the stallion's better qualities is one of the most valuable characteristics a sire can possess.

Above is Ansata Shah Zaman (*Morafic x *Ansata Bint Mabrouka), full brother-sister mating; above right is Ansata Nile Dream (Ansata Ibn Sudan x Ansata Nile Queen), father-daughter mating; at right is Hafiza (Hamdan x Mahfouza), father-daughter mating. Sparagowski photo above.

This confluence of good germ plasm usually springs from a combination of desirable genes - from parents and grandparents of recognized excellence and prepotency. On the other hand, some prepotent individuals are the result of a freak combination. Usually speaking, the great prepotent sires and dams derive from a valued line of prepotent ancestors. Nevertheless, a stallion may have a fabulous pedigree and never sire a decent foal. The explanation is in the inherited genes. Nothing is written in stone. The sire or dam prepotent for dominant characters will stamp them upon its progeny regardless of the other parent. Prepotency is the purity of dominance or the recessiveness of the various sets of genes. Unless selective breeding is continued, thereafter, through careful study of dominant and recessive genes, the ability to develop a strain will disintegrate.

The degree of high breeding required to secure prepotency in a given stallion will evidently depend upon the relative development and breeding lineage of the mares which he services. *The better the mares, and the greater the uniformity in their characteristics, the more prepotent the stallion must be to secure a predominance of his traits in his offspring.* This intensity in the power of transmission can only be produced by still higher breeding. Because of the intense genetic nucleus within the straight Egyptian mares worldwide, it is difficult to create stallions who can dominate over the mares' fixed characteristics especially within particularly powerful families or strains. Many well-bred stallions who may not be dominant over the straight Egyptian mares, may well be dominant over outcross mares (e.g., those of unrelated bloodlines such as Polish, American, Crabbet, or others). For this reason straight Egyptian stallions have been sought after to improve type in other lines. For example, Aswan or Palas was introduced into the Polish herd; El Hilal and Ruminaja Ali, into American bloodlines, and Shaker El Masri into Spanish lines to create outstanding show and performance horses.

The following are some illustrations of developing prepotency through linebreeding:

Prepotency in a sire carries forward:

Ansata Halim Shah (*Ansata Ibn Halima x Ansata Rosetta), a prepotent, homozygous grey sire, Dahman Shahwan - Bukra female line. Escher photo.

Salaa El Dine (Ansata Halim Shah x Hanan) a son of Ansata Halim Shah from the Abayyan Om Jurays - Hanan family. Van Lent photo.

GR Amaretto (Classic Shadwan x Halims Asmara by Ansata Halim Shah) a maternal Halim Shah grandson from the Kuhaylan Rodan - Nazeera family. Escher photo.

(a) The sire of the sire is also the sire of the dam, e.g., Ansata Ibn Sudan (double Nazeer) Ansata Malik Shah (double Ansata Halim Shah) *Tuhotmos (double Shahloul).

(b) The sire of the sire is the grandsire of the dam, e.g., *Ansata Ibn Halima, Aswan, Prince Fa Moniet, Imperial Madori, Imperial Kamar, Ansata Rahmah. This is thought to be a very powerful combination and was used by Von Szandtner particularly by breeding Nazeer to Sheikh El Arab daughters, thus doubling Mansour. However, the dam line usually held the type (e.g., *Ansata Ibn Halima, *Rashad Ibn Nazeer, *Aswan).

(c) Doubling the same female line through the sire and the dam's tail female: Ansata Halim Shah (tail female Dahma through his dam/Bukra line and through his sire/Farida line. El Hilal (tail female Dahma/Farida through both sire and dam), Ansata Nile Nadir (doubling the Dahma/Sabah through sire and dam), Sherifa Tamria (doubling the Dahma/Sabah through *Pharrah in sire and dam's tail female, plus other intense lines to *Ansata Ibn Halima of the Farida family). Ansata Selket (tail female sire and dam tail female Dahma/Bukra/Sabah). *Ibn Moniet El Nefous (tail female Saklawi/Moniet through his sire and dam), The Egyptian Prince (doubling Saklawi/Moniet through his sire and dam), Moniet El Sharaf (all tail female lines to Moniet), MB Sateenha (tail female sire to Serra and dam to Moniet). All of these combinations were designed to effect prepotency, and, in most cases, they did. However, it should also be remembered that prepotency can be effective for positive or negative traits.

Another asset in studying degrees of inbreeding, linebreeding and outcrossing is to color-code pedigrees. One immediately sees the influence of particular families or individuals through similarities of the colors, or names of the individual horses in the pedigrees. Examples of this are noted in illustrations of some of the above pedigrees. Color-coding is often done by using blue for Kuhaylan, red for Saklawi, green for Dahman, and

a) The sire of the sire is also the sire of the dam:

*Tuhotmos (El Sareei by Shahloul x Moniet El Nefous by Shahloul).

Ansata Ibn Sudan (*Ansata Ibn Halima by Nazeer x *Ansata Bint Mabrouka by Nazeer). Sparagowski photo.

Ansata Malik Shah (Ansata Hejazi by Ansata Halim Shah x Ansata Malaka by Ansata Halim Shah). Sparagowski photo.

b) The sire of the sire is also the grandsire of the dam:

c) Doubling the same female line through the sire and the dam's tail female:

*Ansata Ibn Halima (Nazeer by Mansour x Halima by the Mansour son, Sheikh El Arab). Sparagowski photo.

El Hilal (*Ansata Ibn Halima of the Farida female line x *Bint Nefisa of the Farida female line). Sparagowski photo.

Prince Fa Moniet (The Egyptian Prince by *Morafic x Fa Moniet by the *Morafic son *Ibn Moniet El Nefous). Sparagowski photo.

*Ibn Moniet El Nefous (*Morafic of the Moniet El Nefous female line x Moniet El Nefous). Johnny Johnston photo.

Imperial Madori (*Imperial Madheen by Messaoud x Imperial Orianah by the Messaoud son *Orashan).

Sherifa Tamria (Royal Jalliel of the *Pharrah female line x Imperial Daeemah of the *Pharrah female line).

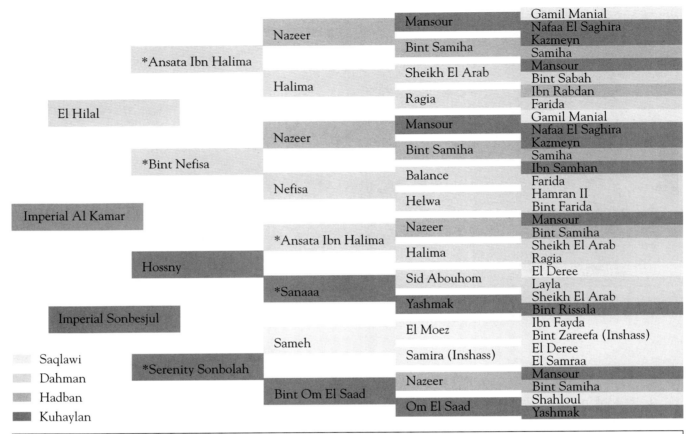

The above sample of a color coded pedigree demonstrates the previously covered points: **a) the sire of the sire is also sire of the dam** [El Hilal/double Nazeer]; **b) the sire of the sire is also the grandsire of the dam** [Imperial Al Kamar/double *Ansata Ibn Halima and *Ansata Ibn Halima/double Mansour]; **c) doubling the same female line through the sire and dam's tail female line** [Imperial Sonbesjul/double Yashmak, and El Hilal/double Farida]. In addition, color coding reveals what strains predominate overall: Dahman and Kuhaylan being nearly equal and the rest divided among Hadban and Saklawi.

shades thereof for specific families within the strain. For example, this type of color-coding could be used to differentiate between Farida and Sabah in the Dahman strain, Helwa and Roga in the Saklawi strain, or Rissala from Riyala in the Kuhaylan Rodan strain.

To ingraft a new or modified character upon those representing a family type without destroying the desirable specific traits of the family, (e.g., within straight Egyptian bloodlines, the Dahman/Farida family, the Saklawi/Moniet family, the Abeyyan/Hanan family) close breeding within the limits of the family can prevent too wide a divergence in the dominant characters. The most obvious objection to close breeding - and it is perhaps the only one of importance - is the difficulty of selecting individuals that are free from constitutional defects, and the danger arising from the tendency of such defects to become dominant in the offspring. Nevertheless it is an important means of improvement when judiciously practiced, and it constitutes the only known method of securing an accumulation of the slight variations, in a particular direction, that one may wish to retain and perpetuate.

Heredity: We often think of heredity in terms of "like begets like" and "like father, like son," or "like mother like daughter;" however, no son is exactly like father, no mother exactly like daughter, and no sister exactly like sister. An example of this would be Ansata Halim Shah who was an absolutely dominant and prepotent sire taking more after his dam line compared to his full brother, Ansata Omar Halim, a handsome and very

Ansata Sinan (Prince Fa Moniet x Ansata Nefara) not reminiscent of either of his parents. Full brother to Ansata Iemhotep at right. Polly Knoll photo.

Ansata Iemhotep (Prince Fa Moniet x Ansata Nefara) resembles his sire. These full brothers are excellent individuals but different in type. Vesty photo.

good sire looking more like his sire. The same holds true for Ansata Abbas Pasha and Ansata Ibn Sudan respectively. More recently Ansata Sinan and Ansata Iemhotep, who are also full brothers, but different in type, with Sinan not resembling either parent, and Iemhotep claiming much after his sire.

We observe that animals sometimes look much like one of the parents, (e.g., *Ansata Bint Bukra "Hosnia" resembled her dam, Bukra, while her sister Bint Bukra, in Egypt, did not), or somewhat like both of the parents (e.g., most of Nazeer's get have good qualities of both sire and dam coming through but predominantly the mare's structure and type). But does "like always beget like"? Not necessarily. In the days before Mendelism, "like begets like" meant that horses look like their parents and grandparents. Sometimes they do resemble them (e.g. *Ansata Ibn Halima closely resembled Sheikh El Arab, his dam's sire). However, it is only when the "like begets like" theory is applied to *genes* that it becomes true. Like horses do not necessarily beget like horses, but *like genes do produce like genes*.

COLORS AND VALUES

Another factor to be taken into consideration is that type often relates to color. In certain families, color generally plays an important part. For example, in the Dahman strain, the greys in the Farida line are different from the bays, again depending on which family they came up from. The chestnuts in the *Bint Bint Sabbah - Ansata Sabiha family are perhaps more beautiful than the greys or mahogany bays, while the bays in the Nile family were initially the most beautiful followed by the greys, and then the chestnuts of that line. In the Moniet El Nefous family, the chestnuts and greys were much more in type to that family than the bays, most specifically because the bay stallion, *Rashad Ibn Nazeer, dominated in type through his bay get, while his chestnut get of that line took more after Moniet in type.

The importance of the grey color can be studied in the old Crabbet herd. Lady Wentworth bred colored horses to colored horses until her mother, Lady Ann Blunt, despaired that she would ever use grey. The con-

Type often relates to color:

Moniet El Nefous [chestnut] (Shahloul [grey] x Wanisa [chestnut])

*Bint Moniet El Nefous [chestnut] (Nazeer [grey] x Moniet El Nefous [chestnut])

Head view of *Bint Moniet El Nefous [chestnut] (Nazeer [grey] x Moniet El Nefous [chestnut])

Rasmoniet RSI [chestnut] (*Rashad Ibn Nazeer [bay] x *Bint Moniet El Nefous [chestnut])

RSI Raya Del Sol [bay] (*Rashad Ibn Nazeer [bay] x *Bint Moniet El Nefous [chestnut])

Bint Bint Moniet [bay] (*Rashad Ibn Nazeer [bay] x *Bint Moniet El Nefous [chestnut])

tinual use of chestnuts to chestnuts produced the high white in the Crabbet herd (primarily through the mare Rodania and the Crabbet stallion Mesaoud), but when the grey Skowronek came along, the entire picture changed and better Arabian type was restored.

Generally speaking, grey is the most consistent color in producing type no matter the family, as can be seen in the Ansata herd, in linebred herds found in Europe (e.g., early Marbach, early Babolna, and some private studs), and the Arab world (private studs in Qatar, Kuwait, and Egypt). Some of the most consistent and prepotent breeding stallions have been grey (Dahman El Azrak, Rabdan, Mansour, Mabrouk Manial, Shahloul, Hamdan, Sheikh El Arab, Nazeer, *Fadl, *Morafic, *Ansata Ibn Halima, The Egyptian Prince, *Imperial Madheen, Ruminaja Ali, etc.). Some influential chestnuts were: Mesaoud (England), Ibn Rabdan, Alaa El Din and Galal (Egypt), Ibn Galal (Germany), and Ibn Galal I (Hungary). The bay stallion El Sareei contributed good daughters in Egypt, and *Tuhotmos, his only son that was used as a sire, was a bay. Neither father nor son were prepotent sires. Of late, bay Egyptian horses have become fashionable in America through The Minstril and Thee Desperado, who both impacted the Egyptian-American breeding programs with a type of their own through linebreeding. Other bays are Anaza El Farid, and his son, Farid Nile Moon, a handsome champion from the *Ansata Bint Zaafarana line which is most frequently found in grey. The former is a proven sire of significance, the latter is just beginning his career at stud.

Blacks are relatively uncommon, the most prolific being Gharib in Germany, who did a credible job primarily due to the high quality mares he serviced at Marbach State Stud, and Fa-Serr in the Babson Farm, who contributed a certain heaviness and shortness of leg to the program. *El Mokhtar, was a stunning black imported to the U.S.A., but he was of minor consequence as a breeding stallion and he died before getting a real chance at stud. Another awe-inspiring black is multi-champion Simeon Sadik, a remarkable indi-

Gharib [black] (Antar x Souhair), sold to Marbach Stud in Germany.

vidual who is endeavoring to make his mark as a sire. Many people are infatuated with black, and H.H. Prince Mohamed Aly was one of them. He made a large figure of a horse out of black paper and placed one in the stall of each of his mares who were in foal, hoping that by looking at it they would be interested and he would get a black stallion. He did this for twenty years, and the last year when he decided to disband the stud, he got a black colt out of Negma by Ibn Rabdan (who sired other blacks for the R.A.S./E.A.O.). Whether it turned into anything decent, he did not mention, but there are no black sires of merit that have been used by the Egyptians in Egypt to this day. Unfortunately high-quality, well-conformed blacks are as rare now as they were in days of yore.

COMPLETING THE PICTURE

Using Mendelism as a tool in selective breeding, and holding firmly in mind one's own vision of the result to be achieved, a breeder is using the law of universal attraction while matching and balancing genes - not merely mating two horses. Using these factors, the breeder sculpts his block of marble into a masterpiece, or a grotesque shape, for in each individual mating a breeder has two tools with which to create: he must choose and utilize two individuals only - the stallion and the mare - as the carriers of the desired genes. It is

*"...holding firmly in mind one's own vision of the result to be achieved, ...one must mate two individuals whose germ plasms will include the genes one chooses to perpetuate." *Ansata Ibn Halima, left and *Ansata Bint Mabrouka, right when chosen in Egypt as yearlings in 1959 as Ansata's foundation.*

not possible to choose genes for one kind of head from one horse, genes for a particular kind of hindquarter from another, genes for a correct shoulder from a third, and genes for a particular shape of legs or neck from a fourth. One must mate two individuals whose germ plasms will include the genes one chooses to perpetuate. If genes from another individual need to be introduced, then it is necessary to wait for another generation. Here is where *patience* and *perseverance* come into the picture.

Mendelism enables one to evaluate an individual not for what it is, but for what it will produce, and it can, thereby, be understood why, out of two full sisters of equal quality, one sister may be a great asset to the breed and the other a dismal failure. However, before decisions can be made, foals must be born. One can no more tell merely by looking at an individual what it will produce than one can determine by looking at a deer how high it can jump. Proof of the pudding is in the progeny. Generally speaking, the genotype (genetic makeup) to a large degree determines the phenotype (outward appearance).

The individual stallion or mare who is prepotent to a high degree of excellence is the greatest asset any breeder can possess. From such prepotent individuals, particularly the mares, a type or strain can be developed and maintained. Dominant mares can produce only a limited number of progeny, but dominant stallions can beget hundreds of foals if given the chance. Breeders recognize that these prepotent individuals exist, and such individuals are often the subject of much discussion. However, it is usually the astute, visionary, and dedicated breeders (large or small) who take advantage of them.

The greatest improvement in the form and qualities of Egyptian horses, or any other livestock, can only be made by those who possess the knowledge and skill to blend and perpetuate all of the desirable variations within the breed through a system of rigorous selection and close breeding, and by minimizing the accumulation of undesirable characters. By the ideals one formulates and by the proximity to the goal one achieves, the breeder brings dreams to fruition and expresses his or her personality. This self-expression is art and "realizing the dream."

"By the ideals one formulates and by the proximity to the goal one achieves, the breeder brings dreams to fruition and expresses his or her personality. This self-expression is art and 'realizing the dream.'" 1971 U.S. National Champion Stallion Ansata Ibn Sudan, the result of mating *Ansata Ibn Halima and *Ansata Bint Mabrouka. Sparagowski photo.

Seti I from the Temple of Karnak. Forbis photo.

CHAPTER III

Dreams of the Past
Realities for the Future

Nazeer (Mansour x Bint Samiha) in old age at the EAO. Forbis photo 1959. "Without knowledge of classic horses of the past, we cannot…come close to achieving our dream of breeding the perfect horse!"

CHAPTER III

Dreams of the Past
Realities for the Future

*The outer world of circumstance
shapes itself to the inner world of thought...
As the plant springs from, and could not be without, the seed,
so every act of man springs from the hidden seeds of thought,
and could not have appeared without them.
Act is the blossom of thought.*

James Allen, *As A Man Thinketh*

PORTRAIT OF A BREEDER

The saying that those who don't learn from history are doomed to repeat it may well be applied to the understanding of pedigrees. Without knowledge of classic horses of the past, we cannot realize the creation of superior bloodstock for the present or the future, nor can we come close to achieving our dream of breeding the perfect horse! Study of the past is an important aspect of a breeder's work as he or she molds and chisels thought before making breeding decisions that result in present realities. What then, might we ask, defines a breeder?

The Arabian Horse Registry of America defines a breeder as the owner of a mare at time of service. However, breeding one or two or a few horses does not necessarily qualify one to be known as a true Breeder; i.e., one of the backbones of the breed. As an official breeder of record, that person should possess the necessary competence and knowledge vital to protecting the breed for the future. A true Breeder is someone who:

Has spent years studying pedigrees and breed types;

Never stops learning about breed history and purpose;

Knows and understands the breed standard;

Is constantly on the lookout for inherent breed problems;

Freely recognizes both the positive and negative points of his or her own animals;

Has an open and objective mind toward competitors' animals;

Has established a breed line (a distinctive type within the breed) much like a master painter has developed a distinctive style that sets his work apart from others;

Has endured years of heartbreak and sacrifice...

All of the above, and more, for the love of the Arabian horse and the preservation of the oldest breed known to man.

In today's society where everything is desired instantaneously with an absolute minimum of blood, sweat, and tears, there is a general lack of commitment to excellence in anything. Settling for second best without endeavoring to learn otherwise is commonplace.

The horses a true Breeder tends and molds are living beings, not paper certificates, and they are predisposed to a variety of stresses just as humans are. The superb show horse, the ideal foal, the magnificent sire, the outstanding broodmare - any or all of which dies

General Tibor Pettkö von Szandtner with Moniet El Nefous.

prematurely (as in the case with the beautiful Hungarian mare, Ibn Galal I-7, "Gala" among others), has caused many a Breeder to wonder why they endure the pain and heartbreak. The ideal mating that did not take place has frustrated countless hopes and plans.

There is no short-cut to success or to becoming known as a Breeder in the Arabian horse world. Only knowledge and understanding combined with years of perseverance and constant study will guide *breeders* into becoming *Breeders* that have a lasting positive impact on upholding the traditional standard of the breed and the classic Arabian horse they have pledged to protect and maintain.

In speaking of a true Breeder, the late General Tibor Pettkö von Szandtner, former Director of the Babolna Stud, Hungary, is one who provides an inspirational example. Love of horse was his guiding principle, made obvious by what he accomplished in his short time at the Egyptian Agricultural Organization's El Zahraa Stud as manager from 1948-1959. He belonged to "the old order of things." Meticulous in detail, the farm and buildings were transformed and scrupulously maintained. He continued to build positively on the foundation of superior bloodstock he inherited from his dedicated predecessors, particularly Dr. Branch, creating a uniform and high-quality herd. He maintained handwritten herdbooks wherein he placed a conformation photograph of the horse, wrote its pedigree, strain and measurements, and concisely evaluated each individual from head to toe, including its gaits. These critiques illustrate the meticulousness with which he went about his task. They also set an example for others to follow and provide a brief insight into ancestral horses which breeders build upon today. Regrettably the complete herdbooks were not copied, but many of the horses that are influential in modern breeding programs are noted in this section.

In 1955 the E.A.O. and the Secretary of Agriculture requested a full accounting of his six years of activities, and he commented: "The stallions which were present in 1949, when I took over the management of the El Zahraa stud farm, were mostly old, partly very old. Many had hereditary faults and were, mostly, unfit for the standard which ought to be set for breeding stallions to raise the general level of horsebreeding in Egypt. During the past six years I was able to eliminate about fifty percent of this faulty material and to replace it by young, new, better stallions of known origin." Is it any wonder that he despaired when he heard Bint Mabrouka, Bint Zaafarana and Ibn Halima had left Egypt? Not to mention the horses that subsequently followed with the Gleannloch and other purchases.

EVALUATING ANCESTRAL HORSES

Authoritative evaluations of ancestral stock are of value if one bears in mind the age and condition of the horse at the time the evaluation is made, that certain eyes are more critical as well as educated than others, that some of these assessments represent absolute facts as well as subjective comments because beauty is in the eye of the beholder. One finds common threads among them, as well as diversity of opinion. The following is a random sampling of evaluations by various authorities; however, some photos of early foundation horses appear with little or no commentary and are included for

One of Von Szandtner's handwritten herdbooks. Sid Abouhom pictured.

reference purposes. It is impossible to list or critique all the foundation stock, but additional information and comments, particularly on the desert-bred stallions such as El Deree, Nabras, Mashaan, El Nasser, Besheir El Ashkar, etc., as well as desert-bred gift mares incorporated in the R.A.S./E.A.O./Inshass programs can be found in *Authentic Arabian Bloodstock* (pp. 133-137), and in other chapters. While more recent E.A.O. horses could have been critiqued, the breeding patterns differed under succeeding management as did the thinking behind the breeding program. Those presented here were the core foundation of modern Egyptians and are of specific value to study even though they are some generations removed.

The majority of the critiques in this section were made by General Von Szandtner, having been hand copied during the 1960's at El Zahraa by the author from his personal herdbooks handwritten in German or Hungarian (there were no photocopy machines at that time!). Some of the handwriting was difficult to read, let alone to copy and decipher, and although these copied notes have been translated by breeding authorities familiar with these languages, and every effort to be accurate has been made, there is some room for other interpretations. Von Szandtner had a consistent pattern to his evaluation of each horse. Some individuals, particularly foals, were appraised without measurements and recorded in the Hungarian language. Mature stock was measured and recorded in German. It is interesting to see his early critiques of young horses that eventually came to America or went to other countries and who made their mark as superior breeding and/or show stock (*Ansata Ibn Halima, *Morafic, *Bint Moniet el Nefous, etc.). Many breeders are still alive who saw these imports of the 50's through 80's and they may have differing opinions having seen the horses at maturity and under totally different environmental conditions (better feed, corrective farrier work, skilled grooming, etc.).

Dr. Mohamed Marsafi's comments were given by him to Bentwood Farms for use in one of their seminar workbooks. As the El Zahraa farm veterinarian under Dr. Ashoub and Von Szandtner, before he was promoted to Director in 1959, he had personally witnessed many of the foundation horses in the 40's and 50's. His comments were made from memory and not from written notes. The remarks by Dr. Ameen Zaher, consultant to the E.A.O. and author of *Arabian Horse Breeding and the Arabians of America*, were spoken to me, which I wrote down during visits to Egypt in the early 60's.

Dr. Ameen Zaher

My personal observations were made after the horses Von Szandtner had evaluated had matured or become aged. Those and comments by other authorities are noted where they add some worthwhile reference to the picture. Some pertinent comments have been quoted from Jack Humphrey's letters to W.R. Brown in 1932, when he selected Brown's stock to be imported to the United States. Especially noteworthy are his statements that ring true in relation to descendants of horses bred by H.H. Prince Mohamed Aly, H.H. Prince

Left, Prince Mohamed Aly in the garden at Manial. Above, *Nasr (Rabdan El Azrak x Bint Yamama) as a young horse at Manial.

Kamel El Dine, and Dr. Branch at the Royal Agricultural Society (although Branch was also a consultant to the other two). Humphrey said, "In speaking of the above three Studs of Arab horses, one might say that all three Studs represent true Arab type and purity of blood. Prince Mohamed Aly gives practically all thought to head; Prince Kamel El Dine seems to have depth of body and heavy quarter muscling strongly in his eye; while Dr. Branch of the Agricultural Society comes near trying to get a horse of all-round balance and excellence." Humphrey's evaluations of horses at that time, and their breeding ability were "right on."

After reviewing all of these photographs and commentaries, one realizes that a Breeder's task is formidable and far more difficult than painting a picture or molding viable clay into a substantial form. We can also observe from these assessments and photographs, and by studying their pedigrees, understand how far modern day Egyptian horses have, or have not, progressed in type, conformation and quality from the early stock and whether certain families have risen to a higher level, or, where and why they may have fallen.

For those who have not yet taken up the challenge of breeding, as well as for those who have already begun - fear not to seek advice, and do not despair at criticism. As Cardinal Newman once said: "Nothing would be done at all if a man waited until he could do it so well that no one could find fault with it." Furthermore, the perfect horse has yet to be born, and who can proclaim a "perfect work of art." One can take inspiration from Carl Sandburg's wisdom: "A satisfied flower is one whose petals are about to fall. The most beautiful rose is one hardly more than a bud wherein the pangs and ecstacies of desire are working for larger and finer growth."

Dr. Mohammed Marsafi with Judith Forbis at the E.A.O.

R.A.S./E.A.O. ANCESTRAL STALLIONS AND MARES

DAHMA SHAHWANIA MARES
Obeya/Sabah Family

Bint Sabah (Kazmeen x Sabah)
<u>Von Szandtner</u>: Moderately wide and deep, left front turned out, sickle-hocked. {This assessment must have been made when she was past 24 years old, ed.}

Bukra (Shahloul x Bint Sabah)
<u>Von Szandtner</u>: Measurements: 155/163/17.5. Fleabitten grey. Very noble. Very noble but a little bit long head, good large eyes, ears well set, but moderately long neck, good withers, very good back, loins and croup, nicely high set on tail, deep and wide, good strong legs, good strong hocks, correct posture, good gaits.
<u>Forbis</u>: A very special and noble mare which Von Szandtner termed *"prima stute."*

Bint Bukra (Nazeer x Bukra) {Full sister to "Hosnia" *Ansata Bint Bukra}
<u>Von Szandtner</u>: Grey. Very noble, noble head, well set on but slightly long ears, rather large lively eyes, well set on arched neck, good withers, slightly forward stretching back, moderately wide and deep, slightly long legs, a bit tied in, slightly toes turned out {front}, good movement.
<u>Forbis</u>: A very attractive mare, shorter-headed than her dam or her full sister, nevertheless she did not have their extreme refinement or quality.

Komeira (Nabras x Layla)
<u>Von Szandtner</u>: Measurements: 161/173/19 Grey. Moderately noble head, short neck, good topline, sufficiently wide and deep, tied in at the knees, left front moderately turned out, sickle-hocked.

Kamar (Nazeer x Komeira)
<u>Von Szandtner</u>: Grey. Very noble. Very noble head, lively beautiful eyes, well set ears, well set but somewhat upside down neck, good withers, back, loins and croup, high set on tail, normally wide and deep, weak foundation {bones}, long cannons, tied in, somewhat steep pasterns, somewhat steep hooves, somewhat toes turned out, good movement.
<u>Forbis</u>: At maturity, an extremely refined head and muzzle, very refined overall. Typical of the earlier assessment, but definite peaked somewhat short croup which often goes with an extreme head.

Bint Obeya (El Halabi x Obeya x El Dahma)

Fardous (Kazmeen x Bint Obeya) dam of Enzahi

Sabah (Mabrouk Manial x Bint Obeya)

Bint Sabah (Kazmeen x Sabah)

Bukra (Shahloul x Bint Sabah)

Bint Bukra (Nazeer x Bukra)

Layla (Ibn Rabdan x Bint Sabah) dam of Komeira and Sid Abouhom.

Kamar / Kamr (Nazeer x Komeira x Layla x Bint Sabah)

DAHMAN SHAHWAN STALLIONS
Obeya/Sabah Family

Awad (Mabrouk Manial x Bint Obeya)
Humphrey: Dr. Branch described as of "Bedouin type". They used him considerably as a sire and he has bred much better than he looks. As an individual, he is a heavy-fronted horse, tending toward Morgan type. He is rather plain throughout and has considerable hair on his legs. He is a horse with one testicle.
Marsafi: A very good stallion, high spirited, good, well-arched reasonably long neck, strong bodied, high tail carriage.
Forbis: Photographs indicate Awad became low in the back as he aged.

Sid Abouhom (El Deree x Layla)
VonSzandtner: Measurements: 156/169/20¼ Grey. Noble, noble head, good expressive eyes, very well set on small ears, very good set on nice long neck, good withers, forward stretching soft back, moderately strong loins, good croup, high set on tail, medium heart region, moderately deep, sufficiently wide, cannons moderately tied in, very little walking on toe on the right front foot, moderately cut out hocks, very good movements, medium drawn in at knee, very little turned to the right hocks, very good gaits, was a very good race horse.
Marsafi: Good size and strength, long neck, not as long-backed as Shahloul. Height 15 hands compared to average size of 14.2; not as good a head as Shahloul; not very high tail carriage; very good in racing. He bred true and produced much better foals bred to Moniet El Nefous than Nazeer.
Forbis: Marsafi's comment above is questionable when one considers *Bint Moniet El Nefous, Maya, *Fakher el Din, who were comparable in quality and better conformed than those by Sid Abouhom; however, he may not have considered them as influential as breeding stock. We found Sid Abouhom to be a handsome very aristocratic horse even though he was in an emaciated state when we saw him just before he died. What impressed us most were his big dark expressive eyes in a handsome classic head with finely chiseled small ears, as well as his fine dry bone, prominent withers and extreme depth and layback of shoulder, He appeared well-balanced even though gaunt. He was more "forehand" and a bit light in the rear (perhaps appearing that way due to age) with tendency to a lower croup, which most of his get had in comparison to those by Nazeer. Sid Abouhom's dam, Layla, was also a maternal half-sister to Sheikh El Arab (x Bint Sabah), another point most likely considered by the General in selecting him.

Above left Awad (Mabrouk Manial x Bint Obeya) Raswan photo. Above right Awad, Humphrey photo courtesy Carol Lyons.

Above and below, Sid Abouhom (El Deree x Layla)

Sheikh El Arab: (Mansour x Bint Sabah)
Von Szandtner: (quoted from his friend Laslo Monostory in Arabian Horse World June 1980): In Szandtner's opinion, the outstanding premier stallion was Sheikh El Arab, born in 1933; he was the most prepotent. {His sire, Mansour, was also the sire of Nazeer, a point obviously considered by the General in choosing him as premier stallion and Sheikh El Arab's successor, ed.}
Marsafi: More elegant than Nazeer; Nazeer's head was more feminine. Sheikh El Arab had the spirit of a stallion. Height 14.3 compared to the average height of 14.2 at that time. Nazeer's head was smaller than Sheikh El Arab's, but Sheikh El Arab had more width between the eyes. Nazeer had shorter ears. Sheikh El Arab had more arch in the neck than either Shahloul or Nazeer. His neck was in harmony with his body; he had good sound legs, level croup, very good tail carriage.
Zaher: Wonderful horse and excellent type. Very good bones, a bit rectangular in shape, but excellent characteristics.

Gazal I : (Nazeer x Bukra)
{Sold to Yemen, 1960, ed.}
Von Szandtner: Grey. Very noble, noble head, beautiful large eyes, well set ears, well set arched neck, good withers, somewhat slightly forward-stretching back, excellent loins, good croup, tail set on high, rather wide but less deep, a bit long legs, toes turned in, good hocks, slightly toes out, its posture, gaits are good. Will be for sale or stay here for breeding.
Forbis: A very classic stallion much as described above, he appeared to be finer headed and better conformed though higher-legged than his full brother, Ghazal, who went to Germany.

DAHMA SHAHWANIA MARES
Farida Family

Bint Farida (Mansour x Farida)
Von Szandtner: Measurements 153/165/18.5. Grey. Noble, sufficiently noble head, good withers, soft back, bad weak loins, sufficiently wide and deep, moderately tied in, moderately toed in, angled hocks.

Halima (Sheikh El Arab x Ragia)
Von Szandtner: Bay. Measurements: 153/ 163/17.5. Noble. Noble head, good lively eyes, well set ears, well set neck, good withers, good back, weak loins, good croup, good high set on tail, wide and deep, angled hocks, good gaits.
Forbis: An attractive mare, not exotic, she was well-balanced typical of the above assessment. She was a superb foundation mare; her three progeny that we saw by Nazeer were better than herself. (Ibn Halima, Gazi and Nawarra), and through her daughter Mohebba (by Sid Abouhom) she founded the Marbach line of distinction.

Above and below left, Sheikh El Arab (Mansour x Bint Sabah)

Above Ghazal I (Nazeer x Bukra)

Above left Bint Farida (Mansour x Farida) R.A.S. photo when young, above right, Bint Farida, photo in old age from the Von Szandtner herdbook.

Above left, Ragia (Ibn Rabdan x Farida), above right, Halima (Sheikh El Arab x Ragia), photos from the Von Szandtner herdbook.

Helwa (Hamran II x Bint Farida)
Von Szandtner: Grey. Noble, very noble head, beautiful expressive eyes, well set on ears, nicely set on flowing neck, good withers, good back, weak loins, good croup. High set on tail. Sufficiently wide and deep, light bone, in front slightly toed in, tied in, sickle hocked and cow hocked, good ground-covering gait.
Forbis: When seen as an aged mare, a classic individual, though very thin, she was typical of the above description.

Nefisa (Balance x Helwa)
Von Szandtner: Measurements: 152/ 166/18. Grey. Sufficiently noble, a little bit small, noble head, well set on not very long neck, good withers, good back, moderately strong loins, good croup and high set on tail, wide and deep, slightly toed in, excellent hocks, front legs moderately massive and tied in, sufficiently good gaits.

*Bint Nefisa (El Sarie x Nefisa) Imported to U.S.A. by Pritzlaff
Von Szandtner: Bay. Rather noble. Rather noble head, lively big eyes, well set ears, well set on neck, good withers, slightly forward stretching back, good loins, and croup, middle high set on tail, not very wide, rather deep, good foundation, somewhat long pasterns (hind right pastern swollen), gaits and posture correct, good movement.
Forbis: The most memorable thing that stands out in recollecting *Bint Nefisa were her wonderful big, dark and intelligent eyes. She was quite typical of the above description, but the eyes were unforgettable.

Faysa II (Sid Abouhom x Nefisa)
Von Szandtner: Grey. Noble, sufficiently noble head, moderately big but lively eyes, well set ears, good moderately long neck, good withers, moderately low back, good loins, good croup, tail set a little bit low, sufficiently wide and deep, sickle-hocked, moderately cut out under knee, correct posture, very good movement.

Kismat (El Sarie x Dahma II) Imported to U.S.A by Pritzlaff as *Bint Dahma)
Von Szandtner: Chestnut. Rather noble, rather noble head, lively big eyes, well set on ears, well set on but slightly ewe neck, good withers, forward stretching back, somewhat slanted croup, middle high set on tail, moderately wide and adequately deep, tied in, slightly sickle-hocked, correct posture and gaits, good movement.
Forbis: At maturity she remained much the same. An exceptionally well-bred mare, she was quite prolific in founding her own family line.

Above left, Helwa (Hamran II x Bint Farida) from the Von Szandtner herdbook. Above right is Nefisa (Nazeer x Helwa), Pritzlaff photo.

*Bint Nefisa (El Sareei x Nefisa)

Fayza II (Sid Abouhom x Nefisa) from the Von Szandtner herdbook

Above left, Dahma II from the Von Szandtner herdbook, above right Kismat/*Bint Dahma (El Sareei x Dahma II)

Gazala (Mashhour x Bint Farida) Imported to U.S.A. by Gleannloch as *Ghazalahh

Von Szandtner: Grey. Noble, large lively eyes, good neck, withers, back, loins, croup, tail, wide and deep, good legs, slightly tied in, slightly sickle-hocked, cow- hocks, posture and gaits are correct. Excellent movement.

Forbis: At maturity she was the epitome of the classic Arabian mare. She was imported by Gleannloch as an aged mare and admired by all who saw her.

*Above left *Ghazalahh (Mashhour x Bint Farida) Forbis photo, above right, *Ghazalahh, photo from the Von Szandtner herdbook.*

DAHMAN SHAHWAN STALLIONS
Farida Family

Balance (Ibn Samhan x Farida)

Von Szandtner: quoted from his friend Laslo Monostory in Arabian Horse World June 1980. "An excellent former race horse. Grey, with good bone structure, long lines, but not too typey, therefore Szandtner gave him only a few mares. He did not consider him a premier stallion. Balance was an old stallion, born in 1927, and Szandtner kept him in the stud farm."

Zaher: Nice neck, good bones and back, but a poor head, though he sired good heads - generally. A horse of good stamina. Most photos of Balance were made when he was in racing trim.

Forbis: Balance inherited some of the plainness of head from his sire, Ibn Samhan, which he passed on with regularity to his get. He sired great race horses, and himself was of racy build and a track record holder for the mile. His daughters, however, particularly Zaafarana and Nefisa, produced some exquisite offspring.

At left and above, Balance (Ibn Samhan x Farida)

Ibn Halima (Nazeer x Halima) - imported to U.S.A. as *Ansata Ibn Halima

Von Szandtner: Grey. Noble, very noble head, lively big eyes, ears well proportioned and well set on. Somewhat short in the neck, good withers, somewhat forward stretching back, good loins, good croup, high set tail, sufficiently wide and deep, somewhat tied in, small degree sickle-hocked, left front to a small degree toes in; gaits and posture good.

Forbis: This evaluation was made when he was a foal. Further comments as he matured are noted in Chapter IV, and his life is chronicled in the book, *Ansata Ibn Halima, The Gift*.

*Above left, Ibn Halima/*Ansata Ibn Halima (Nazeer x Halima) as a weanling in Egypt. Above right, *Ansata Ibn Halima in maturity.*

Marei (Sheikh El Arab x Ragia) full brother to Halima. Von Szandtner herdbook photo.

*Ghazi (Nazeer x Halima) full brother to *Ansata Ibn Halima. Forbis photo.*

DAHMA SHAHWANIA MARES
Bint El Bahreyn/Durra Family

Bint Zareefa (Balance x Zareefa)
Marsafi: A very good mare.

Maisa (Shahloul x Zareefa)
Von Szandner: Measurements: 160/165/17.5. Fleabitten grey. Noble, nice noble head, good large beautiful eyes, well set ears, well set on flowing neck, sufficiently good back, moderate strong loins, good croup, high set on tail, sufficiently wide and deep, weak legs, cannons tied in under the knees, sickle-hocked, moderately toed in, very good ground covering gait.
Forbis: A particularly handsome and big mare she had a special quality about her and was one who always stood out among the herd.

Bint Zareefa (Balance x Zareefa) Von Szandtner herdbook photo

Maisa (Shahloul x Zareefa) Von Szandtner herdbook photo

Bint Maisa (Nazeer x Maisa) Imported to U.S.A. as *Bint Maisa El Saghira by Gleannloch
Von Szandtner: Bay. Noble, rather noble head, beautiful large lively eyes, well set but a bit long ears, well set neck, good withers, back and croup. Tail set on medium high. Adequately wide and deep. A bit long cannons, a bit cut in below the hock, gaits and posture correct, good movement.
Forbis: A very charismatic mare at maturity and outstanding performer. See Chapter VI. One of the mares that helped popularize Egyptian Arabian horses in America.

*Bint Maisa El Saghira (Nazeer x Maisa) in Egypt. Forbis photo

DAHMAN SHAHWAN STALLIONS
Bint El Bahreyn/Durra Family

El Sareei (Shahloul x Zareefa)
{also spelled as El Sarie}
Von Szandtner: Measurements: 156/170/18. Bay. Very noble, very noble head, lively large eyes, well set ears, good well-set neck, good withers, forward stretching soft back, good loins, good croup, high set tail, wide and deep, cannons tied in, pasterns a little bit long, sickle-hocked, very little walking on toes, sound sex organs, good movements.
Marsafi: Structurally better than Shahloul but not as good a head. Height about 15 hands.
Forbis: One of the most splendid and classic Arabians we had the pleasure to see. Beautiful head, well-balanced, prideful carriage, iridescent copper coat; a magnificent specimen of the breed. A good broodmare sire.

El Sareei (Shahloul x Zareefa) Von Szandtner herdbook photo

El Sareei (Shahloul x Zareefa) Forbis photo

Gasser (Kheir x Badia)
{also spelled as Gassir}

Von Szandtner: Measurements: 151/166/20. Fleabitten grey. Sufficiently noble, sufficiently noble head, moderately large but expressive eyes, well set ears, strong a little bit thick moderately long but well set neck, good withers, moderately soft back, good loins, good croup, tail set on high, wide and deep, good legs, good strong hocks, left front foot moderately turned out, sound sex organs, good ground covering gaits

Forbis: A handsome but very heavy stallion; prideful carriage, thick neck but well shaped, compact in body, straight in stifle and very uncomfortable when ridden under saddle. He was Dr. Marsafi's favorite stallion. He was not a premier breeding sire.

Gassir (Kheir x Badia) Von Szandtner herdbook photo

Gassir (Kheir x Badia) Forbis photo

SAKLAWIA JEDRANIA MARES
Radia "Ghadia" Family

Zaafarana (Balance x Samira)

Von Szandtner: Measurements: 155/165/17.5. Grey. Noble, noble head. large lively eyes, well set on ears, well set on but not very long neck, good withers, sufficiently good back, moderately strong loins, good nice croup, tail set on high, unequal hooves, a little bit weak legs, cannons tied in under the knee, wide and deep, moderately turned out hocks, very slightly toed in, good ground covering gait.

Forbis: At maturity, an attractive and refined mare, long rather ordinary head with good expressive eyes, long well-set neck, longlines, good tail carriage, excellent movement and prideful carriage.

Zaafarana (Balance x Samira) Von Szandtner herdbook photo

Zaafarana (Balance x Samira) Forbis photo

Bint Zaafarana (Nazeer x Zaafarana) - imported to U.S.A. as *Ansata Bint Zaafarana

Von Szandtner: Grey. Rather noble head, very nice big eyes, well set on ears, well set on arched neck, good withers, slightly forward stretching back, good loins, good croup, middle high set on tail, properly wide and deep, tied in, slightly sickle hocks, posture and gaits are correct, good movement!

Forbis: At maturity very much the same as the early evaluation with pretty head. A very strong mare with good size, much prettier and better conformed than her dam.

*Near right, *Ansata Bint Zaafarana as a yearling in Egypt. Far right, *Ansata Bint Zaafarana in maturity.*

*At right, Kawsar (Ibn Manial x Zamzam by Gamil III x Bint Radia) from the same dam line as Zaafarana, granddam of Gleannloch's imported mare *Hayam (Mashhour x Tahia by Gassir x Kawsar). Far right, Fadila (Sheikh El Arab x Atlus x Zamzam), dam of Naglaa who went to Germany. Von Szandtner herdbook photos.*

SAKLAWI JEDRAN STALLIONS
Radia Family

Shahloul (Ibn Rabdan x Bint Radia)
Von Szandtner: Measurements: 156/169/19 Grey. Noble, noble head, large eyes, well set ears, well set on neck; good withers, very forward stretching soft back, good loins, good croup with high set tail, moderately strong in the heart area, somewhat drawn up stomach, sufficiently wide, in front a little tied in under the knee and just a little calf-kneed, very strongly sickle-hocked and cow-hocked, good movements.
Marsafi: Shahloul had an elegant head, deep jowl with prominent cheek muscles and good dish, long neck nicely arched; right front leg toed out; comparatively long back, level croup and very high tail carriage, very spirited - the favorite stallion of the royal family, he was very intelligent and calm under saddle (the only time he could be given injections).
Zaher: Best eyes, head and neck which was very arched and lovely; thin skin; rather weak legs.

Hamdan (Ibn Rabdan x Bint Radia)
Zaher: Full brother to Shahloul. About 15 hands tall; generally better than Shahloul except in head, smoother bodied and taller than Shahloul. A true breeding horse.

Antar (Hamdan x Obeya of Inshass)
Forbis: A handsome bright red chestnut stallion of noble bearing, attractive head of medium length, rather high-placed medium eyes set in boney sockets, well set and well-shaped ears, well-shaped nicely arched rather heavy long-enough neck, good withers, somewhat long back which tended to softness as he aged, good croup and high set tail with good tail carriage, good legs and feet but heavy bones; when viewed from the front, his front legs attached to his body differently than other stallions, and one can see this trait in some of his get. Good gaits. He gave the impression of masculinity and power. His disposition was difficult. He was a sire of good daughters.

Shahloul (Ibn Rabdan x Bint Radia) Von Szandtner herdbook photo

Hamdan (Ibn Rabdan x Bint Radia)

Antar (Hamdan x Obeya INS) Forbis photo

*Fadell (Kheir x Bint Radia) imported to the U.S. in 1950 by Queen Mother of Egypt

SAKLAWIA JEDRANIA MARES
Roga El Beda Family

Khafifa (Ibn Samhan x Dalal)
<u>Marsafi</u>: Very much like Medallela; very nice head, excellent dish; good tail carriage.

Medallela (Awad x Khafifa)
<u>Von Szandtner</u>: Measurements: 157/172/17.5. Light bay. Noble, but a little bit bigger head, does not show the noble original Arab descendants, high withers, sufficiently good back, sufficiently wide, tied in, front left foot moderately turned out, very sickle-hocked and cow-hocked (one sees there is also Managie blood in the pedigree.)
<u>Marsafi</u>: Very nice head, excellent dish; very good tail carriage.

El Bataa (Sheikh El Arab x Medallela)
<u>Von Szandtner</u>: Light bay. Measurements: 157/172/17.5 Noble. Good noble head. good neck, very slightly resembling a ewe neck, soft back, loins not strong enough, high croup, tail could be set on higher, deep and wide, light bone, cannons slightly too long and tied in, angled hocks, correct posture, right fore hardly noticeably turned out.
<u>Forbis</u>: An attractive mare, she somewhat resembled Halima but was longer in head, neck and body structure and did not produce strong hindquarters in general.

Amal (Nazeer x El Bataa) - imported to U.S.A. as *Bint El Bataa by Pritzlaff
<u>Von Szandtner</u>: Black. Noble. Very noble head. Lively big eyes, well set on but slightly ..... ears, somewhat ewe neck, good withers, slightly forward stretching back, adequate loins, good croup, middle high set on tail, moderately wide and deep, light foundation (bones), slightly tied in, in front a bit longer pasterns, slightly sickle-hocked, posture and gaits correct, good movement. (It remained small).
<u>Forbis</u>: At maturity a very beautiful head but structurally weak with very poor hindquarters that were worse than her full sisters.

Moniet el Nefous (Shahloul x Wanisa)
<u>Von Szandtner</u>: Chestnut. Measurements: 156/173/18. Extremely noble. Very noble head, large lively eyes, well set on neck, good withers, good back, good loins, the croup just a little bit high, nicely high set tail, correct front feet, moderately cow-hocked, sufficiently wide and deep, correct posture, very good outstretching gait.
<u>Forbis</u>: One of the most exquisite mares of her time. She was Von Szandtner's favorite mare. Her head was exceptionally refined with a teacup muzzle. Structurally she was fine-boned and a bit light in the hindquarters as an aged mare. Her story can be found in *Authentic Arabian Bloodstock*.

Khafifa (Ibn Samhan x Dalal)   *Medallela (Awad x Khafifa)*

*Above, El Bataa (Sheikh x El Arab x Medallela) Von Szandtner herdbook photo. At right, Amal/*Bint El Bataa (Nazeer x El Bataa).*

Moniet El Nefous (Shahloul x Wanisa) in old age. Forbis photo.

*Bint Moniet El Nefous (Nazeer x Moniet el Nefous) - imported to U.S.A. by Pritzlaff
<u>Von Szandtner</u>: Chestnut. Noble. Noble head, very beautiful big eyes, well set and well proportioned ears, well set on neck, good withers, back and loins, nice good croup, highly set on tail, adequately wide and deep, slightly long tied in cannons, slightly sickle-hocked, longer pasterns, quite a bit toes turned out, good movement.
<u>Forbis</u>: At maturity she much resembled her dam, but better in conformation, and she was the finest of the Pritzlaff imports.

Mona (Sid Abouhom x Moniet El Nefous)
<u>Von Szandtner</u>: Chestnut. Very noble, noble head, big lively eyes, well set and proportioned ears, well set neck, good withers, good back, sufficient loins, good croup, middle high set tail, adequately wide and deep, good foundation, somewhat tied in, somewhat sickle hocked, front leg slightly turned out, good movement.
<u>Forbis</u>: At maturity extremely beautiful head, neck a bit cut in in front of withers, long hip but somewhat rafter-hipped. The prettiest head of the three full sisters but poorest conformation.

Mabrouka (Sid Abouhom x Moniet El Nefous)
<u>Von Szandtner</u>: Chestnut. Noble, rather noble head, rather well set neck, good withers, good back, loins, croup, tail set on medium high, moderately wide but deep, a bit tied in below the knee, a bit cut in below the hock, cow hocks, good posture, good gaits.
<u>Forbis</u>: At maturity, head not as beautiful as Mona, but much better than Lubna; long neck, long adequately laid back shoulders, somewhat short croup and a bit straight behind, but strong. Very charismatic and a good mover.

Lubna (Sid Abouhom x Moniet El Nefous)
<u>Von Szandtner</u>: Grey. Noble, noble head, lively eyes, well set ears, well set neck, good withers, good back, adequate loins, good croup, tail set on high, wide and deep, tied in sickle hocks, correct posture, very good movement.
<u>Forbis</u>: At maturity rather plain head and somewhat small eyes, good bodied, not as attractive as her sisters.

Bint Mabrouka (Nazeer x Mabrouka)
Imported to U.S.A. as *Ansata Bint Mabrouka
<u>Von Szandtner</u>: Grey. Noble, very noble, small head, lively very big eyes, well set on ears, well set on neck, good withers, slightly forward stretching back, good loins, beautiful straight croup, high set on tail, wide and deep, slightly tied in, moderately sickle hocked, posture and gaits are correct.
<u>Forbis</u>: At maturity she remained true to this evaluation - but with added pride, carriage and a unique character.

Above left, *Bint Moniet El Nefous as a yearling in Egypt (Nazeer x Moniet El Nefous), above right, *Bint Moniet El Nefous at maturity.

Mona/Mouna (Sid Abouhom x Moniet El Nefous)   **Bint Mona (Nazeer x Mona)*

Mabrouka (Sid Abouhom x Moniet El Nefous)   *Lubna (Sid Abouhom x Moniet El Nefous)*

*Ansata Bint Mabrouka (Nazeer x Mabrouka) as a yearling in Egypt.

SAKLAWI JEDRAN STALLIONS
Roga El Beda Family

Morafic (Nazeer x Mabrouka) - Imported to U.S.A. by Gleannloch
Von Szandtner: Grey. Very noble, noble head, extremely lively big eyes, well set on ears, well set on neck, good withers, back, loins and croup, high set tail, adequately wide and deep, tied in, slightly sickle-hocked, bone spavin, quite a bit toes turned out. Good movement. It remains in the stud as a future breeding stallion!
Forbis: Much has been written about this stallion; he was unique and of special quality - a true Saklawi in type and spirit. His head has never been duplicated as of this writing - and probably will remain unique in Arabian annals. (For his story refer to *Authentic Arabian Bloodstock*.)

Above *Morafic (Nazeer x Mabrouka) in Egypt.

Samhan in old age (Rabdan El Azrak x Om Dalal) of interest because he is a full brother to Dalal I, the dam line of Moniet El Nefous and hence *Morafic. Samhan also sired Ibn Samhan (sire of Balance, Khafifa, Kheir, *Zarife and *Bint Saada), Nadra El Saghira (dam of Farida) and Samiha (dam of Bint Samiha).

SAKLAWI SHIEFI STALLIONS

Kheir (Ibn Samhan x Badouia d.b. {nee Noura of Emir Ali El Hussein of Mecca}
Von Szandtner: Measurements: 155/166/20. Grey. Sufficiently noble, sufficiently noble head, moderately large but expressive eyes, well set ears, strong a little bit thick moderately long but well set neck, good withers, moderately soft back, good loins, good croup, tail set on high, wide and deep, good legs, good strong hocks, left front foot moderately turned out, sound sex organs, good ground covering gaits.
Marsafi: A very good stallion, and much better than El Deree. Thick head, and an average head when compared to Shahloul and Ibn Rabdan. Strong neck, strong overall, very high tail carriage.

El Deree D.B. a Saklawi Shaifi stallion used by the R.A.S.

Kheir (Ibn Samhan x Badouia d.b.) as a young horse.

SAKLAWI STALLIONS /Inshass (DB)
(Said to be of Saklawi strain by Raswan, but there is no strain mentioned in the original Inshass Herdbook.)

Sameh (El Moez x Samira of Inshass)
Grey Stallion
Forbis: The author noted no reference in Von Szandtner's herdbooks but Sameh, like Antar, was chosen by him from the Inshass Herd for retention by the R.A.S./E.A.O. A handsome masculine-type grey stallion of noble bearing, he had an attractive head of medium length, rather average eyes, well set and well-shaped ears, well set on and long enough but slightly heavy neck, good withers, good topline, good croup and hip, and high-set well carried tail. He was wide and deep, had good legs, front slightly toed in, and hocks stood out slightly behind. His legs were extremely fine-boned for the substance of his body. A well-balanced horse overall, he had good gaits and sired excellent performance as well as halter horses. He was especially good at correcting faulty toplines which many of the E.A.O. mares had. He was a sire of superior mares, many of whom became champions and U.S. National winners in America, including the beloved U.S. National Champion Mare, *Serenity Sonbolah. He was a favorite stallion of many American visitors to El Zahraa.

Sameh (El Moez x Samira INS)

HADBAN ENZAHI STALLIONS
Gamila Family

Ibn Rabdan (Rabdan El Azrak x Bint Gamila)
Humphrey: He is a dark chestnut of at least 15.1, rather straight in face line and lacking in depth of jaw. Otherwise, almost perfect. We saw many of his progeny, and he turns them all out either chestnut or dark bay, so dark in some cases as to be almost black. He uniformly transmits more or less of the white strip in face which he has himself. Up to this time he has sired not more than three or four grey horses, in spite of the fact that most of the mares he has mated with are grey.
Marsafi: Very beautiful, very strong, and very high spirited. Height 14.2 hands. Shorter neck than Shahloul. Perfect sound legs. Very good croup and hip. Good tail, but not as high as Shahloul, Sheikh El Arab or Mashhour. Died at age 28.
Zaher: Ibn Rabdan was famous as an all-around good Arabian. A bit longer-backed than Mansour. Slightly large in head. Excellent shoulders and powerful rear. Hocks a bit behind (and this was not generally inherited in his progeny); when viewed from behind he was straight. A horse with great presence, he was not photogenic and few of his photos did him justice. As a sire he was very good.

Above left, above right and at right, Ibn Rabdan (Rabdan El Azrak x Bint Gamila)

HADBA ENZAHI MARES
Bint Rustem Family

Bint Rustem (Rustem x Bint Hadba El Saghira)
Humphrey: ...is coming seven years old, dark bay, plain head in spite of depth of jaw, good withers, medium shoulder and cut in below the knee. I might add here as a general remark that the bulk of the horses at Prince Mohamed Aly's and the Agricultural Society are excellent in shoulders and bone. When one visits the Stud of Prince Kamel El Dine, however, and sees the results of extreme use of Blunt blood, principally Rustem, one is impressed by the fact that this blood has given quality, heavy hindquarter muscling but has straightened shoulders and is almost uniformly cut-in below the knee.

Hind {aka Bint Bint Rustem} (Ibn Rabdan x Bint Rustem)
Humphrey: Bint Bint Rustem is the same type as her mother, but smaller. In action she has a very high show trot, almost hackney, which personally I do not see where she gets.

Above is Bint Rustem (Rustem x Bint Hadba El Saghira).

At right is Hind (Ibn Rabdan x Bint Rustem)

Salwa (Ibn Rabdan x Bint Rustem)
Von Szandtner: Measurements: 157/166/17.5. Moderately noble, sufficiently noble head, moderately big eyes, well set lively ears, well set flat neck, good withers, sufficiently good back, weak loins, short croup with sufficiently well set tail, weak legs, cannons tied in and moderately calf-kneed, front left moderately turned out, sickle-hocked and cow-hocked, good movements.

Latifa/Lateefa (Gamil III x Salwa)
Von Szandtner: Measurements: 152/162/17. Grey. Sufficiently noble, moderately noble head, neck a little bit short, good withers, good back, moderately good loins, short and moderately sloping croup, tail set a little bit low, sufficiently wide and deep, short cannons tied in at the knees, very sickle-hocked and cow-hocked, correct posture and good movements.

Rouda (Sheikh El Arab x Fasiha)
Von Szandtner: Measurements: 160/169/18. Grey. Moderately noble, moderately noble head, large eyes, slightly long ears. A little bit short but very flowing set on neck. Moderately weak back, weak loins, short croup, nicely high set tail. Wide and deep. Weak legs especially in front. Tied in and moderately calf-kneed, moderately sickle-hocked and cow-hocked, moderately walking on toes. Good elegant gaits.

Galila (Sid Abouhom x Rouda)
Von Szandtner: Grey. Noble but with moderately noble head, well set on ears, good large and fiery eyes, good neck, sufficiently good withers, good back, good loins, good croup, middle high tail, good breadth and depth, cannons short but tied in and moderately strong, moderately calf-kneed, very slightly sickle-hocked and cow-hocked, moderately long pasterns, very slightly toed in, very good gaits.
Forbis: At maturity, very much as above, somewhat plain head but beautiful big eyes, not a superior individual but a good broodmare.

Yosreia (Sheikh El Arab x Hind)
Von Szandtner: Measurements: 156/170/17.5. Grey. Very noble, very noble head, large eyes, well set on neck, good withers, good back, good loins, good croup, tail set on a little bit low, wide and deep, moderately sickle-hocked, correct posture, not much ground covering gait.
Forbis: At maturity, an attractive mare with a very classic head, light hindquarters somewhat peaky croup.

Salwa (Ibn Rabdan x Bint Rustem)

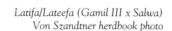

Latifa/Lateefa (Gamil III x Salwa)
Von Szandtner herdbook photo

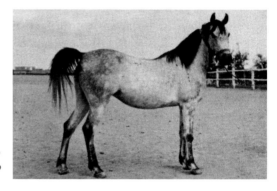

Rouda (Sheikh El Arab x Fasiha)
Von Szandtner herdbook photo

Galila (Sid Abouhom x Rouda)
Forbis photo

Yosreia (Sheikh El Arab x Hind)
Von Szandtner herdbook photo

Farasa (Sid Abouhom x Yosreia) {Farasha}
<u>Von Szandtner</u>: Measurements: 160/170/18. Grey. Sufficiently noble, noble head, well set ears, big lively eyes, well set neck, good withers, moderately weak back, weak loins, short high croup, well set tail, wide and deep, cannons moderately tied in at the knees, moderately sickle-hocked and cow-hocked, correct posture, good movements.
<u>Forbis</u>: At maturity, an attractive mare, similar to her dam in type but not as beautiful in the head, again, a bit light behind.

Frasha/Frashah
(Sid Abouhom x Yosreia)
Von Szandtner herdbook photo

Sheherezade (Nazeer x Yosreia) {Shahrzada}
{Full sister to Aswan, Bint Yosreia, ed.}
<u>Von Szandtner</u>: Grey. Very noble, very noble head, large lively eyes, well set on ears, and well set on arched neck, good withers, a bit forward-stretching back, rather good loins, a bit high croup, rather high set tail, adequately wide and deep, a bit tied in, toes turned out {front}, good movement.
<u>Forbis</u>: At maturity, one of the prettiest-headed and typiest mares at the E.A.O., a desirable combination of Nazeer on Sheikh El Arab daughters. Somewhat steep croup and light behind - the typical marriage of beautiful head and peaked croup.

Sheherezade/Shahrzada
(Nazeer x Yosreia)
Forbis photo

Mohgat (El Sareei x Yosreia)
{Mohga - black}
<u>Von Szandtner</u>: Rather noble, noble head, small but lively eyes, well set on ears, good but a little bit upside down neck, a little bit forward stretching back, good loins, a bit sloping quarters, tail set on medium high, moderately wide, rather deep, light foundation (bones), a bit tied-in below the knee and too long cannon, ... pasterns, sickle hocks, a bit cow hocks, posture and gaits are correct.
<u>Forbis</u>: At maturity, an extremely attractive black mare, much as described, but somewhat weak hindquarters, she was offered to us but we chose the two yearlings instead.

Mohga/Mohgat
(El Sarei x Yosreia)
Forbis photo

Bint Yosreia in old age at Hamdan Stables (Nazeer x Yosreia)
full sister to Aswan and Sheherezade. Forbis photo.

HADBAN ENZAHI STALLIONS
Bint Rustem Family

Mashhour (Shahloul x Bint Rustem)
Von Szandtner: Dark Bay. Measurements: 150/167/18.5. Sufficiently noble head, good expressive eyes, well set on ears, good but little short in the neck, good withers and back, good strong loins, good croup with sufficiently high set on tail, wide and deep. good legs, strong hocks, good posture but left fore slightly toed out. Sound sex organs. Good gaits.
Forbis: A handsome horse, an unusual mahogany bay, long and somewhat plain but pleasing head, he did not give us the impression of quality compared to others of his peer group, some of the old Rustem look coming through. He was not a premier breeding stallion but did sire such prominent daughters as *Ghazalahh, *Hayam, and *SF Bint Mamlouka.

*Ghalii (Nazeer x Galila) Imported to U.S.A. as a gift to The Secretary of Agriculture, who presented him to Michigan State University.
Von Szandtner: Grey. Rather noble. Rather noble head, well set on .... ears, lively big eyes, well set on neck, good withers, slightly forward stretching back, good loins and croup, high set on tail, adequately wide and deep, a bit tied in, sickle-hocked, slightly cow-hocked, right front leg a bit turned out, good movement.

*Mashhour
(Shahloul x Bint Rustem)
Von Szandtner herdbook
photo*

**Ghalii (Nazeer x Galila)
Forbis photo*

HADBA ENZAHI MARES
Samiha Family

Bint Samiha (Kazmeen x Samiha)
Von Szandtner: Shy because of age, very wonderful noble head, very low back due to age, particularly bad hocks, sickle-hocked and cow-hocked. {Laslo Monostory in Arabian Horse World 1980, quotes in Von Szandtner's opinion "Bint Samiha, Nazeer's dam, was the most prepotent broodmare." ed.}
Zaher: One of the most beautiful mares in Egypt. Compared favorably with Bint Dalal and Khafifa. Medium length head, good width between the eyes, but not much dish; large rounded eye.

Shams (Mashaan x Bint Samiha)
Von Szandtner: Measurements: 154/158/17.5 Very noble, especially noble head.

*At right, Bint Samiha
(Kazmeen x Samiha)*

Shams (Mashaan x Bint Samiha)

Maisoona (Kheir x Shams)
Von Szandtner: Measurements: 154/162/17.5. Grey. Very noble, noble head, moderately well set neck, beautiful lively eyes, good withers, sufficiently good back, moderately strong loins, croup a little bit high moderately sloping, middle-set tail, sufficiently wide and deep, cannons a little bit long and tied in at the knees, cow-hocked, correct posture, good movements.

Fatthaya (Sid Abouhom x Shams) {Fathia}
Von Szandtner: Grey. Noble, noble head, big lively eyes, well set ears, well set neck, good withers, good back and loins, good croup, high set on tail, sufficiently wide and deep, good legs, moderately cut out hocks, correct posture, good ground-covering movements.

*Above left Maisouna/Maisoona (Kheir x Shams) Von Szandtner herdbook photo, above right *Bint Maisouna in Egypt at the Police College, Forbis photo.*

Fatthaya/Fathia (Sid Abouhom x Shams) Von Szandtner herdbook photo

Foze (Nazeer x Fathia) Forbis photo

HADBAN ENZAHI STALLIONS
Samiha Family

Nazeer (Mansour x Bint Samiha)
Von Szandtner: Measurements: 152/164/19. Light white horse, very noble, very noble head, extremely big expressive eyes, well set lively ears, well set on sufficiently long neck, good withers, moderately soft back, good loins, nice croup with high set tail, moderately wide and deep, left front foot a little turned out, sickle-hocked, good movements.
Marsafi: A stallion that had to be respected, he had much presence and pride. Fair head, sound legs, thin gaskin and forearm. Height not quite 14.3 hands. Primarily used by General von Szandtner to improve structure.
Zaher: Nazeer got good bones from Mansour; head half from sire and half from dam. Good back from Mansour. Good legs which he transmitted to his get. Short straight pastern, not inherited. Szandtner wanted to work on correcting legs and searched for a stallion from which he could acquire the heritage of good legs. When stallions came back from the depots, Szandtner went to Bahteem {Dr. Zaher told us he went with him, ed.} and Szandtner picked Nazeer.
Forbis: An especially noble stallion, his get inherited his unique quality but he did not generally predominate in structure over the tail female mare line to which he was bred. Photos of this horse showing him with badly trimmed feet and no heels gave rise to some erroneous judgements about his legs. Von Szandtner used him to improve legs.

Nazeer as an old horse (Mansour x Bint Samiha)

KUHAYLAN MIMREH STALLIONS

Ibn Samhan (Samhan x Nafaa El Saghira)
<u>Zaher:</u> In response to Forbis question about Ibn Samhan's head: He did have a good head.
<u>Forbis:</u> It appears he contributed rather boney eye sockets imparting a somewhat unusual triangular shape over the eyes to his get, including his sons *Zarife, Kheir, Balance, and daughters *Bint Saada and Khafifa. This characteristic could be relatively persistent depending on whether it was addressed and selectively bred out.

Mansour (Gamil x Nafaa el Saghira)
<u>Humphrey:</u> I THINK that Nasr is better than Mansour, and I THINK that both the Prince and Dr. Branch value Mansour more as a sire. Seeing the foals of both, I THINK it a toss-up between them, but as you will have Nasr in future use, I thought it better to capture all the blood possible from Mansour...... From various things that Dr. Branch and the Prince have said, I know that they both feel that while this horse is slightly less good in conformation than Manial {*Nasr, ed.} he is a better sire. Mansour has a beautiful head, almost perfect according to their ideal, is sound and well built, with the exception of being slightly low in the back, a quality which he transmits to some of his progeny.
<u>Dr. A.E. Branch:</u> Humphrey, you have made me see several things and I can see my route before me is to use Mansour...
<u>Zaher:</u> Excellent hindquarters and exceptionally good bones. He did not have the finesse Ibn Rabdan had. Was somewhat rectangular in shape, did not have the extra long neck. He had a nice head, which was a type of itself. He had vigor, quality and strength, but lacked some of the light fine adjustments. Very good back. A very powerful horse.
<u>Forbis:</u> The doubling of Mansour through Nazeer and Sheikh El Arab proved a very prepotent combination and frequently resulted in exceptional heads.

**Mabrouk Manial
(Farhan "Saklawi II" x Tarfa)**
<u>Humphrey:</u> ...the sire of Mahrussa {Mahroussa}. He is a grand old horse of 20 years and still in service. His outstanding feature is his wonderful bone, which is still sound and shows considerable quality in spite of age.
<u>Forbis:</u> Mabrouk Manial figured significantly as a sire of superior broodmares.

Above left and right, Mansour (Gamil Manial x Nafaa El Saghira).

Above, Humphrey photo of Mansour showing head type, courtesy of Carol Lyons collection.

Above is Nafaa El Saghira (Meanagi Sebeli x Nafaa El Kebira) included for interest here since she is the dam of both Mansour and Ibn Samhan, prominent sires at the R.A.S.

At left Mabrouk Manial (Farhan/Saklawi II x Tarfa) as a young horse, photo from the H.H. Mohamed Aly scrapbook. Above Mabrouk Manial in old age, Humphrey photo courtesy Carol Lyons.

KUHAYLAN JELLABI STALLIONS

Kazmeen (Sotamm x Kasima)
{Kasmeyn/Kesym}
Humphrey: He is a bay horse of fine quality, descended from Blunt stock. In a general way, one can say that many of the Egyptian horses descended from Blunt stock have extreme quality combined with lightness of bone. This horse has a club foot, but has given some excellent results as a sire.
Marsafi: Extremely beautiful, very fine muzzle, with open wide nostrils. Long beautiful neck.
Zaher: Excellent swan neck and lovely head, but weak croup behind, weak back, and boxy shoes {club feet}.
Forbis: The line of Kazmeen tended to produce weak croups which carried forward, harking back to his dam's extreme weakness in that area. Kazmeen was extremely influential as a breeding stallion, thus this particular trait took time to breed away from. Again, the combination of beautiful head coupled with poor croup.

Above left Kazmeen/Kazmeyn (Sotamm x Kasima) at the RAS, and above right Kazmeen in age. (Jellabi not from Bint Yemama's line.)

Gamil III (Gamil Manial x Aroussa) - Prince Mohamed Aly Stock
Zaher: He was a bit coarse. The Jellabi line didn't stay long enough in the {R.A.S.} stud and the foundation stock was old and died before worthy progeny were dropped for retention.

Gamil III (Gamil Manial x Aroussa) sire of Zamzam and Lateefa. From the Bint Yemama line of Prince Mohamed Aly.

KUHAYLAN RODAN STALLIONS

Rustem (*Astraled x Ridaa) - old Blunt Stock
Humphrey: Bay, the old stud from the Blunts...at the present time he is very low in the back and has a big bone spavin. Dr. Branch had never noticed this before and said he thinks it is the result of old age and heavy use, for he is still the principal stallion in service.

*Rustem (*Astraled x Ridaa) as a young horse at Crabbet stud in England.*

Rustem at maturity.

KUHAYLA RODANIA MARES
Riyala Family

Malaka (Kheir x Bint Bint Riyala)
Von Szandtner: Measurements: 154/164/17.5. Grey. Noble. Noble head, good large eyes, well set on ears, moderately well set short somewhat thick neck, vague withers, moderately soft back, sufficiently strong loins, good croup, high set on tail, wide and deep, correct posture, moderately sickle hocked and cow-hocked, good gaits.
Forbis: At maturity a pretty mare very much as described above and founder of a prominent family.

Malaka (Kheir x Bint Bint Riyala). Von Szandtner herdbook photo.

Samia (Nazeer x Malaka)
<u>Von Szandtner</u>: Measurements: 152/160/18. Grey. Rather noble, rather noble head, well set neck, good withers, back, loins, croup, rather well set on tail. Moderately wide and deep, light foundation (bones), tied in, sickle hocks, cow hocks, a little bit toes turned out, good movement.
<u>Forbis</u>: As a mature mare Samia was a somewhat plain-headed individual and typical of the conformation description. She out-produced herself and founded a good family line.

Samia (Nazeer x Malaka)

KUHAYLA RODANIA MARES
Rissala Family

Kateefa (Shahloul x Bint Rissala)
<u>Von Szandtner:</u> Measurements 152/174/17.5
{no comments written down, ed.}

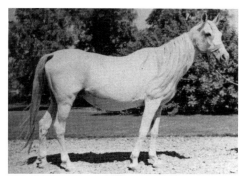

Kateefa (Shahloul x Bint Rissala)
Von Szandtner herdbook photo.

Bint Kateefa (Sid Abouhom x Kateefa)
<u>Von Szandtner:</u> Chestnut. Rather noble, rather noble head, moderately large eyes, well set on ears, slightly upside down neck, good withers, slightly forward-stretching back, good loins, croup, tail set on high, wide and deep, good legs, slightly sickle hocks, left foreleg slightly toes in, good movement.
<u>Forbis:</u> At maturity, a tall long-lined mare, somewhat plain in the head and typical of the above description, she founded a popular family line.

Yashmak (Sheikh El Arab x Bint Rissala)
<u>Von Szandtner:</u> Measurements: 158/172/18. Bay. Very noble, noble head, good large eyes, well set on ears, well set on flowing neck, good withers, moderately soft back, weak loins, good slightly high croup, good high set on tail, wide and deep, slightly weak legs, tied in, slightly long pasterns, sickle-hocked and cow-hocked, posture correct, very good gaits.
<u>Forbis:</u> At maturity a large bay mare with a long well-shaped head with huge black eyes, a long supple neck, long-bodied, short in croup, narrow in stifle. She was much improved in type over her dam. The Saudi Arabian princes during their visit to El Zahraa continually remarked over her size in comparison to many of their desert-bred mares in Arabia.

Bint Kateefa (Sid Abouhom x Kateefa)
Forbis photo.

Yashmak (Sheikh El Arab x Bint Rissala)

Rahma (Mashhour x Yashmak)
Von Szandtner herdbook photo

Om El Saad (Shahloul x Yashmak)
Von Szandtner: Measurements: 158/175/18. Grey. Sufficiently noble. Sufficiently noble head. Nice flowing set on neck, good withers, good back, good loins, good croup and nicely set on tail, sufficiently wide and deep, tied in, very slightly calf-kneed, moderately sickle-hocked and cow-hocked. Barely noticeably toed in. Very good ground-covering gaits.
Forbis: At maturity a big mare, somewhat plain headed, heavy-bodied and relatively long-lined, she was somewhat better conformed than her dam.

Om El Saad (Shahloul x Yashmak) Von Szandtner herdbook photo.

Rafika (Nazeer x Om El Saad)
Von Szandtner: Grey. Noble, noble head, large lively eyes, well set on ears, and well set on arched neck, good withers, slightly forward stretching back, good loins, good croup, tail set on high, wide and deep, slightly back at the knee, tied in, quite a bit sicklehocked, right front leg a little bit turned out, good movement.
Forbis: At maturity a tall mare, plain profile but prettier-headed more refined and more elegant than her mother, enhanced by the quality that Nazeer imparted to his get.

Rafica (Nazeer x Om El Saad) Forbis photo.

KUHAYLAN RODAN STALLIONS
Rissala Family

Alaa El Din (Nazeer x Kateefa)
Von Szandtner: Chestnut. Very noble, noble head, not big but lively eyes, very well set ears, good neck, withers, loins, croup, tail set on middle-high, adequately wide and deep, a bit tied in below the knee, cut in below the hock, movement and posture are regular. Good gaits. Remains a breeding stallion.
Forbis: A tall horse at maturity, very elegant and refined, beautiful head but not large enough eyes, which he tended to transmit. Good length of neck and well shaped. He was well-balanced but tied in at the elbows, narrow in front, had rather small but good dark hooves and he had a disturbing way of walking when viewed from behind, traits which he tended to pass on, especially if inbred.

*At left Alaa El Din (Nazeer x Kateefa) below, *Rashad Ibn Nazeer (Nazeer x Yashmak)*

Rashad (Nazeer x Yashmak) Imported to U.S.A. as *Rashad Ibn Nazeer by Pritzlaff
Von Szandtner: 1955 Bay. Noble, noble head, lively big eyes, well set on but somewhat large ears, well set on neck, rather good withers; only slightly forward stretching back, good loins, high set on tail, wide and deep, light foundation (bones), somewhat sicklehocked, posture is correct, good movement.
Forbis: As a mature horse he was plain-headed, although he had beautiful big dark expressive eyes. He was somewhat straight shouldered, short in croup, narrow in stifle like his dam, and tended to carry his tail to the side. He was trained successfully for dressage at Pritzlaff's ranch by Col. Handler, then Director of the Spanish Riding School.

Pictorial section of additional ancestors

Following is a random selection of ancestors that were not part of the previous reviews in this chapter. These also relate to horses appearing in this book and provide additional reference to comments in Chapter II, e.g., characteristics of strains/families, color relating to type, how photographic poses can be misleading (front versus side views), and how certain "looks" or type does or does not transmit, and through which lines. Through careful study, the serious student will see certain patterns beginning to emerge, despite the quality of these photographs, by comparing the individuals in these photographs with their more recent and better photographed, better posed descendants. *The Pyramid Society's Reference Handbooks* are a continuing source of information and photographs, the first volume published in 1973 and now going into the 10th volume. Other valued pictorial tools are the *Asil Arabian* books published by Olms Press and the Pyramid Society Europe's *Almanach*, both published in Germany. The Egyptian Arabian Horse Breeder's of Egypt are planning to publish reference handbooks to chronicle the Arabian horse in Egypt in this new millennium. Valuable reference photos may also be found in Nagel, Grasso, and other books noted in the bibliography.

DAHMAN SHAHWAN - Sabah family

Enzahi (Nabras x Fardous), was a successful race horse and used pre-Von Szandtner. He sired *Moftakhar (x Kateefa) and *Gamila (x Mamdouha), both imported to U.S.

Ghazal (Nazeer x Bukra). Full brother to *Ansata Bint Bukra. Bred during the Von Szandtner era, Ghazal was exported to Germany. His contribution there was invaluable through several fine daughters (e.g., Ghazala x Hanan, Malikah and Moheba II x Malacha, Ghazalah x Afifa). Regrettably he was never used to his fullest potential.

Saab (Antar x Kamar)

Wahag (Antar x Kamar)

Shadwan (Shaarawi x Kamar)

Akhtal (Amrulla x Hagir x Kamar)

Sons and grandsons of Kamar illustrate the influence she had on post-Von Szandtner breeding. **Saab** (exported to Europe) and **Wahag** were full brothers by Antar, bearing some of his heavier characteristics; **Akhtal** (by Amrulla) and **Shadwan** (by Shaarawi) were more refined with more scope. The latter three were influential at the E.A.O. in Egypt and in subsequent exported bloodstock. None reached superior sire status, however, some impressive and valuable individuals resulted from each of them.

DAHMAN SHAHWAN - Farida family

Inas (*Morafic x *Ghazalahh exported to Gleannloch) a tall good-bodied and well-balanced grey mare, she was the result of mating two of our favorite individuals that we requested Dr. Marsafi to try. It resulted that she was not comparable to her sire or dam in classic type. She produced Aseel (by Sameh), who was used as a sire by the E.A.O., and Bint Inas (by Gassir) who became an influential broodmare in Hungary and transmitted *Ghazalahh's quality.

Inas (*Morafic x *Ghazalahh)

***Deenaa** (Sameh x Dahma II) we selected for the Huebners who exported her to America as a filly. She was owned by various breeders concluding with the St. Clairs. Despite extremely offset knees, she was chosen for her beauty, balance and quality. She became an outstanding broodmare and dam of champions through her daughter, Bint Deena (by *Ansata Ibn Halima). Bint Deena produced well by Ruminaja Ali (e.g., Anaza El Farid, Anaza El Nisr) and was the dam of Anaza Bay Shahh (by Shaikh El Badi), purchased as a breeding stallion by Simeon Stud and exported to Australia.

*Deenaa as a yearling in Egypt (Sameh x Dahma II)

***Ramses Fayek** (Nazeer x Faysa) was bred during Von Szandtner's era. Like many stallions, he was sent to a stallion depot for use as a sire in that area. A classic horse of good balance quality, he was returned later, used briefly at the E.A.O. and Al Badeia Stud, and then imported to America by Martin Loeber where he became the main sire at his Plum Grove Farm in Illinois.

*Ramses Fayek in Egypt (Nazeer x Fayza)

DAHMAN SHAHWAN - Bint El Bahreyn family

Seyf (Mashhour x Elwya) {Seef}, was a half-brother to *Ansata Bint Elwya. A horse of good size and scope and structure of that family line; he was rather plain and long-headed when viewed from the profile, but pleasant from the front with nice black eyes, decent ears, good length of neck and substantial body with correct legs. Seyf appears in numerous E.A.O. pedigrees in Egypt, and through exported sons and daughters who have been frequent winners at halter and performance.

Above left and right, Seef (Seyf)

SAKLAWI JEDRAN

Emad *(El Arabi x Ebeda-Inshass), a black-bay stallion tracing to Radia through Inshass breeding, he was the only son of El Arabi. His sire bequeathed him some refinement over his black dam, but he inherited her plainness of face, somewhat shortness of leg, and length of body. A sire used quite frequently by the E.A.O., he gave black color. One of his prettiest daughters, *Bint Baheera, was exported to the U.S. and eventually purchased by Ansata.*

*Fakher El Din *(Nazeer x Moniet El Nefous), was a classic and well-balanced chestnut son of Egypt's most famous mare. He was imported to the U.S. by Sarah Loken. He stood at public stud and was a successful sire.*

Ameer *(Galal x Moniet El Nefous), born in 1970, a small but well-balanced chestnut stallion with white facial markings, he was Moniet's last foal and was used as a sire at the E.A.O. Neither he nor his half-brother, Soufian (by Alaa El Din) resembled her sons *Ibn Moniet El Nefous or *Fakher El Din, who were of a different style and class.*

Ghalion *(*Morafic x Lubna), a grey stallion the result of doubling the Moniet family, (Lubna being a full sister to *Morafic's dam) he was much improved over his dam in type and quality. He was exported to Hungary and used as a sire. His bloodlines are prominent in Europe.*

*Sultann *{Sultan} (Sameh x Lubna), chestnut stallion, was offered as a mature stallion to Dr. Wentzler of Marbach Stud, who took Gharib instead. He inherited smoothness of body from his sire but the plain profile and eye placement of his dam. He was the sire many champions including multi-halter and performance winner, U.S. National Champion Stallion, *Asadd (x Amani), and *Sakr (x Enayat), a multi-halter champion stallion and winner of U.S. National performance championships. *Sultann is represented back in the pedigree of many excellent individuals.*

Ibtisam *(Nazeer x Mouna), chestnut granddaughter of Moniet El Nefous, an attractive mare better in conformation but not as beautiful as her dam or granddam, her line carries on in Egypt through her beautiful daughter, Bint Ibtisam.*

Emad (El Arabi x Ebeda)

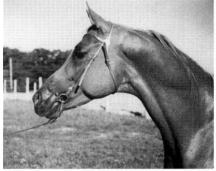
*Fakher El Din (Nazeer x Moniet El Nefous)

Above left and right, Ameer (Galal x Moniet El Nefous)

Ghalion (*Morafic x Lubna)

*Sultann (Sameh x Lubna)

Ibtisam (Nazeer x Mouna)

HADBAN ENZIHI

Hafeed Anter (Wahag by Antar x Basima by Alaa El Din), a small rather unrefined bay stallion who carried himself with pride, much like his paternal grandsire, Antar, but did not have his quality. However, he inherited some of his ancestor's best genes and was a perfect example of genotype not relating to his phenotype. He sired foals much better than himself. He figures prominently in the pedigrees of horses in Egypt as he was used frequently at E.A.O. and by private breeders.

Galal (Nazeer x Farasha), a well-balanced and attractive chestnut stallion, he had a considerable underbite which affected his profile. He was used after Von Szandtner left. He sired some handsome and very influential individuals, although he was not one of the best Nazeer sons. His blood figures prominently in Europe through Ibn Galal "Magdi" (x Mohga) and through use of Ibn Galal I (x Hanan) at Babolna.

Zebeda (El Sareei {El Sarie} x Galila), an attractive grey mare of the Von Szandtner era. When bred to Alaa El Din, the proven successful nick with El Sareei daughters, she produced Alifa, Rasheek and Zohair, and she is influential in many pedigrees worldwide.

Shaker El Masri (*Morafic x Zebeda), chestnut stallion, was an extremely classic and refined *Morafic son when we saw him in Egypt. He figures prominently in the "golden cross" of Egyptian/Spanish pedigrees but it is regrettable that he did not sire many straight Egyptian offspring either in Egypt or after he was exported to Europe.

Shaarawi (*Morafic x Bint Kamla), a grey stallion of impeccable pedigree, he was of good size and scope, head somewhat long and heavy foreface, good eyes and ears, nice length of neck, stronger body than his dam, but he did not inherit her refinement of bones nor his sire's quality. Nevertheless, "Blood will tell" and he was a valuable sire for the E.A.O., outproducing himself.

***Nabda** (Wahag x Neamat), grey mare, was an extremely beautiful individual foaled in 1974 and imported by Stanley White at the time of Don Ford's Lancer Arabians importations. She became influential in American programs particularly through her son, Makhsous.

Hafeed Anter (Wahag x Basima)

Galal (Nazeer x Farasha)

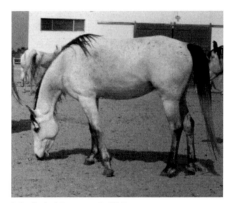
Zebeda (El Sareei x Galila)

Shaker El Masri (*Morafic x Zebeda)

Shaarawi (*Morafic x Bint Kamla)

*Nabda (Wahag x Neamat)

'ABEYYAN OM JURAYS

Ibn Hafiza *(Sameh x Hafiza)*, bay stallion, was at a stallion depot before he was imported by Gleannloch. A good-moving, handsome smooth-bodied horse with a pleasing though not exotic head typical of his sire's and better than his dam's, it was hoped he would cross well with *Morafic bloodlines and provide a look similar to El Arabi. He was bred frequently to non-Nazeer blood! He sired some prominent winners in America.

El Arabi *(*Morafic x Hafiza)*, bay stallion, was the most beautiful Egyptian stallion bred during Marsafi's time and one of the most classic stallions ever bred in Egypt. He was given a limited book of good and some not good mares his first season at stud, before he ran down the driveway into a bus and was killed. His line carried on through his son, Emad, and through his few daughters, one of which was the 1969 bay mare, *Habeebaa (x *Bint Hanaa) imported to U.S.A.

Bilal *(*Morfafic x Mona Inshass)*, was an exceptionally classic and refined tall white stallion, a perfect combination of these two bloodlines inheriting the best of both parents. He became the herd sire for the well-known Shams El Asil Stud in Egypt.

*Above and below El Arabi (*Morafic x Hafiza)*

Ibn Hafiza (Sameh x Hafiza)

*Above Bilal (*Morafic x Mona)*

KUYHAYLAN STRAINS

KUHAYLAN JELLABI (SAQLAWI JEDRAN?)
(Bint Yemama line of Prince Mohamed Aly)

Atfa *(*Morafic x Ghazala-Inshass)*, one of the few mares of this family from the old Ghazala Inshass, she was sold from the E.A.O. and acquired by the Sherei family, who later sold her back to the Marei's Al Badeia Stud.

KUHAYLAN KRUSH
(From Inshass El Kehila – 1921 desert-bred bay mare given to King Fouad by King Abdul Aziz Ibn Saud – considered to be a Kuhaylan strain because of her name and her daughter is entered in the Inshass herdbook as Kuhaylan Krush).

Shahbaa *(Hamdan x Shahd)*, one of the Inshass mares retained by Von Szandtner. An attractive fleabitten mare, she produced two good daughters by Gassir that have been influential in Egypt and other countries.

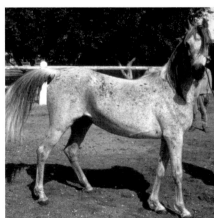
*At left, Atfa (*Morafic x Ghazala-Inshass), in Egypt the only remaining female line to Prince Mohamed Aly's Bint Yamama. Above, Shahbaa, the only female line of the Kuhaylan Krush.*

KUYHAYLAN STRAINS continued

KUHAYLAN RODAN

Amina {Ameena} (Hamdan x Yaman) grey mare retained from the Inshass herd, she was a long-lined refined and good size broodmatron of the Rissala family. She returned to the E.A.O. under Von Szandtner. She bred well with the *Morafic lines producing Enayat (dam of *Sakr), and *Omnia (by Alaa El Din), both owned by Gleannloch.

Ikhnatoon (*Farazdac x Bint Om El Saad), an impressive white stallion used heavily by the E.A.O. as a sire. He had a very refined head with somewhat high eye placement, long foreface nicely tapered and a refined muzzle; good mitbah and good length of neck; good depth of shoulder and high withers, soft back and rather long underline; decent croup and good tail set, good carriage and good movement. His bloodlines are in numerous pedigrees in Egypt and internationally.

Kayed (*Morafic x Kaydahom), grey stallion, remindful in structure and size of his dam, he was one of the *Morafic sons used in Egypt and a foundation sire for the Marei's Al Badeia Stud. A tall bold-moving horse, his blood is represented worldwide through many of the Marei exports.

*Lancer's Asmara (Seyf x Hebah by *Ibn Hafiza), was a substantial very attractive well-conformed Nazeera granddaughter. Imported by Lancer Arabians, she was shown to many halter championships including Regional and U.S. Top Ten, and she was many times a Western Pleasure champion. A fine and substantial broodmare, her bloodlines are influential in international breeding programs.

Nawaf (Amrulla x Nazeera), grey stallion, was an attractive son of Nazeera inheriting some of the good qualities from both his sire and dam. He was brought back from a stallion depot to be used at El Zahraa. He sired Enayah (x Adaweya) who was bred to Ikhnatoon and produced Adl (used as a stallion at the E.A.O. and by private breeders), and his full sister, *Nageia, imported to America and used in the Ansata program.

Ameena/Amina (Hamdan x Yaman)

Ikhnatoon (*Farazdac x Bint Om El Saad)

Kayed (*Morafic x Kaydahom)

Above, Nawaf (Amrulla x Nazeera). At right, *Lancers Asmara (Seef x Heba x Nazeera)

Ahmed Pasha Hamza with his beloved Hamdan

Having examined closely the material elements of certain individuals, it seems fitting to close this chapter by dedicating it to a horse whose life story, as told by Sarah Loken after she snatched him from the jaws of death, is a tribute to the spirit and the soul of the Arabian horse - those sensed but unseen attributes which do not appear in physical evaluations, but which have endeared this breed to mankind from time immemorial:

"Hamdan was to me magnificent and awesome, a creature of a moment in time in a land where time is measured in thousands of years. He was like the breath-taking Karnak Temple of Luxor, in that they were both created in the form and proportion, the ability and function, the dignity and integrity – and hence the beauty – to have survived time and the misuses of man. Whether in their prime or in ruins, each of us is an undeniable creation. Such is Karnak; such was Hamdan."

Dahman, 19th century lithograph by Victor Adam

CHAPTER IV

Realizing the Dream

Ansata Ibn Halima's signature photo taken in Oklahoma by Sparagowski.

CHAPTER IV
Realizing the Dream

*Cherish your visions; cherish your ideals
cherish the music that stirs in your heart,
the beauty that forms in your mind,
the loveliness that drapes your purest thoughts,
for out of them will grow all delightful conditions,
all heavenly environment;
of these, if you but remain true to them,
your world will at last be built.*

James Allen - As A Man Thinketh

THE ANSATA BREEDING PROGRAM

When dreams become realities, the agonies and the ecstacies begin. The dream world allows freedom without restrictions; however, the real world requires boundaries and presents physical challenges. The conceptualizing has been accomplished; the material reality, in this instance - *The Ansata Breeding Program* - has begun.

Which came first, the chicken or the egg, the stallion or the mare? Given that one cannot exist without the other, I have nevertheless arranged the Ansata Herdbook according to tail female lines as we know them today, just as the Arabs have done throughout traditional history, and as the emissaries of Abbas Pasha did in compiling *The Abbas Pasha Manuscript*. While changes may occur in the light of new DNA information, or unearthing of contradictory records, they will not change the reality of what physically stands before us today as "families" and the inherited qualities of the individuals we select from whether or not they actually belong to that "strain." This is why it is of great importance to study individual horses as well the ancestral photos, make picture pedigrees, and note whether they represent what they should, and to learn, if possible, where the change in the line occurred and what caused it.

In weaving the Ansata tapestry over the past 45 years, the selection of mares representing certain families has been paramount; however, the use of ideal sires has been equally important. While many stallions have been used to complete the picture, certain key individuals have provided the framework. These are depicted, along with selected art, to illustrate the vision held when conceptualizing our breeding program.

One of the great challenges in breeding is making right decisions to achieve desired results as frequently and rapidly as possible. Trial and error are costly and patience-producing aspects of every program; beautiful individuals do not always guarantee productive value. Nevertheless, even when one strives hard to hit the mark, we learn valuable lessons when we miss.

Contemplate the key stallions illustrated at the beginning of this chapter. Any successful breeder knows that superior breeding stallions are few and far between. For one farm to have bred or acquired such a collection is a rarity. It is clear in viewing their pictures that there is a continuity in quality, and although there are some differences in type and structure, necessary to maintaining balance within a herd, there is an essence that is unmistakably classic Arabian.

The mare section, arranged by strains and families, is a pictorial treasure. It portrays most of the individuals candidly and *au naturel*, with little or no enhancement by over-stretched posing, artificial contrivances, excessive clipping or outrageous make up. These are mostly "for the record" photos. Few are artistic. Due to the number of horses and allotted space for this chapter, personal commentary has generally been limited to ancestors we have seen, to imports and their first generation offspring, and to acquired individuals purchased in North America. This commentary together with photographs provides a good overview of the program, even though each horse is not commented on.

Not all mares were intended to be major cornerstones of the stud. Some were chosen to "breed up" a line that was in danger of being lost, or was not being bred to its highest potential. There are personal favorites and those less favored, but each strain and family has played a role in the overall Egyptian picture and is worth being maintained to its highest potential. This we have endeavored to do. Additionally, there are a few progeny out of mares we sold who remained in our A.H.R.A. recorded ownership until the sales contract was final. Although we were listed as the breeders, the new owners may not have used the Ansata prefix and may have selected the stallions without our consultation. As Ansata was still listed as breeder on the Registry's books, because of contractual agreements, most of the resulting foals are still noted in this section.

It would have been a pleasure to create an artistic "coffee table book" selecting the finest pictures of the best horses to make everything look wonderful; however, that would have been an easy task, but would have served little instructive purpose. Instead, the horses pictured - from the super to the average to the less favored - as well as the personal opinion commentary, are intended to be teaching and learning tools. While some people may disagree with our observations and opinions, everyone will surely agree that, although the perfect horse exists only in the dream world, there is much pleasure on earth in attempting to realize the impossible dream.

KEY STALLIONS

The following key stallions stood at public and private stud at the Ansata farms. They have been selected as a visual reference and appear in the order in which they were incorporated within the program - not specifically by strain as compared to the mare section. Some photos have been reversed for comparative purposes (*Jamilll, Ansata Sinan, Ansata Malik Shah). The sequence of incorporation in itself tells a story as to traits being selected for at the time each was used.

A short personal commentary about these key stallions is included for useful reference along with the photographs. It is important to remember that in the early days of the Ansata program Egyptian mares were limited in number. The Ansata stallions of yesteryear would have even greater impact today because of the increased variety and quantity of mares within the entire Egyptian community, and with frozen and shipped semen now available.

In selecting stallions for the stud, of course one strives to maintain the breed standard and consequently a correct horse, but it is always necessary to give up something and make choices as to what is most important in one's own program. Our criteria, with rare exceptions, have been: Classic purebred Arabian type (i.e., a good to exotic head and an overall picture of balance, harmony and refinement), well-shaped long enough neck, deep well-angled shoulders, good topline with strong loins, good hip and croup and natural high tail set and high tail carriage, broad enough chest, front legs well-shaped with well-defined forearms, and set on at the corners; relatively fine bones, emphasis on well-let down strong hindquarters with good stifle and gaskins, free movement in front with strong drive from behind. Color is immaterial – a good horse can be any color, but we prefer minimal white markings. Grey is our favorite color as we believe it to be most consistent for transmitting classic Arabian type. Last, but assuredly not least, tractable dispositions and intelligence are requisite. Ansata stallions have been known for good dispositions, a trait contributed by *Ansata Ibn Halima and passed on from generation to generation.

Nazeer blood has been emphasized and we have never strayed far from its intense use. Linebreeding and inbreeding to Nazeer (and consequently to Mansour), *Morafic and *Ansata Ibn Halima have been important to the Ansata program along with close crosses to certain mare families, as will be evident to anyone who studies the pedigrees of Ansata-bred horses.

Nazeer (Mansour x Bint Samiha – Hadban Enzahi/ Samiha). The only time we saw Nazeer was in 1959. He was 25 years old and still in good condition. Very noble and regal was our first impression of him. A stallion of adequate size, stature and refinement, very handsome head of medium length; large expressive dark eyes, well-set lively alert ears, good length of neck, good withers, deep and well laid-back shoulder; good topline for his age with adequate hip, nice croup and high-set, high-carried tail. His front legs were straight; somewhat short and slightly straight pasterns, good hooves, fine clean bones. He was wide and deep enough. His hind legs stood out a bit behind him due to old age. He moved lightly beside his groom with grace befitting his years. A successful race horse, with good movement, he was chosen as a sire by Von Szandtner to improve structure and legs, which he did with regularity. His get inherited his unique quality that set them apart in a herd, but he did not generally change the type of the mare or mare line to which he was bred (witness the differences between *Ansata Ibn Halima, *Rashad Ibn Nazeer, *Aswan, *Morafic, *Talal). He did, however, transmit many of the best qualities of his ancestors. The classic head characteristics of Mansour and Bint Samiha bred forward. Mansour's strong structure also predominated and helped to overcome the weak croup and club foot influence from the Crabbet-bred Kazmeen, sire of many valuable R.A.S. broodmares. Nazeer was the sire of his era, and the consistent quality of his get was evident at our first, and succeeding, visits. Dr. Branch's comment to Humphrey that he had shown him the importance of using Mansour, was reflected in Nazeer and Sheikh El Arab's get.

Nazeer as a young stallion at the E.A.O.

*Ansata Ibn Halima (Nazeer x Halima - Dahman Shahwan/Farida). Selected as a yearling in Egypt, he embodied the ideal classic Arabian we had been searching for. Excellent balance, beautiful short dished head with large dark expressive eyes, small well-set and well-shaped ears, good mitbah, average length but well shaped neck in balance with his short-coupled strong body and good topline. Good tail carriage, good angulation behind, wide and deep enough; not perfect front legs, strong well-defined hindquarters, gaskins well let down; wonderful way of going in front and driving well off the hocks; and, very importantly, a good disposition that bred on from generation to generation. A sire of all colors, including black, he was a prepotent breeding stallion and "The Halima Look" solidified his name in history. He was known for siring short heads, big dark eyes, little to no white markings, good balance and movement, well-shaped but not always long enough necks, and occasional offset knees. He gave good shoulders and broad chests, but the long extreme laid-back shoulders came through incorporation of *Morafic. The combination of *Ansata Ibn Halima with *Morafic blood was the "golden nick" of the era. As a show horse *Ansata Ibn Halima won U.S. Top Ten honors three times and, although shown only one year in performance, he was a multiple Class A and Regional Champion in Park and English pleasure categories. He sired U.S. National Champions in halter and performance, race winners, and international champions, and he was one of the very few stallions to ever sire a U.S. National Champion son and daughter (Ansata Ibn Sudan and Fa Halima).

Ansata Ibn Sudan (*Ansata Ibn Halima x *Ansata Bint Mabrouka - Saklawi Jedran/Moniet). The result of breeding two of the first imports together, Sudan was an ideal blending of the Dahman and Saklawi strains and a true representative of what "double Nazeer" stood for. He possessed a very classic well-shaped head with nostrils capable of great expansion; very expressive eyes though not as large as his sire's or dam's; very pricked, well-set and well-shaped relatively small ears; wonderful mitbah with exceptionally well-shaped long neck set onto a good body; excellent topline, high set tail and extremely high tail carriage. Wide and deep enough; correct legs front and back, front legs set at the corners; good moving but not equal to his sire or dam in that department. He was the epitome of a classic Arabian stallion - a prideful, spirited, natural show horse, loving people and happiest when being admired. When he walked into the show ring, he dominated it completely, and it was no surprise that he was crowned the 1971 U.S. National Champion Stallion in a class of 83 entries, and first on two of the three judges' cards. As a sire, he gave many of his good traits, including his people-loving disposition, and he sired equal quality in all colors, begetting good show and breeding stock and U.S. National halter and performance winners of both sexes. In addition, he was a very important grandparent in a pedigree.

Ansata Shah Zaman (*Morafic x *Ansata Bint Mabrouka - Saklawi Jedran/Moniet). This was an inbreeding - full brother to full sister - that worked. Shah Zaman was a very dramatic horse, a handsome masculine head, rather long in foreface, some dish and a prominent nosebone; big dark expressive fiery eyes, good well-set well-shaped ears; long neck, but somewhat angular mitbah, similar to but better than *Morafic's; exceptionally long and well-laid back shoulders and prominent withers, short coupling with strong topline and superb tail set and carriage; wide and deep enough; good legs but a slight tendency to toe out, like some of the Moniet line; front legs set at the corners; long gaskins and well let down behind, he was exceedingly fast and could literally spin on a dime. His disposition was fiery, but not mean, and he gave great presence and most of his good qualities to his get. He was invaluable for siring prettier heads than his own, big dark eyes, long necks, smooth strong short-coupled bodies, high tail set and high tail carriage, and he gave color but no black. He was a multiple Class A Halter Champion, Most Classic Arabian winner, and a sire of National and International winners. The combination of *Ansata Ibn Halima, Ansata Ibn Sudan and Ansata Shah Zaman resulted in some of Ansata's best.

Mansour
      Nazeer
               Bint Samiha
*Ansata Ibn Halima
                Sheikh El Arab
      Halima
                Ragia

                Nazeer
      *Ansata Ibn Halima
                Halima
Ansata Ibn Sudan
                Nazeer
      *Ansata Bint Mabrouka
                Mabrouka

                Nazeer
      *Morafic
                Mabrouka
Ansata Shah Zaman
                Nazeer
      *Ansata Bint Mabrouka
                Mabrouka

Ansata Abu Sudan (Ansata Ibn Sudan x *Ansata Bintmisuna - Hadban Enzahi/Samiha line). Intensely Nazeer-bred, he descended from Nazeer's dam line in tail female and represented what his pedigree said he should be. He was an extremely classic individual; beautiful head, good eyes but a slight tendency to some white typical of the Hadban line; nice well-set and nicely sculptured ears, long well-shaped neck and good mitbah, good overall balance but a bit weak in the loins like many of the Hadbans; good tail set and excellent tail carriage; correct legs all around; front legs set at the corners; well let down behind and exceptionally free movement in front while driving well from behind - traits which he transmitted regularly to his get. He sired some color, no blacks, but mostly greys. His use was somewhat limited due to the small size of the herd at that time, but nevertheless, his contribution to the stud was most important, particularly through his daughter, Ansata Sudarra. He was a Class A Halter Champion and Most Classic Arabian winner with much potential for performance, had he been shown further.

Ansata Ibn Shah (Ansata Shah Zaman x Ansata Jezebel - Dahman Shahwan/Futna). A perfectly balanced horse, Ibn Shah was a very classic individual, beautiful head, wonderful eyes and expression, well-shaped ears but set a bit wide and low; excellent length and well-shaped neck and mitbah, strong body and topline, good high tail set and carriage; wide and deep enough, and good legs that set out at the corners, well let down behind, he had good movement. He too was limited at use in the stud, but he contributed important daughters; he sired color but no black. He was seven times a Class A Champion at halter, missing U.S. National Top Ten honors (placing 11th) by a hair, and still had a show career ahead of him. Regrettably, he died after returning from being on lease before there was time to use him to fullest advantage.

Ansata Halim Shah (*Ansata Ibn Halima x Ansata Rosetta - Dahman Shahwan/Bukra). An archetype. A perfect combination in pedigree and genetic inheritance. Nobility personified. Exceptionally beautiful very refined head, good dark lively eyes, well-set and well-shaped ears, well-shaped long enough neck, well-laid back shoulders and good withers, excellent topline, excellent tail set and high tail carriage, strong hindquarters and well let down behind; wide and deep enough; not perfect front legs; front legs that set out at the corners; a good mover and he could have been an outstanding performance horse. He was shown to U.S. Top Ten Futurity Stallion and to Res. Jr. World Champion, Salon du Cheval, Paris, remaining in Europe to stand at stud where he was used by Dr. Nagel, Marbach State Stud, and Babolna State Stud. He achieved highest stallion licensing honors and recognition in Germany, and literally changed the face of Egyptian breeding in Europe and subsequently in the Arab world after his purchase by the Emir of Qatar's Al Shaqab Farm. Stallions of his beauty and quality are rare in any century, and sires of his caliber even rarer. He was his "own self" an absolutely distinguishable type that set him apart from any others of his era. As a sire he was a homozygous grey, and all of his get bore his unmistakable stamp. He sired equally good stallions and mares. He crossed especially well when bred to *Jamilll daughters or into the Hanan line, nicking admirably with Alaa El Din blood. The double of Halim Shah blood (Halim Shah stallion to Halim Shah daughter) has been very successful no matter the strain/family.

*Ansata Ibn Halima
    Ansata Ibn Sudan
       *Ansata Bint Mabrouka
Ansata Abu Sudan
       Nazeer
    *Ansata Bintmisuna
       Maysouna

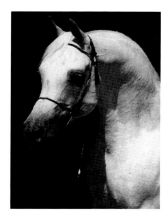 

       *Morafic
    Ansata Shah Zaman
       *Ansata Bint Mabrouka
Ansata Ibn Shah
       *Ansata Ibn Halima
    Ansata Jezebel
       *Ansata Bint Sameh

       Nazeer
    *Ansata Ibn Halima
       Halima
Ansata Halim Shah
       Ansata Shah Zaman
    Ansata Rosetta
       *Ansata Bint Bukra

*Jamilll (Madkour I x Hanan - Abeyyan Um Jurays/ Inshass Mona). A unique very refined elegant stallion; very beautiful short wedgy head with deep jowls, somewhat short mouth and rather small nostrils; huge black expressive eyes, tiny pricked ears, long well-shaped neck, good topline but could have been stronger in hindquarters (a touch of Hadban Enzahi coming through)- a slight rafter hip, which he tended to pass on; good tail carriage, good shoulders; slightly narrow in chest, his front legs attached differently to his body than those of the Ansata horses; somewhat small but good hard dark hooves, good straight front legs which he passed on with regularity; fine bones, a good mover with an excellent German Stallion Licensing record. *Jamilll had his own "look" - a very classic type overall, he was shown to several Class A Halter championships and was a U.S. National Top Ten Stallion the same year Ruminaja Ali was U.S. Reserve National Champion Stallion. He sired bay and chestnut colors as well as grey, and he contributed significant mares to the stud as well as handsome sons, but they were not comparable as breeding stock to his daughters. Unfortunately, he died soon after he was returned to Germany, having been on lease to Ansata. Halim Shah too had a bad health experience after returning to America from being on lease to Nagel in Germany. It seems as if the "evil eye" had been beamed at both of them.

Prince Fa Moniet (The Egyptian Prince x Fa Moniet - Kuhaylan Jellabi/ Maaroufa). A very elegant well-balanced horse, handsome relatively short head, somewhat small nostrils, huge jowls, dark average size very expressive eyes; well-set but not extremely sculptured ears - the shape of which he inherited from *Ibn Moniet El Nefous; long well-shaped neck, good somewhat longer-lined body and topline, excellent tail set and tail carriage, well let down hindquarters, wide and deep enough; good straight legs, front legs set out at the corners; good movement. Prince Fa had been injured twice at Bentwood before we purchased him at their dispersal; first he put a front leg through a glass window, the scars disfiguring it badly; next he stepped in a hole and broke a pastern, resulting in major surgery, screws, plates, and thereafter a permanently turned out foot. Had he not been disfigured, he had the potential of reaching highest show ring honors, both in performance and halter. He was an excellent sire, a homozygous grey, passing on height, stretch and straight legs. His get have won Class A, National Top Tens, and international halter championships, and his daughters and sons have become sought after breeding stock. Prince Fa Moniet's pedigree, like *Ansata Ibn Halima's, represented the breeding adage "let the sire of the sire be the grandsire of the dam." He spent his final days in Australia on lease to Forest Hill Arabian Stud, where he was laid to rest before he could be returned home.

Ansata Hejazi (Ansata Halim Shah x Ansata Sudarra - Dahman Shahwan/Bukra) A regal, refined and well-balanced stallion, much resembling his sire, Hejazi possesses a very classic dry head, large dark expressive eyes, good well-shaped and well-set ears, long well-shaped clean neck set on extremely deep well-laid back shoulders; short-coupled, good topline, high tail set and high tail carriage; well-let down behind; wide and deep; front legs set at the corners, fine boned; exceptionally free movement in front and driving well off his hocks - capable of a breathtaking elevated trot when "turned on." A homozygous grey, Hejazi has proven to sire excellent quality in both sexes, his sons and daughters becoming much sought after as show and breeding stock. A proven successful nick has been Hejazi on Prince Fa Moniet daughters. Hejazi was shown to Res. Grand Champion Stallion at the Egyptian Event, and also at Class A shows. Having contributed significant sons and daughters to Ansata, many being retained and some sold abroad as show and foundation stock for other stud farms, he was sold to Mohammed J.K. Al Marzouk, as the major breeding stallion for his Ajmal Stud. Marzouk has generously shared him with other private breeders and the Arabian Horse Center of Kuwait, and the influence of this well-named stallion is now felt throughout the Arab world.

Hadban Enzahi
    Madkour I
        Moheba II
*Jamilll
        Alaa El Din
    Hanan
        Mona (Inshass)

        *Morafic
    The Egyptian Prince
        *Bint Mona
Prince Fa Moniet
        *Ibn Moniet El Nefous
    Fa Moniet
        Fada

        *Ansata Ibn Halima
    Ansata Halim Shah
        Ansata Rosetta
Ansata Hejazi
        Ansata Abu Sudan
    Ansata Sudarra
        Ansata Delilah

Ansata Iemhotep (Prince Fa Moniet x Ansata Nefara - Dahman Shahwan/Bukra) This horse is a reincarnation of Ansata Ibn Sudan. Tall, elegant, handsome masculine head which fits him, with large dark expressive eyes that look right at you, good well-set well-shaped ears, exceptionally long, upstanding, well-shaped arched neck with fine mitbah, well laid-back shoulders, good withers, strong topline, good tail set and very high tail carriage, correct legs both front and hind, wide and deep, with front legs setting out at the corners; free and bold moving. Like Ibn Sudan, he is a "people horse," loves to show off, and has a very distinctive personality and "look" all his own which he imparts to his get. He was shown to many Class A championships at halter and English performance, being an excellent athlete, and he was also a 1996 U.S. National Futurity Stallion Top Ten winner. A homozygous grey, like his sire and his dam's sire, Hotep inherited the best traits of his sire and dam and has been consistent in siring size and scope as well as good straight legs. His get have been exported to numerous countries, and have won significant honors in the show ring. He sires equal quality in both sexes.

Ansata Sinan (Prince Fa Moniet x Ansata Nefara - Dahman Shahwan/Bukra) A very classic and well-balanced individual, with extremely beautiful head, good lively dark eyes, good well-set and nicely formed ears, well-shaped long enough neck and good mitbah, good shoulders, short coupled, strong enough loins; good croup, tail set and tail carriage; well let down behind; wide and deep enough; good clean legs, front legs set on at the corners; correct movement. Sinan is the consummate show horse; pure white and exceptionally popular in Europe, he was an instant show winner, winning championships in America and then in Paris at the Salon du Cheval, and the Nation's Cup at Aachen, among other wins when he was owned by Veruska Arabians. He was reimported to U.S.A. by his owners and stood at Ansata in 2000. During the time he was used in Mena he sired good colts and fillies. Also a homozygous grey, he passes on his beauty and many of his good traits. His get in Europe are very popular winners at major shows, and they are sought after as breeding stock. Several of his get bred by Ansata were exported to Europe and the Arab world; some remained in America.

Ansata Sirius (Ansata Iemhotep x Ansata Sekhmet - Dahman Shahwan/ Bukra). A perfectly balanced pure white stallion, he resembles Ansata Halim Shah in refinement and body type. He has a unique extremely short head, good expressive dark eyes, well-set and very small pricked ears (inherited from *Jamilll and Hadban Enzahi), huge jowls, good dish, small muzzle; a well-shaped upstanding arched neck of proper length in proportion to his body; good mitbah; good shoulders, excellent body and topline, high set tail and high tail carriage, well let down behind; sufficiently wide and deep; good front legs set out at the corners, and very good movement despite the fact he shattered his left front pastern as a yearling. His tremendous courage and spirit, and thanks to excellent surgery performed at Oklahoma State University College of Veterinary Medicine, allowed his survival when many other horses would have given up. Despite a year's stall rest, which initially hindered his growth, he is now perfectly sound and moves freely with good ground covering gaits. His extreme type and quality are evident in his first foal crops, and he probably will be a homozygous grey sire. Shown only once after his injury, he was Top Ten at the 2002 Egyptian Event in the stallions age 5-10 World Class Stallion competition.

The Egyptian Prince
    Prince Fa Moniet
        Fa Moniet
Ansata Iemhotep
        Ansata Halim Shah
    Ansata Nefara
        Ansata Sudarra

        The Egyptian Prince
    Prince Fa Moniet
        Fa Moniet
Ansata Sinan
        Ansata Halim Shah
    Ansata Nefara
        Ansata Sudarra

        Prince Fa Moniet
    Ansata Iemhotep
        Ansata Nefara
Ansata Sirius
        Ansata Halim Shah
    Ansata Sekhmet
        Ansata Samarra

Ansata Sokar (*Imperial Madheen x Ansata Samantha - Dahman Shahwan/ Bukra). A stallion of good size and impressive demeanor, Sokar was bred to provide some outcross to the intense Ansata bloodlines. The Madheen lineage has nicked well with Ansata breeding, and it also is a source of color. Sokar took much after his dam, inheriting her rich chestnut coat, white star on the forehead and exceptionally strong level topline. He much resembles her in the face, with broad forehead, big dark expressive eyes, well-set well-shaped ears, deep jowls, average length of head; long enough neck well set on with good mitbah, but slightly heavier at the base somewhat typical of Madheen get. Good withers, with strong back, loins and croup and a tail that comes right out and up - carried like a plume - traits that he has passed on consistently. Strong foundation, not perfect front legs, well let down behind, powerful movement; he gives the appearance of strength and masculinity. He has sired all colors, including black, and his get have been exported to Europe and the Middle East. His chestnut get seem to relate to him in type. His elegant chestnut daughter, Ansata Bint Sokar, a very refined intensely-bred Saklawia Jedrania, won Top Ten at the 2002 Egyptian Event at the age of 7 months in a class of 32 yearlings. Sokar's pedigree reflects strong lines to the Saklawi Jedran and Dahman strains, with the Dahman most prominent in his outward appearance. He provides a useful combination of related European and Ansata bloodlines.

Ansata Malik Shah (Ansata Hejazi x Ansata Malaka - Dahman Shahwan/ Halima). Retained as a replacement for his sire, Malik Shah is a very charismatic Dahman-looking white stallion of average size and extreme classic type. A slow horse to mature, which is typical of the Halima line, he has a very classic well-shaped dry head of medium length, large lively black eyes, well-set and well-shaped ears, a good mitbah, nicely arched and well-shaped long enough neck, good withers, somewhat forward-stretching short back, good loins, long croup and very deep hip with high-set and high-carried tail. He is fine-boned, front legs set at the corners, well let down behind, very good movement and gaits. He bears Halim Shah's regal and somewhat aloof attitude, and he makes an immediate impression when he enters the show ring. Although he was very immature at age three, he was received favorably by spectators and judges and appeared more mature than he was. A Res. Grand Champion the only time shown at Class A Halter, he won the Extended Specialty Futurity at the 2002 Egyptian Event. His first foals are due in 2003.

A study of Malik Shah's pedigree is worthwhile, as it reflects many of the important close breeding/linebreeding combinations discussed in Chapter II. He is double Ansata Halim Shah (Halim Shah son to Halim Shah daughter – sire from the Sabah and dam from the Farida/ Dahman tail female), and he also includes crosses to Ansata Shah Zaman, *Ansata Ibn Halima, *Jamilll, Ansata Ibn Sudan and Ansata Abu Sudan – all major sires used by Ansata. Six of his 8 great-grandparents are of the Dahman Shahwan strain, and they reflect tail female lines of his two parents and four grandparents. Nazeer appears only twice in the 16 line, yet the influence of this stallion cannot be denied in the background. Malik Shah perfectly reflects what his pedigree says he should be. Whether his genotype will be the same as his phenotype will not be known until his foals arrive in 2003 and beyond.

Messaoud
*Imperial Madheen
Madinah
Ansata Sokar
*Ansata Ibn Halima
Ansata Samantha
Ansata Delilah

Ansata Halim Shah
Ansata Hejazi
Ansata Sudarra
Ansata Malik Shah
Ansata Halim Shah
Ansata Malaka
Ansata Malika

ADDITIONAL ANSATA STALLIONS USED IN THE PROGRAM

Additional Ansata-bred stallions have also contributed to the program, but may or may not have stood at the farm. Some sired progeny before being sold, some were of limited use, and others were later represented through their get purchased or acquired by Ansata. All were Dahman Shahwan except Ansata Abbas Pasha and Ansata Ibn Aziza who were Saklawi Jedran.

The following sired offspring at Ansata:

Stallion	Family Line
*Ansata El Wazir	(Farida/Futna)
Ansata Mourad Bey	(Bukra/Regina)
Ansata Manasseh	(Bukra/Regina)
Ansata Halima Son	(Bukra/Damietta)
Ansata Haji Halim	(Bukra/Delilah)
Ansata Omar Halim	(Bukra/Rosetta)
Ansata Osiron	(Bukra/Delilah)

The following were sires of stock acquired:

Ansata Imperial	(Bukra/Delilah)
Ansata El Naseri	(Bukra/Bukra)
Ansata Ibn Aziza	(Samira/Zaafarana)
Ansata Abbas Pasha	(Moniet/Mabrouka)

THE INFLUENCE OF *MORAFIC IN THE ANSATA PROGRAM.

In addition to the previous stallions, it is important to acknowledge *Morafic's influence, past, present and future, in the Ansata program, as he was a stallion we admired and believed in.

While in Egypt, we requested Dr. Marsafi to breed Moniet El Nefous to *Morafic, which he did, and *Ibn Moniet El Nefous was the result. He was imported to the U.S. as a yearling by Jay Stream. As a mature stallion he was purchased by Jarrell McCracken, who shocked the world by syndicating him for $1,000,000.

In 1959 we tried to buy the exquisite *Bint Mouna (Mona), full sister in blood to *Morafic and *Ansata Bint Mabrouka, but we had to make a choice between the two fillies. We chose the latter. When *Morafic was imported to Gleannloch, we bred him to *Ansata Bint Mabrouka (his full sister), creating Ansata Shah Zaman, who became an integral Ansata sire. Gleannloch imported *Bint Mouna and bred her to *Morafic, (her full blood brother) producing The Egyptian Prince, an invaluable sire. Bentwood Farms later purchased The Egyptian Prince and bred him to Fa Moniet (by *Ibn Moniet El Nefous), and the magnificent stallion, Prince Fa Moniet resulted.

When inbred or linebred very carefully, (also considering type in relation to color and the heritable faults of the ancestors), this Moniet family can be a major asset to a breeding program before some degree of outcrossing will be required. The *Morafic/*Ansata Ibn Halima combination of bloodlines proved to be a "golden nick" for Ansata and for many other breeders…and especially for those who incorporated Ansata Halim Shah into their program.

*Above left, *Morafic at Gleannloch.*

*Above *Morafic's son, *Ibn Moniet El Nefous at Bentwood.*

*At right, *Bint Mona and her son, The Egyptian Prince, by *Morafic.*

The Ansata type carries forward from generation to generation.

Above left, Ansata Ibn Sudan, above right, *Ansata Ibn Halima. Sparagowski photo

Above left, Ansata Iemhotep, above right, Ansata Sirius. Judith Wich photo

Ansata stallions – inherently noble, spirited and classic!

Ansata Shah Zaman
(*Morafic x *Ansata Bint Mabrouka)

Sparagowski photos

Prince Fa Moniet
(The Egyptian Prince x Fa Moniet). Bred by Bentwood Farms.

Ansata Ibn Shah (Ansata Shah Zaman x Ansata Jezebel)

Ansata Halim Shah (*Ansata Ibn Halima x Ansata Rosetta), in Qatar. Van Lent Jr. photo.

Ansata Hejazi (Ansata Halim Shah x Ansata Sudarra), in Arkansas before moving to Kuwait. Sparagowski photo.

*Jamilll (Madkour I x Hanan), on lease from Dr. Nagel to Ansata in Arkansas. Sparagowski photo.

Ansata Abbas Pasha (*Ansata Ibn Halima x *Ansata Bint Mabrouka), in Texas before returning to spend his final days in Arkansas. J. Johnston photo.

Ansata Iemhotep (Prince Fa Moniet x Ansata Nefara). Sparagowski photo.

Ansata stallions – inherently noble, spirited and classic!

Ansata Sirius (Ansata Iemhotep x Ansata Sekhmet). Sparagowski photo.

Ansata stallions – inherently noble, spirited and classic!

Ansata Sokar (*Imperial Madheen x Ansata Samantha). Sparagowski photo.

Ansata Malik Shah (Ansata Hejazi x Ansata Malaka). Sparagowski photo.

Al Duhaymat

Dahman Shahwan

Aquarelle, 1840, Juliusz Kossak. Museum Okregowe de Tarnow.

The Dahman Shahwan Strain

As history records, the legendary horses of King Solomon were incomparable in beauty, and his collection rivalled all others. According to the Abbas Pasha emissaries, the ancestors of the Dahman strain were in Solomon's possession. Like connoisseurs and early breeders, we too were attracted to this strain because of its overall harmony, balance, and beauty. When we first visited the E.A.O. in 1959, at the time we chose *Ansata Ibn Halima, we were not very conversant with the history or characteristics of strains and/or families in Egyptian bloodlines; however, we saw in him something that transcended the physical. Further study of photographs, art, and literature provided additional clarification of this attraction, but actually breeding this strain gave a better understanding of its continuing form and spirit.

Abbas Pasha I placed the Dahman strain first in *The Abbas Pasha Manuscript*, his emissaries having traveled far and wide to ascertain its history that is recorded from pages 247-283. Some excerpts are as follows:

> "When Mohammed ibn-Qarmalah, Sheikh of Qahtan, was asked about the history of Al Dahmah - the family strain - he replied: "Al Dahmah belongs to Shahwan, and she is from the horses which belonged to our Lord Suleiman (King Solomon), peace be upon him. And the Kuhaylah was called Al Dahmah because of her dark color. And the eyes of the Kuhaylah were as if rimmed with kohl. And all the present day pure horses existing are descended from the female line of the aforementioned."

Shahwan wrote several inspirational poems about his favorite mare. One is noted below. Another particularly descriptive ode introduces the Dahman strain in Chapter VI.

Oh you hunters who shoot with arrows,
I, Shahwan, am the best at shooting prey with arrows
And I am the most skilled marksman of all.

And some hunters may let their prey escape,
But I and my horse can catch up with it and bring it back.

And the hunters who let their prey escape are weak,
And no one can catch up with it
Except one which has the strength and swiftness of my horse.
And people have great admiration for this swiftness.

The most beautiful type of horse
Is the one who stands on three legs and raises the fourth,
And my horse is of this kind.
And she does more than what is expected of her during battle.

During the attacks, she strikes with her forefeet and hind feet,
And the rider strikes with his sword and arrows.
And the higher the pitch of battle,
The more aroused the mare becomes....

She defends her rider
Even unto putting herself in peril,
And does not give the enemy a chance
To wound her rider....

Horses are what the heart most cherishes,
And the Arabs use no other means
For fighting and riding except horses.

Although the first mares we imported in 1959 were of the Saklawi Jedran strain, the primary Ansata foundation mares came from the Dahman Shahwan strain descending from El Dahma of Ali Pasha Sherif. Both branches through Sabah and Farida were important to the stud; the Bint Sabah line through her Babson-bred granddaughters, Fay-Sabbah, Fa-Habba, and our 1965 import, *Ansata Bint Bukra, and later through the Kamar line added in 1999. We acquired the Farida line through *Ansata Bint Sameh (Amal), imported from Egypt, *JKB Masouda, on lease from J. Kayser, and the purchase of Dal Macharia at a sale in the U.S. Others descending from the various Antar/Abla daughters were acquired later. The Bint El Bahreyn family was also represented at Ansata through lines to Elwya and Maisa.

From this Dahman strain many outstanding show horses, as well as racing, endurance winners and superior breeding stock have descended.

The Bint Sabah Family

The Bint Sabah family through Bukra.

Bint Sabah was not alive when we visited Egypt; however, her photos show her to be a bay mare of reasonable quality and pleasing head and better conformed than her mother, pictures of whom are not very flattering. Bint Sabah contributed much as a broodmare to the R.A.S./E.A.O. program. One of her finest daughters was Bukra (by Shahloul), a true matriarch of the breed, the "prima stute" as General von Szandtner called her. A classic fleabitten grey mare of good size and overall balance, her head was exceptionally beautiful having the long refined foreface and fine muzzle. She had the somewhat short peaky croup, like many of the Sabah's, a trait that often seems married to a beautiful head.

*Ansata Bint Bukra (by Nazeer), was very similar in type to her dam, with an even more beautiful head, stronger bodied, and with that certain indefinable essence and quality that marked all the Nazeer get, yet left the mare line intact. She was an exquisite gentle soul, and people from around the world greatly admired her when they came to visit Ansata. Crippled shortly after birth, through an accident that knocked down her hip and damaged her front legs, she never complained and went on to produce her foals with ease. Her photograph decorated the cover of magazines, and she became known as the "prima stute" foundation mare of Ansata. Although we tried to buy her as a yearling, the African Horse Sickness had prevented any horses from being imported to America from Egypt, and we were forced to wait until this ban was lifted. By that time she was a mature mare and had already been serviced by Sameh. At the time of purchase, she was not guaranteed to be in foal, or to be able to carry a foal due to her pelvic injury. Traveling all the way to America on a ship, and in a crate by herself, she arrived at Ansata and was safely in foal to Sameh much to everyone's surprise and our good luck for having taken the risk.

*Ansata Bint Misr (by Sameh), imported in utero, arrived as a bay wearing a large white star that illumined her broad forehead and accentuated her big dark eyes. Her head was shorter and squarer than her dam's, but better than her sires' and she inherited her sire's excellent body structure and topline. *Ansata Bint Misr (which means daughter of Egypt) didn't have her dam's refinement, but Sameh ironed out the short peaky croup typical of that line, and good toplines carried on through most of her descendants. She became a keystone in the Ansata program, and lines to her have produced some of the most beautiful horses in the Bukra family.

Ansata El Sherif (by *Ansata Ibn Halima), was a most classic individual combining much of his sire and dam in head and in overall type, but needed extra length of neck. He was a winning show horse in halter and performance and became a good sire for his owners.

Ansata Rhodora, the full sister to Sherif, had a very pretty head and big eyes, short neck, very short-coupled, a bit small, needed more length of leg; a good mover.

Ansata Sherifa, a bay full sister to Rhodora, was most exquisite but she died as a weanling after being sold to Jarrell McCracken.

Ansata Rosetta (by Ansata Shah Zaman), was a marvelous grey mare of good size, longer head and foreface like her dam, long well-laid back shoulder, strong body, good hip and croup, good tail set and carriage.

Ansata Shahrazada, a bay full sister to Rosetta, and even more beautiful, unfortunately ran into a fence pole, broke her neck and died.

Ansata El Naseri (by Ansata Ibn Sudan), looked much like his sire, though not quite as strong in topline. He was shown very successfully and became a valued sire for his owners, the Flicks, who had purchased Ansata Ali Pasha, the first foal ever bred by Ansata.

Ansata Bint Sudan (by Ansata Ibn Sudan), a tall handsome mare, pleasing head, big eyes, somewhat wide ears, long neck, good balance and structure. She was a good broodmare, but was not as refined as one would have expected from that mating.

Ansata Regina (by *Ansata El Wazir), was a tall extremely good-bodied and very correct mare, somewhat plain in the head early in her development but it eventually dried out and became very pretty. As an overall harmonious picture, she was the best of all the daughters.

Ansata Abu Jamal (by Ansata Abu Sudan), was a good colt, but not as spectacular as the full brother who followed him.

Ansata Ibn Bukra (by Ansata Abu Sudan), turned heads everywhere. One of the most exquisite stallions ever bred at Ansata, he was purchased by Count Federico Zichy-Thyssen and eventually exported to South America. He did not prove very fertile and was gelded even though he was capable of siring foals.

Each one of these first generation *Ansata Bint Bukra daughters bred on at Ansata and in subsequent breeding programs. The most prolific was the line of *Ansata Bint Misr. Each made its own definable family that has contributed immeasurably to the worldwide Egyptian scene. Her sons Ansata El Naseri and Ansata El Sherif were both champions and made significant contributions as breeding stallions.

The subsequent generations of the Bukra family were generally of high quality, but differences within the families were dependent on the degree of linebreeding or inbreeding to the stallions used. Offset knees and a tendency to short necks, poor mitbah, and a somewhat short peaky croup (harking back to Kazmeen) were traits that needed to be overcome. This line has continually produced classic type and well-balanced horses that have proven themselves in various kinds of performance including racing in Egypt. Without question, the most spectacular individual from this family was Ansata Halim Shah, a stallion much like a fine porcelain who was almost a mutant, having his own identity in type, and stamping his get with that "Halim Shah look." He was in type and pedigree the ideal representative of the Sabah/Farida cross, a true renaissance horse, and the introduction of his blood revolutionized the breeding of Egyptian horses in Europe and the Middle East.

*Ansata Bint Bukra (Hosnia) in Egypt with Don Forbis

El Dahma through Bint Sabah
Bukra Branch

Bint Obeya

Sabah

El Dahma c. 1880 Dahmah Shahwania mare from the stud of Ali Pasha Sherif

Obeya 1894 x **Koheilan El Mossen**

Bint Obeya x **El Halabi**

Sabah x **Mabrouk Manial**

Bint Sabah x **Kazmeyn**

Bukra x **Shahloul**

*Ansata Bint Bukra x **Nazeer**

*Ansata Bint Misr x **Sameh**

Bint Sabah

Bukra

*Ansata Bint Misr

*Ansata Bint Bukra in Egypt

*Ansata Bint Bukra and her caretaker, Linda Sain, in Lufkin, Texas.

*Ansata Bint Bukra #33487

Grey Mare; August 23, 1959 - October 21, 1981
Sire: Nazeer   Dam: Bukra
Bred by E.A.O., Cairo, Egypt
E.A.O #268, Volume II
Registered as **Hosnia** in Egypt
Imported by Mr. and Mrs. Donald L. Forbis, Chickasha, Oklahoma, July 8, 1965.

PRODUCE

*Ansata Bint Misr #36452   April 5, 1966
Bay mare by Sameh. Imported in utero.

Ansata El Sherif #41580   April 5, 1967
Grey stallion by *Ansata Ibn Halima.
Sold to Clarence and Daisy Hardin, Santa Ynez, California.

Ansata Rhodora #49569 March 29, 1968.
Grey mare by *Ansata Ibn Halima.

Ansata El Naseri #56303 April 7, 1969
Grey stallion by Ansata Ibn Sudan. Sold to
Willis Flick, Miami, Florida.

Ansata Sherifa #61934 April 12, 1970
Bay mare by *Ansata Ibn Halima. Sold to
Jarrell McCracken, Waco, Texas. Died.

Ansata Rosetta #70167 April 6, 1971
Grey mare by Ansata Shah Zaman.

Ansata Bint Sudan #94537 February 16, 1973
Grey mare by Ansata Ibn Sudan.

Ansata Regina #107422 April 19, 1974
Grey mare by Ansata El Wazir.

Ansata Shahrazada #122312 July 1, 1975
Bay mare by Ansata Shah Zaman. Died
June 18, 1976.

Ansata Abu Jamal #196860 February 19, 1979
Grey stallion by Ansata Abu Sudan.
Sold to Ed and Dorothy Cote, Williston, North Dakota.

Ansata Ibn Bukra #219545 February 20, 1980
Grey stallion by Ansata Abu Sudan. Sold to Count Federico
Zichy-Thyssen; exported to Argentina, S.A.

Ansata Bint Bukra. Photos in Texas by Polly Knoll

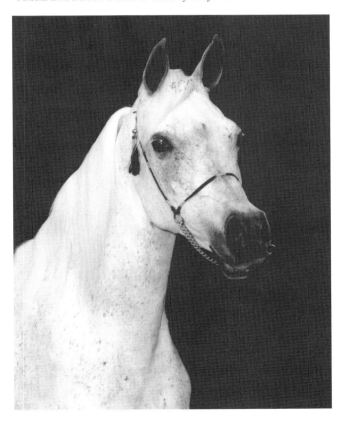

Progeny of *Ansata Bint Bukra

*Ansata Bint Misr

Ansata El Naseri

Ansata El Sherif

Ansata Bint Sudan

Ansata Regina

Ansata Rosetta

Ansata Rhodora

Ansata Shahrazada

Ansata Sherifa

Ansata Ibn Bukra

Ansata Abu Jamal

*Ansata Bint Misr

Anssata Damietta

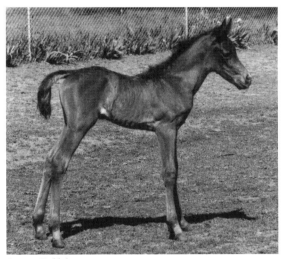
Ansata Delilah

Ansata Abu Tai

The *Bint Misr Family

**Ansata Bint Bukra/*Ansata Bint Misr*

*Ansata Bint Misr #36452
Bay Mare; April 6, 1966 – January 15, 1991
Sire: Sameh Dam: *Ansata Bint Bukra
Bred by E.A.O., Cairo, Egypt.
Imported in utero by Mr. and Mrs. Donald L. Forbis, Chickasha, Oklahoma

PRODUCE

Ansata Damietta #70168 April 7, 1971
Grey mare by Ansata Shah Zaman.

Ansata Delilah #80322 April 10, 1972
Grey mare by Ansata Shah Zaman.

Ansata Abu Tai #094543 July 10, 1973
Bay stallion by Ansata Ibn Sudan. Sold to
Jerry Oppenheimer, Cromwell, Connecticut,
February 1976.

Ansata El Nasrany #143107 September 7, 1976
Grey stallion by *Ansata Ibn Halima. Sold to Sabbath
Arabians, Rev. R.E. Tarter, Oak Ridge, Tennessee,
October 1980.

Ansata Magid Shah #170222 January 29, 1978
Chestnut stallion by Ansata Shah Zaman. Sold to
Gibson Arabians, Loomis, California, October 1980.

Ansata Alexandria unregistered, September 22, 1979.
Grey mare by *Ansata Ibn Halima. Died November 1, 1979.

Ansata Aya Halima #220099 November 9, 1980
Bay mare by *Ansata Ibn Halima.

Ansata Halim Bay #259629 March 15, 1982
Bay stallion by *Ansata Ibn Halima. Sold to
Naguib Audi. Exported to Brazil, South America.

Ansata Cairo Bey #278158 March 18, 1983
Bay stallion by Ansata Shah Zaman. Donated
to Fern Leigh Foundation in 1993.

Ansata Misriya #0355911 April 19, 1986
Grey mare by *Jamilll.

Ansata Nadir Shah #0410780 April 6, 1988
Grey stallion by Ansata Halim Shah. Gelded and
sold to John Lossing, Fayetteville, Arkansas.

Ansata Darius #440720 June 24, 1989
Grey stallion by Ansata Halim Shah. Gelded.

Progeny of *Ansata Bint Misr

Ansata El Nasrany

Ansata Cairo Bey

Ansata Misriya

Ansata Majid Shah

Ansata Halim Bay

Ansata Aya Halima and foal, Ansata Aya Maria. JEF photo

Ansata Damietta in Oklahoma. Sparagowski photo

Ansata Halim Bey

Ansata Halim Son

The Damietta Line of the Bint Misr Family

**Ansata Bint Bukra/Ansata Bint Misr/Ansata Damietta*

Ansata Damietta #70168
Grey Mare; April 7, 1971
Sire: Ansata Shah Zaman   Dam: *Ansata Bint Misr
Sold to Mr. and Mrs. Don Saluti, Birchknoll Arabians, February 1985. Sold by Saluti to Paolo Gucci. February 1989.

PRODUCE

Ansata Halim Bey #137118 July 28, 1976
Grey stallion by *Ansata Ibn Halima. Sold to Rev. and Mrs. T.H. Vallee, Woodstown, New Jersey, August 1978.

Ansata Halima Son #163793 September 3, 1977
Grey stallion by *Ansata Ibn Halima. Sold to Halima Son Partnership, Cleveland, Texas, August 1991. Died 1993.

Ansata El Mabrouk #179149 August 29, 1978
Grey stallion by *Ansata Ibn Halima. Sold to Zichy-Thyssen Investment Corporation, November 1983. Exported to Argentina.

Unnamed grey mare by *Ansata Ibn Halima, April 23, 1980 Died May 18, 1980.

Ansata Dia Halima #237265 May 15, 1981
Chestnut mare by *Ansata Ibn Halima.

Ansata Abu Halim #259988 June 4, 1982
Grey stallion by *Ansata Ibn Halima. Sold to Seven Fields Arabian Bloodstock, Beaver, Pennsylvania.

Ansata Marietta #279330 June 21, 1983
Grey mare by Ansata Ibn Sudan.

Ansata Marvella #303398 June 15, 1984
Bay mare by *Jamilll. Died November 18, 1984.

BKA Damilll #331132 July 1, 1985.
Grey stallion by *Jamilll.
Owned by Don Saluti, Birchknoll Arabians, Malvern, Pennsylvania.

Bred at Ansata:
G Dendera #435767 August 7, 1988
Grey mare by Ansata Halim Shah. Bred and owned by Paolo Gucci, Millfield Stables, Yorktown, New York.

G Princess Fadamiet #0459243 April 21, 1990
Grey mare by Prince Fa Moniet. Bred and owned by Paolo Gucci, Millfield, Stables, Yorktown, New York.

Progeny of Ansata Damietta

Ansata Dia Halima

G Dendera

Ansata Marietta

Ansata Abu Halim. Scott Trees photo.

Ansata El Mabrouk

BKA Damilll

Ansata Bint Misr/Ansata Damietta/Ansata Dia Halima

Ansata Dia Halima #0237265

Chestnut Mare; May 15, 1981
Sire: *Ansata Ibn Halima   Dam: Ansata Damietta
Sold to Dr. Chess Hudson, Lawrenceville, Georgia, February 1992.

PRODUCE

Ansata Damitha #0340688 October 10, 1985
Grey mare by *Jamilll. Sold to Dr. Chess Hudson, Lawrenceville, Georgia.

Ansata Damitha

Ansata Starletta

PRODUCE *of Ansata Dia Halima continued*

Ansata Amen Hotep #0389368 February 3, 1987.
Chestnut stallion by Ruminaja Bahjat.
Gelded and transferred to Robin Thompson, Sheridan, Arkansas, July 1989.

Ansata Radia #0431988 April 26, 1989
Grey mare by Ansata Mourad Bey. Died 1990.

Ansata Omar Kayam #0453473 May 15, 1990
Grey stallion by Prince Fa Moniet. Transferred to Jerry White, Mena, Arkansas. Gelded.

Ansata Monalima #0472856 May 18, 1991
Grey mare by Prince Fa Moniet.
Sold to a Canadian Partnership, January 1994.

Sold in foal to Ansata Manasseh to
Dr. Chess Hudson, Lawrenceville, Georgia.

Ansata Bint Misr/Ansata Damietta/Ansata Dia Halima/Ansata Marietta

Ansata Marietta #0279330

Grey Mare; June 21, 1983
Sire: Ansata Ibn Sudan   Dam: Ansata Damietta
Sold to Anita Wallin. Exported to Sweden.
Resold to Roger Spruyt, Belgium.

PRODUCE

Ansata El Shams #0396390 May 29, 1987
Bay stallion by Dorian Shah El Shams.
Gelded and transferred to Peggy Stanz, Marshall, Texas.

Ansata Abu Simbel #0424827 May 20, 1988
Grey stallion by Ansata Halim Shah.
Gelded and transferred to Randy Peale, Chesapeake, West Virginia, April 1989.

Ansata Starletta #0430213 June 6, 1989
Grey mare by Ansata Halim Shah.
Sold to Cynthia Culbertson, White, Georgia, December 1990.

Ansata El Farhan #0495485 March 2, 1991
Grey stallion by Prince Fa Moniet. Transferred To Mark and Candace Otto, Unadilla, Georgia.

Ansata Delilah in Texas. JEF photo

The Delilah Line of the Bint Misr Family

**Ansata Bint Misr/Ansata Delilah*

Ansata Delilah #80322
Grey Mare; April 10, 1972
Sire: Ansata Shah Zaman Dam: *Ansata Bint Misr

PRODUCE

Ansata Imperial #137119 March 18, 1976
Grey stallion by Ansata Ibn Sudan. Sold to Douglas and Barbara Griffith, Imperial Egyptian Stud, Parkton, Maryland, July 1978.

Ansata Amon Ra #157650 March 25, 1977
Chestnut stallion by Ansata Ibn Sudan. Sold to Alan and Karen Spitzer, Chino, California, February 1980.

Ansata Emir Halim #170224 April 12, 1978
Grey stallion by *Ansata Ibn Halima. Sold to Dana Arabian Stud, Roy, Washington, October 1980.

Ansata Samantha #196854 September 19, 1979
Chestnut mare by *Ansata Ibn Halima.

Ansata Haji Halim #219543 October 6, 1980
Grey stallion by *Ansata Ibn Halima. Sold to Giullermo Ribeiro, July 1989. Exported to Brazil.

Ansata Binthalima #237263 September 25, 1981
Grey mare by *Ansata Ibn Halima.

Ansata Sudarra #259990 September 21, 1982
Grey mare by Ansata Abu Sudan.

Ansata Deborah #302737 May 2, 1984
Grey mare by Ansata Halim Shah. Sold to Paolo Gucci, Millfield Stables, Yorktown, New York, January 1988 in foal to Ansata Abu Sudan. Foaled colt; unregistered.

Ansata Haji Jamil #327993 May 27, 1985
Grey stallion by *Jamilll. Sold to the Royal Stables of H.M. King Hassan II, Morocco. Exported 1989.

Ansata El Sami #355990 May 27, 1986
Bay stallion by *Jamilll. Gelded and transferred to Betty Jones, Texarkana, Arkansas.

Ansata Imperial

Ansata Amon Ra

Shams El Nejd #394567 May 26, 1987
Bay stallion by Dorian Shah El Shams. Gelded and transferred to Yvonne Head, Mena, Arkansas, December 1987.

Unnamed grey stallion by Ansata Halim Shah, May 27, 1988 Died August 3, 1988 before registration.

Ansata Sarai #453441 February 20, 1990.
Grey mare by Prince Fa Moniet.

Progeny of Ansata Delilah

Ansata Emir Halim

Ansata Haji Halim

Ansata Binthalima

Ansata Sudarra

Ansata Samantha

Ansata Deborah

Ansata Haji Jamil

Ansata Bint Misr/Ansata Delilah/Ansata Samantha

Ansata Samantha #196854
Chestnut Mare; September 19, 1979 -1999
Sire: *Ansata Ibn Halima   Dam: Ansata Delilah

PRODUCE
Ansata Samarra #303389 July 5, 1984
Bay mare by *Jamilll.

Ansata Samaria #327855 June 26, 1985
Bay mare by *Jamilll.

Unnamed bay stallion by *Jamilll, June 12, 1986.
Died June 15, 1986.

Ansata Sharifa #392691 August 25, 1987
Grey mare by Ansata Ibn Shah.

Ansata El Salaam #0421591 August 15, 1988
Chestnut stallion by Ansata Ibn Shah. Sold to
Heinrich and Margaret Kretschmann. Exported to Germany.

Ansata Samsara #0440751 September 8, 1989
Grey mare by Prince Fa Moniet.

Ansata Samai #0471998 April 7, 1991
Grey mare by Prince Fa Moniet. Sold to Dr.
Chess Hudson, Lawrenceville, Georgia, February 1992.

Ansata Samaria

Ansata Samantha in Arkansas. JEF photo

Progeny of Ansata Samantha

Ansata Samaria

Ansata El Salaam

Ansata Samai

Ansata Sharifa

Ansata Samarra

Ansata Samsara

Ansata Samiha

PRODUCE of Ansata Samantha continued

Ansata Samiha #481160 April 8, 1992
Grey mare by Prince Fa Moniet. Sold to Veruska Arabians,
Exported to Europe. Re-acquired by Ansata in 2000.

Ansata Shalimar #0493684 April 24, 1993
Grey mare by Prince Fa Moniet.

Ansata Sokar #0518767 March 19, 1995
Chestnut stallion by *Imperial Madheen.

Ansata Bint Misr/Ansata Delilah/Ansata Samantha/Ansata Samarra

Ansata Samarra #0303389
Bay Mare; July 5, 1984
Sire: *Jamilll   Dam: Ansata Samantha
Sold to H.R.H. Prince Khalid S.K. Al Saud, Nejd Stud,
December 1998.
Exported to Saudi Arabia.

PRODUCE

Ansata Selket #0418521 May 16, 1988
Grey mare by Ansata Halim Shah.

Ansata Sekhmet #0430225 May 26, 1989
Grey mare by Ansata Halim Shah.

Ansata Orion #0472042 May 14, 1991
Grey stallion by Prince Fa Moniet. Transferred
to Gail Halvorsen, Provo, Utah. Gelded.

Ansata Memnon #0481158 May 16, 1992
Grey stallion by Prince Fa Moniet.
Gelded. Sold to Joseph & Mary Brown, Tampa, Florida,
June 1996.

Ansata Safari August 23, 1993
Grey stallion by Prince Fa Moniet. Died prior to registration.

Ansata Sammoura #0529508 May 29, 1996
Grey mare by Ansata Hejazi. Sold to H.R.H. Prince Khalid
S.K. Al Saud, Nejd Stud. Exported to Saudi Arabia, October
1996.

Ansata Samari #0542103 June 3, 1997
Grey mare by Ansata Hejazi.
Sold to Mohammed J.K. Al Marzouk,
Al Ajmal Stud, Kuwait, and exported.

Ansata Saamir #0556012 September 1, 1998
Grey stallion by Ansata Iemhotep. Owned by
H.R.H. Prince Khalid S.K. Al Saud, Nejd Stud, Saudi Arabia.

Ansata Shalimar in Kuwait. Filsinger photo

Ansata Sokar

Ansata Samarra

Progeny of Ansata Samarra

Ansata Sekhmet

Ansata Orion

Ansata Saamir

Ansata Sammoura

Ansata Samari   Ron Shimer photo

Ansata Selket as a filly in Arkansas with Ansata's trainer, David Bradbury. Sparagowski photo.

*Ansata Bint Misr/Ansata Delilah/Ansata Samantha/Ansata Samarra/Ansata Selket

Ansata Selket #0418521
Grey Mare; May 16, 1988
Sire: Ansata Halim Shah Dam: Ansata Samarra
Co-owned with Claudia Quentin. Sold to Sheikh Abdulaziz Al Thani, Al Rayyan Farm, Qatar. Exported in foal to Ansata Iemhotep.

PRODUCE
Ansata Amir Sarin #0482307 May 10, 1992
Grey stallion by Prince Fa Moniet. Died 1995.

Unnamed premature grey stallion by Prince Fa Moniet, May 20, 1993. Died.

Ansata Safwan #0313232 September 13, 1994
Grey stallion by Arabest Kalid. Sold to Jay Smith, Eureka Springs, Arkansas, October 1996. Gelded.

Unnamed mare by Thee Desperado, February 26, 1996
Died March 1996.

Ansata Serqit #0542111 May 31, 1997
Grey mare by Ansata Iemhotep. Owned by Claudia Quentin.

Ansata Osiron #0557007 August 26, 1998
Grey stallion by Ansata Iemhotep. Sold in 2001 to Sheikha Sarah Al Sabah. Exported to Kuwait.

Exported in foal to Ansata Iemhotep. Foaled filly in Qatar.

Ansata Serqit

*Ansata Bint Misr/Ansata Delilah/AnsataSamantha/Ansata Samarra/Ansata Selkhmet

Ansata Sekhmet #0430225
Grey Mare; May 26, 1989
Sire: Ansata Halim Shah Dam: Ansata Samarra
Sold to H.R.H. Prince Khalid S.K. Al Saud, December 1997.
Exported to Saudi Arabia.
Died on arrival.

PRODUCE
Ansata Sabika #0494080 January 16, 1993
Grey mare by Prince Fa Moniet.
Sold to Canadian Partnership January 1994.

Ansata Shahnaz #0505493 February 15, 1994
Grey mare by Prince Fa Moniet. Sold to Omar Sakr. Exported to Egypt.

Ansata Osiron

Ansata Sabika

Ansata Shahnaz

PRODUCE of Ansata Sekhmet continued

Ansata Sirius #0537019 August 23, 1996
Grey stallion by Ansata Iemhotep.

Ansata Al Bassam #0551163 February 13, 1998.
Grey stallion by Ansata Iemhotep.
Sold to H.R.H. Prince Khalid S.K. Al Saud, Nejd Stud.
Exported to Saudi Arabia.

AnsataBintMisr/AnsataDelilah/AnsataSamantha/Ansata Samaria

Ansata Samaria #0327855
Bay Mare; June 26, 1985
Sire: *Jamilll   Dam: Ansata Samantha
Sold to Sheikh Abdulaziz Al Thani, Al Rayyan Farm.
Exported to Qatar, March 1994

Ansata Sirius

PRODUCE

Unnamed grey stallion by Ansata Halim Shah, April 9, 1991
Transferred to Rick Hawkins, Howard, Colorado. Gelded.

Ansata El Shahraf #0479623 May 28, 1992
Grey stallion by Ansata Halim Shah. Leased to
Paul and Anne Walker, Montebello Farms, Canada,
January 1994. Returned to Ansata.
Sold to Barbara and Tyrone Lewis, Baraka Farm,
Cove, Arkansas.

1993 serviced by Fa Daalim #091824
Foaled bay filly at Al Rayyan Farm in Qatar, 1994.

Ansata Al Bassam

Left, Ansata Samarra
with her daughters.
Ansata Sekhmet, below right.

Sparagowski photos.

Ansata Selket photographed
in Qatar by Judith Wich.

Ansata Sharifa. Sparagowski photo.

Ansata El Shahraf

G Shafaria

Ansata Shahriyar

*AnsataBintMisr/AnsataDelilah/AnsataSamantha/Ansata Sharifa

Ansata Sharifa #0392691
Grey Mare; August 25, 1987
Sire: Ansata Ibn Shah   Dam: Ansata Samantha
Sold half interest to Paolo Gucci, Millfield Stables, New York, January 1, 1988. Sold by Forbis and Gucci to Sheikh Abulaziz Al Thani, Al Rayyan Farm, October 1993. Exported to Qatar.

PRODUCE

G Shafaria #0488155 February 19, 1992
Grey mare by Prince Fa Moniet. Co-owned with P. Gucci.
Full interest purchased by Ansata 1993.

Ansata Shahriyar #0493780 February 25, 1993.
Grey stallion by Prince Fa Moniet.
Full interest purchased from Gucci by Ansata 1993. Gelded.

1993 serviced by Prince Fa Moniet. Exported to Qatar.
Foaled 1994 filly.

Ansata Selma in Kuwait

Ansata Selman in Qatar

Ansata Safeer

*AnsataBintMisr/AnsataDelilah/AnsataSamantha/Ansata Sharifa/G Shafaria

G Shafaria #0488155
Grey Mare; February 19, 1992
Sire: Prince Fa Moniet Dam: Ansata Sharifa
Sold to Sheikh Abdulaziz Bin Khalid Al Thani, Al Rayyan Farm, Qatar.

PRODUCE

Ansata Selma #0552499 June 14, 1998
Grey mare by Ansata Hejazi.
Sold to Sh. Abdulaziz Bin Khalid Al Thani, Al Rayyan Farm, Qatar. Exported.

Ansata Selman #0566312 September 18, 1999
Grey stallion by Ansata Hejazi. Sold to Sh. Abdulaziz Bin Khalid Al Thani, Al Rayyan Farm, Qatar. Exported.

Exported in foal to Ansata Hejazi. Foaled filly in Qatar.

*AnsataBintMisr/AnsataDelilah/AnsataSamantha/Ansata Samsara

Ansata Samsara #0440751
Grey Mare; September 8, 1989 - July 1998
Sire: Prince Fa Moniet Dam: Ansata Samantha
Sold to Kuwait Equestrian Club (Kuwait Arabian Horse Center), May 1998. Died before export.

PRODUCE

Ansata El Samir #0493688 June 6, 1993
Grey stallion by Ansata Halim Shah. Gelded.
Transferred to Jeremy and Chad Harper, Prattsville, Arkansas.

Ansata Samiri #0510996 August 28, 1994
Grey mare by Ansata Halim Shah.
Sold to Virgilio Sadnik, Italy, November 1995. Exported.

Ansata Safeer #0529509 February 16, 1996
Grey stallion by Ansata Hejazi. Sold to Jan and Arja Lancee. Exported to the Netherlands, November 1996. Later acquired by H.R.H. Princess Zeyn Bint Hussein and Majdi Al Saleh, Jaafar Stud, Jordan.

Ansata Al Murtajiz #0540868 February 14, 1997
Grey stallion by Ansata Hejazi. Sold to Usamah Al Kazemi, July 1997. Exported to Kuwait.

Ansata Suleyma #0551161 February 1998
Grey mare by Ansata Hejazi. Sold to Usamah Al Kazemi and exported to Kuwait.

Ansata Samiri

Ansata Al Murtajiz

Ansata Suleyma

*AnsataBintMisr/AnsataDelilah/AnsataSamantha/Ansata Samiha

Ansata Samiha #0481160
Grey Mare; April 8, 1992
Sire: Prince Fa Moniet Dam: Ansata Samantha
Sold to Veruska Arabians and exported to Europe.
Reimported by Veruska; Re-acquired by Ansata in 2000.

*AnsataBintMisr/AnsataDelilah/AnsataSamantha/Ansata Shalimar

Ansata Shalimar #0493684
Grey Mare; April 24, 1993
Sire: Prince Fa Moniet Dam: Ansata Samantha
Sold to Mohammed J.K. Al Marzouk, Al Ajmal Stud, July 1999. Exported to Kuwait in foal to Ansata Hejazi.

PRODUCE
Ansata Sherrara #0552198 June 1, 1998.
Grey mare by Ansata Hejazi. Sold to Kuwait Equestrian Club (Kuwait Arabian Horse Center) February 1999. Exported to Kuwait in foal to Ansata Hejazi; foaled filly in Kuwait.

Ansata Sherrara in Kuwait

Ansata Bint Halima. JEF photo.

*AnsataBintMisr/AnsataDelilah/Bint Halima

Ansata Binthalima #0237263
Grey Mare; September 25, 1981
Sire: *Ansata Ibn Halima   Dam: Ansata Delilah
Sold to H.H. Sheikh Hamad bin Khalifa Al Thani, Al Shaqab Farm, Doha, Qatar, January 1994.

PRODUCE

Ansata El Jalil #327994 June 26, 1985
Grey stallion by *Jamilll.   Gelded and
transferred to J. and A. Parks, Pilot Point, Texas.

Ansata El Hamis #0355953 June 5, 1986
Grey stallion by *Jamilll.
Gelded and transferred to Carrie Hatcher, Palestine, Texas.

Ansata Pristina #0392716 August 28, 1987
Grey mare by Ansata Ibn Shah. Sold to Paolo Gucci, Millfield Stables, New York, November 1987. Died at Ansata.

Ansata Ramose #0419889 September 4, 1988
Grey stallion by Ansata Ibn Shah.
Gelded and transferred to Dawn Hotubee, Mena, Arkansas, June 1989.

Ansata Prince Hal #0440750 September 10, 1989
Grey stallion by Prince Fa Moniet. Transferred to Jerry White, Mena, Arkansas. Gelded.

Ansata Nawarra #0472004 February 27, 1991
Grey mare by Prince Fa Moniet.

Ansata Amir Fahim #0482309 April 2, 1992
Grey stallion by Prince Fa Moniet.
Sold to Teresa Lyons, Panama, Oklahoma, October 1993. Gelded.

Ansata Neoma #0493649 April 10, 1993
Grey mare by Prince Fa Moniet.

Exported to Al Shaqab Stud, Qatar, January 1994 in foal to Ansata Manasseh. Foaled filly in Qatar.

Ansata Binthalima in Qatar. Van Lent Jr. photo

Ansata Nawarra

At left and right, Ansata Neoma

AnsataBintMisr/AnsataDelilah/Bint Halima/Ansata Nawarra

Ansata Nawarra #0472004
Grey Mare; February 21, 1991
Sire: Prince Fa Moniet   Dam: Ansata Bint Halima
Sold to Sh. Abdulaziz Al Thani, Al Rayyan Stud, Qatar.

PRODUCE
Elmiladi Zaynah # 0571382 March 21, 1998
Grey mare by Zedann.

Unnamed grey stallion by Zedann, May 19, 1999.
Gelded.

Exported to Al Rayyan Farm, Qatar.

Ansata Neamet

AnsataBintMisr/AnsataDelilah/Bint Halima/Ansata Neoma

Ansata Neoma #0493649
Grey Mare; April 10, 1993
Sire: Prince Fa Moniet   Dam: Ansata Bint Halima
Sold to Ray and Jamie Roberts, Excelsior Springs, Missouri.

PRODUCE
Ansata Neamet #0552500 May 28, 1998
Grey mare by Ansata Hejazi.
Sold to Claudia Quentin, Buenos Aires, Argentina, S.A.

Ansata Nadra #0566463 August 28 1999
Grey mare by Ansata Hejazi.

Ansata Nariya #0579273 August 12, 2000
Grey mare by Ansata Hejazi.

NF Naeema #585660 August 18, 2001
Grey mare by Ansata Sirius. Owned by Ray and Jamie Roberts, Excelsior Springs, Missouri.

Ansata Nadra

Ansata Nariya   *Elmiladi Zaynah*   *NF Naeema*

Ansata Sudarra in Qatar at the Emir's Stables, Al Shaqab Stud. JEF photo

AnsataBintMisr/AnsataDelilah/Ansata Sudarra

Ansata Sudarra #0259990

Grey Mare; September 21, 1982 - July 24, 1995
Sire: Ansata Abu Sudan   Dam: Ansata Delilah
Sold to H.H. Sheikh Hamid Bin Khalifa Al Thani, Al Shaqab Farm, Doha, Qatar. Exported to Qatar, January 1994.

PRODUCE

Ansata Etherea #0355950 March 8, 1986.
Grey mare by *Jamilll #268275. Died 1987.

Ansata Nefertiti #0389421 March 1987
Grey mare by Ansata Halim Shah.

Ansata Nefara #0410754 March 13, 1988
Grey mare by Ansata Halim Shah.

Ansata Nefertari #0453399, April 2, 1990
Grey mare by Prince Fa Moniet.

Ansata El Suhayli #0476891 May 8, 1991
Grey stallion by Prince Fa Moniet. Donated to Fernleigh Equine Foundation.

Ansata Hejazi #0479624 May 8, 1992
Grey stallion by Ansata Halim Shah.

Ansata Nefri #0497407 June 3, 1993
Grey mare by Ansata Manasseh.

1994 exported in foal to Ansata Halim Shah. Foaled filly in Qatar.

Ansata Hejazi in Kuwait. Filsinger photo.

Ansata Nefara

Ansata Nefertiti

Ansata Nefri

*AnsataBintMisr/AnsataDelilah/Ansata Sudarra/Ansata Nefertiti

Ansata Nefertiti #0389421
Grey Mare; March 24, 1987
Sire: Ansata Halim Shah   Dam: Ansata Sudarra
Sold to Mohamed J.K. Al Marzouk, Al Ajmal Stud, Kuwait. Exported.

PRODUCE
Ansata Nefer Isis #0274048 August 11, 1991
Grey mare by Prince Fa Moniet. Sold to the Dr. F. Santoro, La Frasera Arabian Stud. Exported to Italy, July 1993.

Ansata El Faheem #0473236 August 15, 1992
Grey stallion by Prince Fa Moniet. Donated to Fern Leigh Foundation, December 1993.

Ansata Queen Nefr #0498356 September 1, 1993
Grey mare by Prince Fa Moniet. Sold to Arja Lancee, Netherlands. Exported.

Ansata Tousson #0519647 February 27, 1995
Grey stallion by Prince Fa Moniet. Gelded and sold to Linda Hacker, Texarkana, Texas, March 1997.

Unnamed grey stallion by Ansata Manasseh May 12, 1996 Transferred locally. Unregistered.

Ansata Nefer #0542102 May 7, 1997
Grey mare by Prince Fa Moniet.

Ansata Nashmia #057305 February 20, 1999
Grey mare by Ansata Sokar. Sold to Sheikh Abdulaziz Al Thani, Al Rayyan Farm, Doha, Qatar. Exported.

Exported in foal to Ansata Sokar. Foaled filly in Kuwait at Al Ajmal Stud.

Ansata Nefertiti in Kuwait at Ajmal Stud. Filsinger photo.

Ansata Queen Nefr

Ansata Nefer Isis

Ansata Nefer

Ansata Tousson

Ansata Queen Nefr

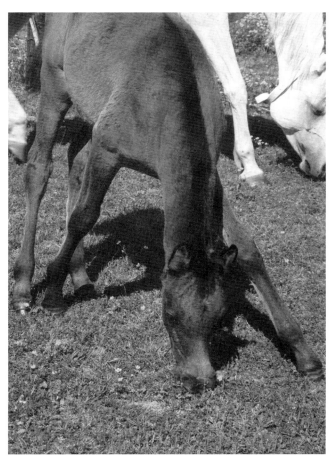
Ansata Nashmia

Ansata Iemhotep just before winning 1996 U.S. Top Ten Futurity Stallion. Vesty photo.

Ansata Iemhotep winning his yearling class, Egyptian Event.

Ansata Sinan and Ansata's trainer, Richard Sanders.

AnsataBintMisr/AnsataDelilah/Ansata Sudarra/Ansata Nefara

Ansata Nefara #0410754

Grey Mare; March 24, 1988 – December 14, 1996
Sire: Ansata Halim Shah   Dam: Ansata Sudarra

PRODUCE

Ansata Sinan #0482308 February 16, 1992
Grey stallion by Prince Fa Moniet. Sold to
Peter and Vera Stoessel-Vocka, Switzerland and
Spain. Exported to Europe, September 1994.

Ansata Iemhotep #0493704 March 4, 1993
Grey stallion by Prince Fa Moniet.

Unnamed grey stallion by Prince Fa Moniet, March 13, 1994.
Died September 15, 1994.

Ansata Nafisa #0518781 March 22, 1995
Grey mare by Prince Fa Moniet. Sold to
Omar Sakr, November 1995. Exported to Egypt.

Unnamed grey stallion by Prince Fa Moniet,
September 16, 1996.
Died.

Ansata Nafisa

Ansata Nefara as a filly in Arkansas. Sparagowski photo

AnsataBintMisr/AnsataDelilah/Ansata Sudarra/Ansata Nefertari

Ansata Nefertari #0453399
Grey Mare; April 2, 1990 - July 1998

PRODUCE
Ansata Shalim #0511213 August 4, 1994
Grey stallion by Ansata Halim Shah. Sold to
Al Naif Stud, Sheikh Abdullah Bin Nasir Al Thani.
Exported to Qatar, February 1997.

Unnamed stallion May 21, 1996. Grey colt by Ansata
Manasseh. Gelded. Transferred to Y. Head, Mena.

Ansata Sirdar #0556196 May 27, 1998
Grey stallion by Ansata Sokar. Gelded. Sold to
Dr. and Mrs. Andrew David, Mena, Arkansas.

AnsataBintMisr/AnsataDelilah/Ansata Sudarra/Ansata Nefri

Ansata Nefri #0497407
Grey Mare; June 4, 1993
Sire: Ansata Manasseh Dam: Ansata Sudarra

PRODUCE
Ansata Shammar #0566534 May 27, 1999
Grey stallion by Ansata Iemhotep.

Ansata Bint Nefri #0579272 August 27, 2000
Grey mare by Ansata Iemhotep.

Ansata Emir Sinan #0589161 November 4, 2001
Grey stallion by Ansata Sinan.

Ansata Shalim

Above and below, Ansata Shammar

Ansata Sirdar

Ansata Bint Nefri

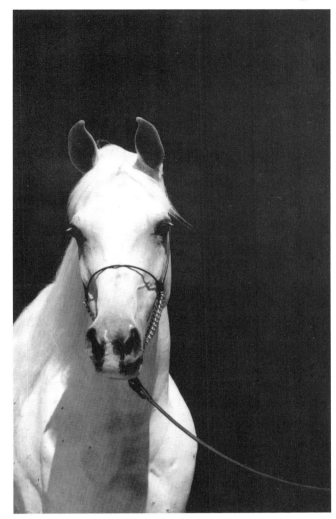
Ansata Sarai

Ansata Emir Sinan

*<u>*AnsataBintMisr/AnsataDelilah/Ansata Sarai</u>

Ansata Sarai #0453441
Grey Mare; February 20, 1990
Sire: Prince Fa Moniet   Dam: Ansata Delilah

PRODUCE
Ansata Al Kadir #0313333 August 16, 1994
Grey stallion by Arabest Kalid. Transferred to Gail Halvorsen, Provo, Utah. Gelded.

Ansata Ramzy #0563252 March 22, 1996
Grey stallion by Ansata Ramazan. Gelded. Sold to L.J. Detter of Oklahoma.

Ansata Aya Halima
Sparagowski photo

AnsataBintMisr/AnsataDelilah/Ansata Aya Halima

■ Ansata Aya Halima #0220099
Bay Mare; November 11, 1980
Sire: *Ansata Ibn Halima   Dam: *Ansata Bint Misr
Sold to Paolo Gucci, Millfield Stables, New York, July 8, 1988. Exported to England.

PRODUCE

Ansata Aya Maria #327826 May 3, 1985
Chestnut mare by Ansata Shah Zaman.

Ansata Aya Nadira #0353591 May 23, 1986
Grey mare by *Jamilll. Sold to Philippe and Yvonne de Bourbon, Haras Jinnah Al Tayr, July 16, 1986. Exported to Uruguay, South America.

Ansata Aya Nabila #0390309 May 17, 1987
Bay mare by Dorian Shah El Shams #206955. Transferred to Ronald J. Hutchings. Exported to England.

Bred at Ansata:
G Halisha #0433560 February 16, 1989
Grey mare by DAJ Abbas Pasha.
Owned by Paolo Gucci, Millfield Stables. N.Y.

G Princess Halima #0466935 March 10, 1990
Grey mare by Prince Fa Moniet.
Owned by Paulo Gucci, Millfield Stables, N.Y. Sold to Corinna and Wolfgang Bindl.
Exported to Germany.

G Ayyah #0482552 March 27, 1991.
Bay mare by AK Sirhalima. Owned by Millfield Stables, N.Y.

*AnsataBintMisr/AnsataDelilah/Ansata Aya Halima/Ansata Aya Maria

Ansata Aya Maria #0327826
Chestnut Mare; May 3, 1985
Sire: Ansata Shah Zaman   Dam: Ansata Aya Halima
Sold to Erwin and Annette, Escher. Exported to Germany, February 1994.

PRODUCE

Ansata Aya Adora #0430201   April 21, 1989
Grey mare by Ansata Abu Sudan. Sold to
Steve and Debbie Brunhild of Texas, December 1990.

Ansata Aya Maria leased to Dr. Chess Hudson, Lawrenceville, Georgia, 1990-1994.

*AnsataBintMisr/Ansata Misriya

Ansata Misriya #0355911
Bay Mare; April 19, 1986
Sire: *Jamilll   Dam: *Ansata Bint Misr
Sold to Joe Young, Plainfield, New Jersey, June 1991, in foal to Ansata Halim Shah.

PRODUCE

Ansata Allegra #0471999   April 9, 1991
Grey mare by Ansata Halim Shah.
Sold to Virgilio Sadnik, Italy, July 1995. Exported in foal to Ansata Hejazi.

ESA Halim #0489810   April 20, 1992
Grey stallion by Ansata Halim Shah. Owned by Joe Young. Gelded.

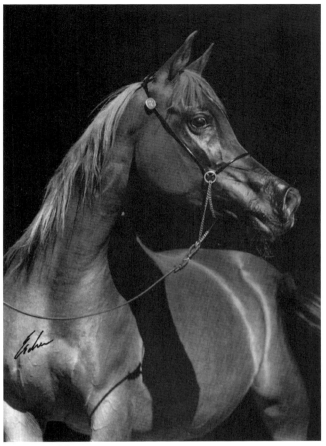

Ansata Aya Maria.

Ansata Allegra. Filsinger photo.

Ansata Aya Nadira winning Champion Mare in Uruguay.

Ansata Rhodora (left) with Ansata Bint Sudan in Arkansas. Sparagowski photo

The Rhodora Family

Ansata Bint Bukra/Ansata Rhodora

Ansata Rhodora #49569
Grey Mare; March 29, 1968
Sire: *Ansata Ibn Halima   Dam: *Ansata Bint Bukra
Transferred to Dr. Chess Hudson, Lawrenceville, Georgia, March 1991.

PRODUCE

Ansata El Shahwan #94541 June 16, 1973
Bay stallion by Ansata Shah Zaman. Sold to Alan Pond, Forest Hills Arabian Stud, October 1983. Exported to Australia.

Ansata Rhozira #122314 February 18, 1975
Grey mare by *Ansata El Wazir.
Sold to Cathy Downing, Chino, California, January 1978.

Ansata Ibn Rashid #157651 February 6, 1977
Bay stallion by Nabiel. Sold to Judy Ayres, Toronto, Canada, October 1977.

Ansata Kamal Bey #170221 February 12, 1978
Grey stallion by Ansata Ibn Sudan. Sold to Robert and Joanne Thorndike, Orville, Washington, February 1980.

Ansata El Shahwan

Ansata Rhozira

Ansata Kamal Bey

Ansata Ibn Rashid

Ansata Ramona. Sparagowski photo

Ansata Mari Isis

Ansata Rebecca

PRODUCE of Ansata Rhodora continued

Ansata Rebecca #196869 April 2, 1979
Grey mare by Ansata Abu Sudan.

Ansata El Rahim #278156 February 15, 1983
Bay stallion by Ansata Halima Son.
Died December 9, 1983.

Ansata Ramona #0208572 October 6, 1984
Bay mare by *Jamilll. Sold 1989 to Mourad
El Cassabugui, Athens, Texas.

Ansata Rhoda #355992 May 29, 1986
Grey mare by *Jamilll. Sold 1989 to Terry and
Nancy Alleman, San Antonio, Texas.

Unnamed stallion June 12, 1987, by Dorian Shah
El Shams. Died prior to registration.

*Ansata Bint Bukra/Ansata Rhodora/Ansata Rebecca

■ Ansata Rebecca #0196869
Grey Mare; April 2, 1979
Sire: Ansata Abu Sudan   Dam: Ansata Rhodora
Sold to Frau Liesel Vetter, December 1990. Exported to
Germany.

PRODUCE

Ansata Mari Isis #278163 July 29, 1983
Grey mare by Ansata Ibn Shah. Sold to
Dr. Aloysio Faria, Delta Arabians,
November 1984. Exported to Brazil, South America.

Ansata El Shahir #0308013 July 3, 1984
Grey stallion by Ansata Ibn Sudan. Gelded and
transferred to Bill and Karen Barnes, Mountain
Harbor, Arkansas.

PRODUCE of Ansata Rebecca continued

Ansata Taj Malik #0327963 June 15, 1985
Grey stallion by *Jamilll. Gelded and
transferred to Lynn Carle, Springfield, Missouri.

Ansata Ranita #0389382 April 24, 1987
Grey mare by Ansata Halim Shah. Sold
To Paolo Gucci, Millfield Stables, New York.
Died 1990.

Ansata El Ibriz #0418442 April 11, 1988
Grey stallion by Ansata Halim Shah. Transferred to
Andrew and Dorian Weil, Dorian Farm, Goshen,
New York, February 1989.

Ansata Ramazan # 0430202 April 7, 1989
Grey stallion by Ansata Halim Shah. Sold to
Deborah Wood, Littleton, Colorado. Gelded.

Ansata Ramesses #0453442 April 20, 1990
Grey stallion by Prince Fa Moniet. Sold to
Royal Stables of HM King Hassan II. Exported to
Morocco.

Ansata Bint Bukra/Ansata Rhodora/Ansata Ramona

Ansata Ramona #308572
Bay Mare; October 7, 1984
Sire: *Jamilll Dam: Ansata Rhodora
Sold in 1987 to Mourad El Cassabagui, Athens, Texas
Sold by El Cassabagui to Majdi Al Saleh.
Exported to Jordan November 1996.

PRODUCE

El Nabil #0420769 August 12, 1988
Grey stallion by Ansata Halim Shah. Owned
By Mourad El Cassabagui, Athens, Texas.

Ansata Ranita

Ansata Ramazan

Ansata El Ibriz

Ansata Ramesses

Ansata Rosetta being exhibited at an Ansata seminar in Arkansas. JEF photo

The Rosetta Family

**Ansata Bint Bukra/Ansata Rosetta*

Ansata Rosetta #70167
Grey Mare; April 6, 1971 – Deceased
Sire: Ansata Shah Zaman   Dam: *Ansata Bint Bukra

PRODUCE

Ansata Ghazala #137116 April 19, 1976
Grey mare by Ansata Ibn Sudan.

Ansata El Ghazal #170223 March 25, 1978
Grey stallion by Ansata Ibn Sudan. Sold to
Mary and Garner Tullis, Days Creek, Oregon,
April 1980.

Ansata Omar Halim #196861 April 25, 1979
Grey stallion by *Ansata Ibn Halima. Sold to
Lyle and Virginia Bertsch, Fort Wayne, Indiana.
Sold by Bertsch to Peter and Vera Stoessel-Vocka
of Switzerland and Spain. Exported April 1994 to Belgium
and died after arrival.

Ansata Halim Shah #219546 April 28, 1980
Grey stallion by *Ansata Ibn Halima. Sold to H.H.
Sheikh Hamad Bin Khalifa Al Thani, Al Shaqab Stud, Doha,
January 1994. Exported to Qatar. Died September 1994.

Ansata Raja Halim #259984 January 28, 1982
Grey stallion by *Ansata Ibn Halima. Gelded and
transferred to Ed Shinn, London, Arkansas.

Ansata Rose Queen #278155 February 10, 1983
Grey mare by Ansata Halima Son. Died
August 11, 1994.

Ansata El Ghazal

Ansata Omar Halim

Ansata Raja Halim

Ansata Halim Shah

Ansata Prima Rose

Ansata Ghazala

PRODUCE of Ansata Rhodora continued

Ansata Prima Rose #327992 April 13, 1985
Grey mare by *Jamilll.

Ansata Haji Hamid #419472 February 2, 1988
Grey stallion by Ansata Haji Halim. Gelded.
Tansferred to Elaine Watkins, October 1988.

Ansata Bint Bukra/Ansata Rosetta/Ansata Ghazala

Ansata Ghazala #137116
Grey Mare; June 10, 1976 – January 10, 1991
Sire: Ansata Ibn Sudan   Dam: Ansata Rosetta
Sold to Jim Wells, Germantown, Tennessee, November 1989

Ansata Rose Queen

PRODUCE

Ansata Ghazia #237262 September 4, 1981
Grey mare by *Ansata Ibn Halima.

Ansata Ghazi Pasha #278154 February 6, 1983
Grey stallion by Ansata Halima Son. Gelded.
Transferred to Tilman Gosnell, Mena, Arkansas.

Ansata Ghazi Bey #303390 March 1, 1984
Grey stallion by *Jamilll. Gelded.

Ansata Exotica #0355992 February 23, 1986
Grey mare by *Jamilll.

Ansata Galia #0389318 May 22, 1987
Bay mare by Dorian Shah El Shams
Sold to Jim Wells, Germantown, Tennessee,
October 1989.

Ansata Exotica

*Ansata Bint Bukra/Ansata Rosetta/Ansata Ghazala/Ansata Ghazia

Ansata Ghazia #237262
Grey Mare; September 4, 1981 – June 6, 1987
Sire: *Ansata Ibn Halima   Dam: Ansata Ghazala

PRODUCE

Ansata Gloriana #0356019 March 21, 1986
Grey mare by *Jamilll.
Leased to Dr. Hans Nagel. Exported to Germany,
September 1990.

Ansata El Azrak #0390185 April 9, 1987
Grey stallion by Dorian Shah El Shams
Gelded. Transferred to Charley Hayes, Waldron,
Arkansas, September 1990.

*Ansata Bint Bukra/Ansata Rosetta/Ansata Ghazala/Ansata Ghazia/
Ansata Gloriana

Ansata Gloriana #0356019
Grey Mare; March 21, 1986 – July 30, 1995
Sire: *Jamilll   Dam: Ansata Ghazia
Leased to Dr. Hans Nagel. Exported to Germany 1990
Died July 30, 1995 in Germany while on lease.

PRODUCE

Helala GASB #13084 March 29, 1992.
Bay mare by Salaa El Dine. Owned by Dr. Nagel,
Germany.

El Mousheer GASB March 28, 1994
Grey stallion by Salaa El Dine.

Ansata Helwa GASB 24730 May 3, 1995
Grey mare by Salaa El Dine, sold to Omar Sakr, Egypt.

Ansata Ghazia

Ansata Gloriana

Ansata Helwa in Egypt. Toichel photo

Helala in Germany at Katharinenhof.

Ansata Bint Bukra/Ansata Rosetta/Ansata Ghazala/Ansata Ghazia/ Ansata Gloriana/Ansata Shahrezade

Ansata Shahrezade #0479620

Grey Mare; April 4, 1992
Sire: Ansata Halim Shah Dam: Ansata Exotica

PRODUCE

Unnamed stallion by Ansata Hejazi, June 2, 1996
Died June 5, 1996.

Ansata Sahir #0566465 May 10, 1999
Grey stallion by Ansata Iemhotep. Sold to Sheikh Abdullah Bin Nasir Al Thani, Al Naif Stud, Doha, Qatar. Exported.

Ansata AlGhazzali #0579274 August 30, 2000
Grey stallion by Ansata Iemhotep.
Sold to Ian Woodward, England. Exported 2001.

Ansata Bint Shahrezad #0597185 April 1, 2002
Grey mare by Farres.

Ansata Sahir

Ansata AlGhazzali

Ansata Bint Bukra/Ansata Rosetta/Ansata Prima Rose

Ansata Prima Rose #0327992

Grey Mare; April 13, 1986
Sire: *Jamilll   Dam: Ansata Rosetta
Sold half interest to Wolfgang and Sylvie Eberhardt, Katr El Nada Stud Farm, Germany, December 15, 1990. Exported to Germany in 1991. Resold to Al Shaqab Stud, H.H. Sheikh Hamad bin Khalifa Al Thani, Doha, Qatar. Exported from Germany to Qatar 1993.

PRODUCE

Ansata El Iman #0430159 March 22, 1989
Grey stallion by Imperial Imdal. Gelded.

Ansata Al Barak #0453430 March 31, 1990
Grey stallion by Prince Fa Moniet. Gelded.
Transferred to Sue Hillman, Mayflower, Arkansas.

Ansata Ken Ranya #0511904 January 21, 1992
Grey mare by Salaa El Dine. Co-owned with Katr El Nada, Germany. Imported by Ansata 1994.

Ansata Ken Rashik #0511905 April 28, 1993
Grey stallion by Salaa El Dine
Co-owned with Katr El Nada. Imported by Ansata 1994.
Sold 1998 to Gary and Tracy Davis, Ozark, Missouri.

Ansata Bint Shahrezad

Ansata Prima Rose

Ansata Ken Ranya

Ansata Ken Rashik

*Ansata Bint Bukra/Ansata Rosetta/Ansata Prima Rose/Ansata Ken Ranya

Ansata Ken Ranya #0511964
Grey Mare; January 21, 1992
Sire: Salaa El Dine   Dam: Ansata Prima Rose
Bred and co-owned with Sylvie and Wolfgang Eberhardt, Germany.
Imported by Mr. and Mrs. Donald L. Forbis, Mena, Arkansas 1994. Exported to Dr. Hans Nagel, Germany, 1998.

PRODUCE
Ansata Rafi #0534408 March 9, 1996
Grey stallion by Prince Fa Moniet. Transferred to a charitable foundation, Houston, Texas.

Ansata Rahmah #0542439 March 22, 1997
Grey mare by Ansata Hejazi. Owned by
Sylvie and Wolfgang Eberhardt of Germany, transferred 1998 to Donald and Judith Forbis.

*Ansata Bint Bukra/Ansata Rosetta/Ansata Prima Rose/Ansata Ken Ranya/Ansata Rahmah

Ansata Rahmah #0542439
Grey Mare; July 2, 1996
Sire: Anata Hejazi   Dam: Ansata Ken Ranya
Bred by Mr. and Mrs. Donald L. Forbis, Mena, Arkansas.

Ansata Rahotep #0587267 March 22, 2001
Grey stallion by Ansata Iemhotep.
Sold 2001 to Joseph and Kimberly Murgola, Mena, Arkansas.

Ansata Al Rahim #059950 September 10, 2002
Grey stallion by Ansata Iemhotep.

Ansata Rafi

Ansata Rahotep

Ansata Rahmah

Above and below, Ansata Bint Sudan

The Bint Sudan Family

Ansata Bint Bukra/Ansata Bint Sudan

■ Ansata Bint Sudan #94537
Grey mare; February 26, 1973 – Died in Germany 1993
Sire: Ansata Ibn Sudan   Dam: *Ansata Bint Bukra
Sold to Heinrich and Margaret Kretschmann, July 1989.
Exported to Germany.

PRODUCE
Ansata King Tut #153295, March 24, 1977
Originally named Ansata Tashrif. Grey stallion
By Nabiel.   Sold to William and Karen
Scott, Sunnymead, California, February 1980.

Ansata El Tareef #183873 January 31, 1979
Grey stallion by *Ansata Ibn Halima. Sold to
Dr. Aloyso Faria, Fazenda Fortaleza. Exported to
Brazil.

Premature grey stallion by *Ansata Ibn Halima,
February 25, 1980. Died.

Ansata Nahema #237264 September 21, 1981.
Grey mare by Ansata Abu Sudan. Sold at Pyramid Society
Auction.

Ansata Amir Zaman #308116 February 21, 1984
Grey stallion by Ansata Shah Zaman. Exported to
Germany on lease to Kretschmann. Sold to
Gunter Seidlitz family, Germany.

Ansata Chandra #355951 May 28, 1986
Grey mare by *Jamilll. Sold to
Guillermo Henin, December 1990. Exported to Argentina.

Shams Al Sahara #394543 June 26, 1987
Bay stallion by Dorian Shah El Shams.
Gelded and transferred to Susan Vaught, Mena, Arkansas,
February 1988.

Ansata Shahira #410782 June 25, 1988
Grey mare by Ansata Halim Shah. Sold to
Heinrich and Margaret Kretschmann. Exported to Germany.

Ansata Moniet Sudan G.A.S.B. #21334 June 4, 1990
born in Germany.
Grey mare by Prince Fa Moniet. Leased to Dr. Hans Nagel.

Ansata El Tareef

Ansata King Tut

Ansata Amir Zaman

*Ansata Bint Bukra/Ansata Bint Sudan/Ansata Nahema

Ansata Nahema #0237264
Grey Mare; September 12, 1981
Sire: Ansata Abu Sudan   Dam: Ansata Bint Sudan
Sold to Barbara Reilly, Pyramid Society Sale II
Resold to Miguel Muzquiz, Mexico City, Mexico.

PRODUCE

Ansata Angela #0363140 May 12, 1986
Grey mare by ET Crown Prince. Died.

Ansata El Ajeeb #0410762 March 22, 1988
Grey stallion by Ansata Haji Halim. Gelded and sold to Dr. Chess Hudson, Lawrenceville, Georgia.

Ansata Almira #0453474 March 18, 1990
Grey mare by Prince Fa Moniet. Sold to Miguel Muzquiz, Mexico City, Mexico.

Ansata Latifa #0480535 April 6, 1991
Grey mare by Ansata Halim Shah. Owned by Miguel Muzquiz, Mexico City, Mexico.

*Ansata Bint Bukra/Ansata Bint Sudan/Ansata Chandra

Ansata Chandra #0355951
Grey Mare; May 28, 1986
Sire: *Jamilll   Dam: Ansata Bint Sudan
Sold to Giullermo Henin, December 1990.
Exported 1991 to Argentina, South America.

PRODUCE

Ansata Al Shahab #0458058 May 15, 1990
Grey stallion by Prince Fa Moniet. Gelded.

Ansata Nahema

Ansata El Ajeeb

Ansata Latifa

Ansata Regina in Texas. JEF photo

The Regina Family

Ansata BintBukra/Ansata Regina

Ansata Regina #107422
Grey Mare; April 19, 1974 – November 16, 1985
Sire: *Ansata El Wazir   Dam: *Ansata Bint Bukra

PRODUCE

Ansata Mourad Bey #196858 May 4, 1979
Grey stallion by Ansata Ibn Sudan.

Ansata Raqessa #219547 May 2, 1980
Grey mare by *Ansata Ibn Halima.

Ansata Serena #240426 April 28, 1981
Grey mare by *Ansata Ibn Halima.

Unnamed bay stallion by *Ansata Ibn Halima May 2, 1982
Died prior to registration.

Ansata Shah Osman #278162 June 15, 1983
Chestnut stallion by Ansata Shah Zaman.
Transferred to Victoria Varley, Mena, Arkansas.

Ansata Aliha #308574 June 9, 1984
Grey mare by *Jamilll

Ansata Serena

Ansata Mourad Bey

Ansata Aliha

Ansata Raqessa

**Ansata Bint Bukra/Ansata Regina/Ansata Raqessa*

Ansata Raqessa #219547
Grey Mare; May 2, 1980
Sire: *Ansata Ibn Halima   Dam: Ansata Regina
Sold to Jan Van Ness, Flaxman Arabians, December 1990.
Exported to the Netherlands.

PRODUCE

Unnamed bay mare by *Jamilll February 15, 1984
Died August 24, 1984.

Ansata El Sirhan #0327856 February 11, 1985
Bay stallion by *Jamilll. Transferred to
Chad Goodner, Mena, Arkansas, July 1987. Gelded.

Ansata Kamriya #0355954 April 12, 1986
Bay mare by *Jamilll. Donated to
Society for the Preservation of the Egyptian
Arabian Horse, Lexington, Kentucky.

Ansata Athena #0392423 April 27, 1987
Grey mare by Ansata Halim Shah. Sold to
Paolo Gucci, Millfield Stables, Yorktown, New York.

Ansata Alyssa #0410804 July 26, 1988.
Grey mare by Ansata Halim Shah. Sold to
John Bacot, Tucson, Arizona, December 1990.

1990 serviced by Prince Fa Moniet. Sold mare in
foal. Foaled filly in the Netherlands.

Ansata Kamriya

derivedtext *Ansata Bint Bukra/Ansata Regina/Ansata Serena*

Ansata Serena #240426
Grey Mare; April 28, 1981
Sire: *Ansata Ibn Halima   Dam: Ansata Regina
Sold to Anita Wallin, December 1990. Exported to Sweden.

PRODUCE

Ansata Serenada #327888 August 4, 1985
Grey mare by *Jamilll. Transferred to Dorian Farm,
Goshen, New York, February 1989.

Ansata El Sadiqy #0356021 July 3, 1986
Grey stallion by Jamilll. Gelded and transferred to
Gene Lyle, Wickes, Arkansas, May 1987.

Ansata El Sarin #0394606 May 26, 1987
Grey stallion by Ansata Halim Shah.
Gelded and transferred to John Shepherd, Mena, Arkansas,
December 1987.

Ansata Haji Said #0417449 May 26, 1988
Grey stallion by Ansata Haji Halim. Transferred to
Cindy Mize, Cadiz, Kentucky, October 1988.

Ansata El Naji #0430160 May 16, 1989
Grey stallion by Ansata Halim Shah. Gelded and transferred
to Ruth Krause, Barrington, Illinois, March 1990.

RA Rebecca #0466044 March 8, 1991
Chestnut mare by Prince Fa Moniet, owned by
Anita Wallin, Sweden.

Ansata Athena

Ansata Bint Bukra/Ansata Regina/Ansata Aliha

Ansata Aliha #0308574
Grey Mare; June 9, 1984
Sire: *Jamilll   Dam: Ansata Regina
Sold to H.H. Sheikh Hamad Bin Khalifa Al Thani,
Al Shaqab Stud, Doha, Qatar. Exported, January 1994.

PRODUCE

Ansata Manasseh #0440770 April 25, 1989
Grey stallion by Ansata Halim Shah.
Sold in 2002 to Cynthia Culbertson, Carrizozo, New Mexico.

Ansata Manasseh

Ansata Bint Aliha #0472021 March 23, 1991
Grey mare by Ansata Halim Shah.
Sold to Peter and Vera Stoessel-Vocka, Switzerland.
and Spain. Exported to Europe, September 1994.

Ansata Atallah #0479622 March 9, 1992
Grey mare by Ansata Halim Shah. Died.

Ansata Shah Abbas #0493685 May 6, 1993
Grey stallion by Ansata Halim Shah. Sold to
Kuwait Equestrian Club (Kuwait Arabian Horse
Center), May 1998. Exported to Kuwait.

Ansata Bint Aliha

Ansata Bint Bukra/Ansata Regina/Ansata Aliha/Ansata Bint Aliha/VA Ahlam

VA Ahlam # 0568249
Grey Mare; July 2, 1996
Sire: Ansata Sinan Dam: Ansata Bint Aliha
Bred by Veruska Arabians, Switzerland and Spain.
Acquired by Ansata in 2000.

Ansata Desert Rose #0596948 February 9, 2002
Grey mare by Ansata Iemhotep.

Ansata Shah Abbas

VA Ahlam

Ansata Desert Rose

Falima, founder of the Nile family. (*Ansata Ibn Halima x Fa-Habba x *Bint Bint Sabbah). JEF photo.

The Bint Sabah family through *Bint Bint Sabbah

We did not see *Bint Bint Sabbah (by Baiyyad) and there is only one published picture of her being ridden by George Cason, the Babson manager. She also appears in a home movie taken by visitors at the Babson Farm. She looked to be similar to her dam, Bint Sabah, and nowhere near as exquisite in the head as her half-sister, Bukra (sired by Shahloul). We did, however, see most of *Bint Bint Sabbah's first generation descendants.

Fay-Sabbah (by Fay El Dine, one of our very favorite Saklawi stallions) was a tall and refined bay mare, somewhat plain headed with long foreface, and longer bodied and weaker in croup than her bevy of sisters by *Fadl. We leased her as an old mare, knowing it was risky, and were unable to get a foal from her.

Sabrah (x Serrasab x Fay-Sabbah), a black mare, was eventually acquired in trade for Ansata Abbas Pasha. The intensification of *Fadl blood, especially through Fa-Serr, and through black progeny, tended to get heavier horses in general, and we limited acquiring intense breeding to Fadl, particularly through Fa-Serr. Sabrah (by Fabah - who was not known for producing beautiful heads) was a tall maiden mare when she arrived at Ansata, and she was - kindly put - very homely and unharmonious. She was strong-bodied but somewhat rafter-hipped and short in the croup. However, "blood will tell" and when bred to *Ansata Ibn Halima, she produced two outstanding mares, Ansata Sabiha, an attractive foundation broodmare and dam of multiple champions, and Fa Halima, a charismatic beautifully balanced mare who became Reserve European Champion at the Salon du Cheval, Paris and 1981 U.S. National Champion Mare. Subsequent generations produced top champions and breeding stock. Sabrah was a perfect example of "never judge a book by its cover." She taught us a good lesson about breeding potential despite outward appearances.

Fa-Habba (by *Fadl), was a well-balanced bay mare; pleasing head, somewhat straight neck lacking a good mitbah; strong body, level croup and good tail carriage; good-legged and good-moving. Bred to *Ansata Ibn Halima she produced the very classic bay mare, Falima, beautiful of head, strong of body and an excellent mover, who in turn founded the Nile Family through breeding her to *Morafic, Ansata Ibn Sudan, and Ansata Shah Zaman. This family has been linebred and inbred, and continues to produce beautiful heads, good bodies, good legs, good tail carriage, and exceptionally strong-moving individuals who can "trot a hole in the wind." Rarely did we practice breeding father to daughter, but we did so in this family. Breeding Ansata Ibn Sudan to his exquisite bay daughter, Ansata Nile Queen resulted in a most classic bay mare, Ansata Nile Dream and a pleasing but plainer, heavier-bodied chestnut mare, Ansata Nile Charm. Both were excellent producers, but as Murphy's Law would have it, Charm outlived her sister and was more prolific. In general, the bays and greys were more refined and elegant in this Nile family, while the Ansata Sabiha/Sabrah/Fay-Sabbah line produced exceptionally pretty chestnuts (e.g. Ansata Sabrina and Ansata Splendora) as well as beauty in all colors. The Nile family has been popular in Europe and the Middle East, and some of Ansata's most beautiful stallions and mares are from it.

Asal Sirabba (x Habba, a full sister to Fa-Habba), was acquired on lease because we wanted a line to her dam, Habba, and because her sire, Sirecho, was by the grand old stallion, *Nasr, who unfortunately did not get many straight Egyptian progeny in America. We saw Habba when she was owned by the Krausnicks. She was a very classic well-balanced grey mare, the most beautiful of the *Bint Bint Sabbah daughters we saw by *Fadl. Asal Sirabba was heavier than her dam, and not as pretty in the head, although she had a longer better shaped neck. She was bred exclusively to *Ansata Ibn Halima, producing several good daughters and sons who have carried on in other breeding programs worldwide.

Aana (by Fay El Dine x Fa Saana [who was a full sister to Habba and Fa-Habba]). We did not see Fa Saana, but she appears to have been extremely beautiful. In fact, the greys seemed to outclass the bays in type and quality from this *Bint Bint Sabbah line. Aana was an exceptionally fine and beautiful mare of wonderful type, and although she had been productive, she had become a breeding problem and we took a chance in leasing her. We could not keep her in foal, and she was not productive for the Babson farm after being returned.

Rose of Egypt (x Fa-Abba, a full sister to Habba, Fa-Habba and Fa Saana), provided a line to the bay Fa-Abba, who was prettier and more refined than her full sister, Fa-Habba, but not as compact. Rose of Egypt (by Fa-Serr) was a dark bay whom we acquired from Walter Schimanski when she was old and full of tumors. We had seen her when she was a young mare at Milt Thompson's and liked her very much, but she took on the heavier Fa-Serr look as she aged. Due to her condition, she could not get in foal at Ansata and died without further production.

The *Bint Bint Sabbah line through Fay-Sabbah carried on at Ansata in tail female and provided excellent foundation stock for other breeders. However, as of 2002, only the Fa-Habba/Nile family tail female line remains at Ansata. In generalizing, the Fay-Sabbah family was most beautiful in the chestnut and grey colors, the *Fadl daughters in grey, and the Nile line in copper bay and grey. The mahogany bays and blacks appeared heavier and not as elegant. The value of this line to Egyptian breeding programs has been immeasurable, both in the production of successful show and performance stock.

*Fa-Habba (*Fadl x *Bint Bint Sabbah)*

Ansata Blue Nile (Prince Fa Moniet x Ansata Stari Nile). Filsinger photo.

El Dahma through Bint Sabah
*Bint Bint Sabbah and Layla branches

*Bint Bint Sabbah

Bint Sabah

El Dahma c. 1880 Dahmah Shahwania mare from the stud of Ali Pasha Sherif

Obeya 1894 x **Koheilan El Mossen**

Bint Obeya x **El Halabi**

Sabah x **Mabrouk Manial**

Bint Sabah x **Kazmeyn**

Sabah

Layla

*Bint Bint Sabbah x **Baiyad**

Fay-Sabbah x **Fay El Dine**	Fa-Habba x ***Fadl**	Habba x ***Fadl**		Layla x **Ibn Rabdan**
Serrasab x **Fa-Serr**	Falima x ***Ansata Ibn Halima**	Asal Sirabba x **Sirecho**	Fa-Abba x ***Fadl**	Komeira x **Nabras**
Sabrah x **Fabah**			Rose Of Egypt x **Fa-Serr**	Kamar (Kamar) x **Nazeer**

Kamar

Tamria x ***Tuhotmos**

*Pharrah x **Farag**

Fa-Saana x ***Fadl**

Aana x **Fay El Dine**

*Pharrah

Hagir x **El Sareei**

*Hegrah x **Alaa El Din**

AK Rishafa x **AK Shah Moniet**

Fa-Habba

Falima

Fay Sabbah

Serrasab

Fa-Abba

Sabrah

Asal Sirabba

Aana

Hagir

*Hegrah

183

*Bint Bint Sabbah Family

The Fay-Sabbah Family

<u>El Dahma/Obeya/Bint Obeya/Sabah/Bint Sabah/*BintBint Sabbah/ FaySabbah</u>

Fay-Sabbah #1513
Bay Mare; May 15, 1938
Sire: Fay-El-Dine   Dam: *Bint Bint Sabbah
Bred by Henry B. Babson, Dixon, Illinois
Leased by Ansata, Chickasha, Oklahoma.

PRODUCE
1963 and 1964 serviced by *Ansata Ibn Halima. Open.
Returned to Babson. Died without further production.

Fay Sabbah

<u>*BintBint Sabbah/FaySabbah/Serrasab/Sabrah</u>

Sabrah #28350
Black Mare; May 24, 1964
Sire: Fabah   Dam: Serrasab
Bred by Henry B. Babson, Dixon, Illinois
Obtained by Ansata from the Babson farm
in exchange for Ansata Abbas Pasha.
Sold to Robert Cowling, Houston, Texas, May 1969.

PRODUCE
Ansata Sabiha #47966 March 1, 1968
Grey mare by *Ansata Ibn Halima.

Ansata El Reyhan #55154 April 7, 1969
Grey stallion by Ansata Ibn Sudan.
Sold with dam to Robert Cowling, Houston, Texas.

Sabrah

Ansata Sabiha

Ansata El Reyhan

Fa Halima

PRODUCE of Sabrah continued

Fa Halima #67401 April 13, 1970
Grey mare by *Ansata Ibn Halima.
Owned by Robert Cowling.

*BintBintSabbah/FaySabbah/Serrasab/Sabrah/Ansata Sabiha

Ansata Sabiha #47966
Grey Mare; March 1, 1968
Sire: *Ansata Ibn Halima   Dam: Sabrah
Sold to Mike and Kiki Case, Glorieta Ranch,
Lufkin, Texas June 1976
in foal to Ansata Ibn Sudan.

PRODUCE

Ansata El Emir #050324 February 20, 1972
Grey stallion by *Morafic. Sold to Jay Jones,
Arabesque Farm, Illinois.

Ansata Sabrina #04542 February 21, 1973
Chestnut mare by Ansata Ibn Sudan.

Ansata Shah Zam #105818 February 28, 1974
Grey stallion by Ansata Shah Zaman.
Sold to Cathy Downing, Chino, California.

Ansata Reza Shah #125642 February 26, 1975
Grey stallion by Ansata Shah Zaman. Sold at dam's
side in 1976 to Glorieta Ranch, Lufkin, Texas.

Ansata Sabrina

Ansata El Emir

Ansata Reza Shah

Ansata Shah Zam

Glorieta Sabdana

Sundar Sabbahalim

Sundar Alisha

PRODUCE *of Ansata Sabiha continued*

Glorieta Sabdana #143943 March 2, 1976
Grey mare by Ansata Ibn Sudan. Owned by Glorieta Ranch, Lufkin, Texas.

Glorieta Ali Sudan #0160078 March 6, 1979
Grey stallion by Ansata Ibn Sudan. Owned by Glorieta Ranch, Lufkin, Texas.

BintBintSabbah/FaySabbah/Serrasab/SabrahAnsataSabiha/ GlorietaSabdana/BKA Ali Sabbah/Sundar Sabbahalim

Sundar Sabbahalim #0424555

Grey Mare: January 15, 1989
Sire: Ansata Halim Shah   Dam: BKA Alisabbah
Bred by V. Don and Margaret Saluti, Malvern, Pennsylvania
Purchased from Rainer Kristinus by Ansata, February 1990.
Sold to Dr. Chess Hudson, Lawrenceville, Georgia
November 1993.

PRODUCE

Ansata Sabbara #0493646 March 6, 1993
Grey mare by Prince Fa Moniet. Sold to a Canadian Partnership, January 1994.

Zandai Fa Sabbah #0502680 March 3, 1994
Grey mare by Prince Fa Moniet. Owned by Dr. Chess Hudson, Lawrenceville, Georgia.

BintBintSabbah/FaySabbah/Serrasab/SabrahAnsataSabiha/ GlorietaSabdana/BKAAli Sabbah/SundarAlisha

Sundar Alisha #0465493

Grey Mare; February 28, 1990
Sire: Ansata Halim Shah   Dam: BKA Alisabbah
Bred by Mr. Rainer Kristinus, Pennsylvania
Purchased from Miguel Muzquiz, Mexico,
sold to Guiseppe Fontanella, Italy. Exported to Italy.

PRODUCE

Ansata Alima #0518762 May 3, 1995.
Grey mare by Prince Fa Moniet
Sold to H.R.H. Prince Khaled S.K. Al Saud, Nejd Stud,
Exported to Saudi Arabia.

Ansata Aida #0529848 April 29, 1996.
Grey mare by Prince Fa Moniet. Sold to
Sheikh Abdullah bin Nasir Al Thani, Al Naif Stud.
Exported to Qatar, February 1997.

Ansata Halisha #0556198 September 5, 1998.
Grey mare by Ansata Iemhotep. Owned by
Guiseppe Fontanella, Italy. Exported to
Italy, March 1999.

Ansata Alima

Ansata Sabrina

BintBintSabbah/FaySabbah/Serrasab/Sabrah/Ansata Sabiha/Ansata Sabrina

Ansata Sabrina #94542
Chestnut Mare; February 21, 1973 - September 3, 1993
Sire: Ansata Ibn Sudan   Dam: Ansata Sabiha

PRODUCE

Unnamed grey stallion by Hossny. February 16, 1978
Died prior to registration.

Ansata Ali Halim #196855 October 10, 1979
Chestnut stallion by *Ansata Ibn Halima.

Ansata Astarra #237261 September 6, 1981
Grey mare by Ansata Shah Zaman.

Ansata Asmarra #278157 March 3, 1983
Chestnut mare by Ansata Shah Zaman.
Sold to Jim Wells, Germantown, Tennessee, March 13, 1987.

Ansata Aly Jamil #307000 August 25, 1984
Chestnut stallion by *Jamilll. Sold to
Claudia Quentin, Haras las Cortaderas October 3, 1985.
Exported to Argentina.

Ansata Aida

Ansata Halisha

Ansata Aly Jamil

Ansata El Muteyri

Ansata Sabrina produce continued

Ansata Splendora #355890 February 17, 1986
Chestnut mare by *Jamilll. Sold February 199
to Sheikh Abdulaziz Al Thani, Al Rayyan Stud, and exported to Doha, Qatar.

Ansata Haji Tahir #394590 August 17, 1987
Grey stallion by Ansata Haji Halim. Gelded and
Transferred to Ray Goodner, Mena, Arkansas,
July 1988.

Ansata Abu Mansur #418467 August 4, 1988
Grey stallion by Ansata Halim Shah.
Gelded and transferred to Phil Caldwell, Fort Smith,
Arkansas, November 1990.

Ansata El Muteyri #0471990 February 26, 1991
Grey stallion by Prince Fa Moniet.
Transferred to Jeremy & Chad Harper, Prattsville, Arkansas.
Gelded.

Ansata Splendora in Qatar at Al Rayan Farm. Filsinger photo

Ansata Star O

Ansata Asmarra. Sparagowski photo

BintBintSabbah/FaySabbah/Serrasab/SabrahAnsataSabiha/Ansata Sabrina/Ansata Astarra

Ansata Astarra #237621
Grey Mare; September 6, 1981 – December 3, 1995
Sire: Ansata Shah Zaman   Dam: Ansata Sabrina
Sold to Dr. Chess Hudson, Lawrenceville, Georgia, November 1993.

PRODUCE
Ansata Ali Baba #0355892, March 2, 1986
Bay stallion by *Jamilll. Gelded and
Transferred to Kristy Ward, Marionville, Missouri, October 1987.

Unnamed chestnut colt by Ansata Haji Halim, February10, 1988. Transferred to Ray Stokley of Arkansas, August 1988.

Ansata Mara #0430226 March 28, 1989
Grey mare by DAJ Abbas Pasha.
Sold to Cornerstone Partnership, Covington, Louisiana, December 1990.

Ansata Al Hamid #0456414 April 1, 1990
Grey stallion by DAJ Abbas Pasha.
Gelded.

Ansata Star O #0472013 April 2, 1991
Grey mare by Prince Fa Moniet.
Sold to a Canadian Partnership, January 1994.

Ansata Fakhir unregistered January 17, 1993
Grey stallion by Prince Fa Moniet. Transferred to Brian Edwards, Hot Springs, Arkansas. Died 1994.

BintBintSabbah/FaySabbah/Serrasab/Sabrah/Ansata Sabiha/Ansata Sabrina/Ansata Asmarra

Ansata Asmarra #278157
Chestnut Mare; March 3, 1983
Sire: Ansata Shah Zaman   Dam: Ansata Sabrina
Sold to Jim Wells, Evergreen Arabians, Germantown, Tennessee, March 1987. Sold to Kim McGill of Switzerland, February 1992.

PRODUCE
Ansata El Simitar #0389943 June 17, 1987
Grey stallion by Ansata Halim Shah. Owned by Evergreen Arabians.

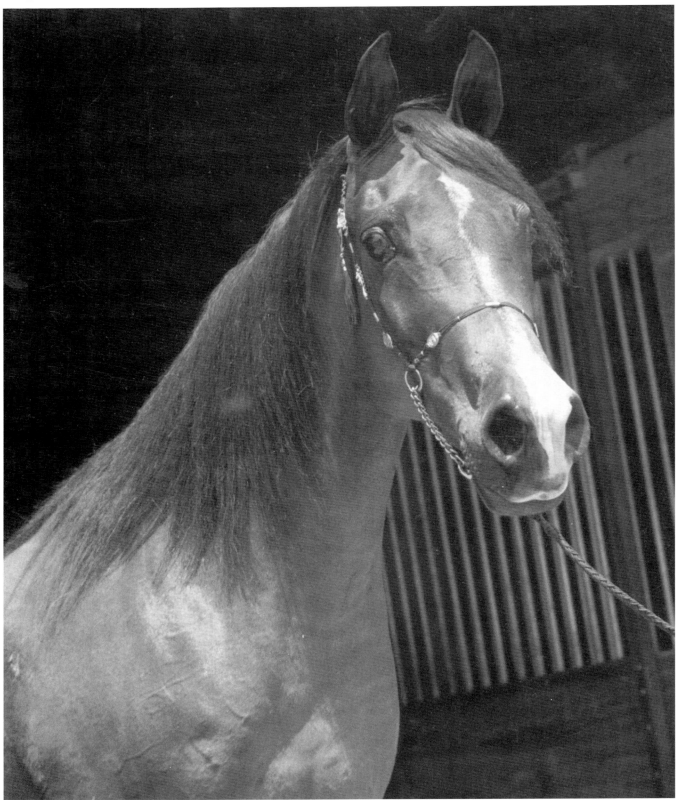
Falima (*Ansata Ibn Halima x Fa-Habba) Sparagowski photo.

The Nile Family

BintBintSabbah/Fa-Habba

Fa-Habba #4483
Bay Mare; April 26, 1947
Sire: *Fadl   Dam: *Bint Bint Sabbah
Bred by Henry B. Babson, Dixon, Illinois.
Leased by Ansata, Chickasha, Oklahoma.

PRODUCE
Falima #31956 July 22, 1965
Bay mare by *Ansata Ibn Halima.

1965 aborted colt by *Ansata Ibn Halima. Returned to Babson where she produced her last foal, Serr Habba, a filly by Ibn Fa Serr.

Fa Habba

BintBint Sabbah/Fa-Habba/Falima

Falima #31956
Bay Mare; July 22, 1965 – June 16, 1988
Sire: *Ansata Ibn Halima   Dam: Fa-Habba
Bred by Mr. and Mrs. Donald L. Forbis, Chickasha, Oklahoma.

PRODUCE
Ansata Nile Queen #59483 April 4, 1969
Bay mare by Ansata Ibn Sudan.

Ansata Nile Jewel #61935 April 20, 1970
Grey Mare by *Morafic. Sold to John Hacklander, Janesville, Wisconsin, January 1976.

Ansata Nile King #70165 April 13, 1971
Grey stallion by Ansata Shah Zaman. Sold to Douglas Griffith, Imperial Stud, Parkton, Maryland.

Falima

Ansata Nile King

Ansata Nile Queen

PRODUCE of Falima continued

Ansata Nile Star #80323 April 11, 1972
Grey mare by Ansata Shah Zaman. Died 1979.

Ansata Nile Mist #94535 April 4, 1973
Bay mare by Ansata Ibn Sudan. Sold to
Douglas Griffith, Imperial Egyptian Stud,
Parkton, Maryland.

Ansata Nile Pharo #105817 April 4, 1974
Bay stallion by Ansata Ibn Sudan. Sold to Don
And Joan Johnson of California. Died May 20, 1975.

Ansata Nile Sun #237266 March 24, 1981
Grey stallion by Ansata Ibn Sudan. Sold to
Philippe de Bourbon, Haras Jinnah Al Tayr,
March 1982. Exported to Uruguay, South America.

Ansata Nile Mist

Ansata Nile Jewel

Ansata Nile Star

Ansata Nile Sun

Ansata Nile Pharo

BintBintSabbah/Fa-Habba/Falima/Ansata Nile Queen

Ansata Nile Queen #59483
Bay Mare; April 4, 1969 – March 27, 1980
Sire: Ansata Ibn Sudan   Dam: Falima

PRODUCE

Ansata Nile Storm #94536 April 3, 1973
Grey stallion by Ansata Shah Zaman. Gelded. Sold to Marc Stacer, Vandervoort, Arkansas March 1983.

Ansata Nile Ruler #122315 February 25, 1975
Bay stallion by *Ansata El Wazir. Sold to Sabbath Arabians, Oakridge, Tennessee. Died February 16, 1976.

Ansata Nile Dream #137117 March 17, 1976
Bay mare by Ansata Ibn Sudan.

Ansata Nile Charm #153298 July 2, 1977
Chestnut mare by Ansata Ibn Sudan.

Ansata Nile Dawn #196864 May 26, 1979
Grey mare by Nabiel.

Ansata Nile Ruler

Ansata Nile Storm   *Ansata Nile Charm*

Ansata Nile Dream

Ansata Nile Dawn

**Bint BintSabbah/Fa-Habba/Falima/Ansata Nile Queen/Ansata Nile Dream*

■ Ansata Nile Dream #137117
Bay Mare; March 17, 1976 – August 13, 1987
Sire: Ansata Ibn Sudan   Dam: Ansata Nile Queen

PRODUCE

Ansata Nile Glory #0237267 January 24, 1981
Grey mare by Ansata Halima Son. Sold to Leanor Romney, Somerset Farms, Santa Ynez, California.

Ansata Nile Flame #259985 February 20, 1982
Chestnut stallion by Ansata Ibn Shah. Sold to Ronald J. Hutchings, Meranti Arabian Farm. Exported to England.

Ansata Riva Nile #278161 May 27, 1983
Grey mare by Ansata Ibn Shah. Sold to Leanor Romney, Somerset Farms, Santa Ynez, California March 1, 1986. Later sold by Romney Estate and exported to Brazil.

Ansata Nile Pasha #304261 June 5, 1984
Grey stallion by *Jamilll. Sold to Dr. Patterson and Sherry Moseley, Sherbrook Egyptian Stud, Winter Springs, Florida. Exported to England on lease to P.W. Maxwell; sold to F. Huemer, Austria.

Ansata Nile Spark #0355857 July 6, 1986
Chestnut stallion by *Jamilll. Gelded. Transferred to Carrie Fletcher, Palestine, Texas.

Ansata Nile Fame #0390051 August 12, 1987
Chestnut mare by Dorian Shah El Shams. Sold to Dr. Ann Campbell, Oklahoma City, Oklahoma, December 1990.

Ansata Nile Glory

Ansata Riva Nile

Ansata Nile Flame

Ansata Nile Fame

Ansata Nile Pasha

BintBint Sabbah/FaHabba/Falima/Ansata Nile Queen/Ansata Nile Charm

■ Ansata Nile Charm #0153298
Chestnut Mare; July 2, 1977 – October 31, 1991
Sire: Ansata Ibn Sudan   Dam: Ansata Nile Queen

PRODUCE

Ansata Nile Rose #259986 March 21, 1982
Chestnut mare by Ansata Ibn Shah.

Ansata Stari Nile #302848 April 12, 1984
Bay mare by Ansata Ibn Shah.

Ansata Nile Wine #327887 April 20, 1985
Bay mare by *Jamilll. Sold to Dr. and
Mrs. Les Nunnally, Ocala, Florida.

Ansata Nile Sheik #0356045 April 9, 1986
Grey stallion *Jamilll. Sold to Larry and Roberta Kirberger, Blanchard, Oklahoma, January 1991.

Ansata Regal Nile #0390079 April 16, 1987
Bay stallion by Dorian Shah El Shams.
Transferred to Richard Marshall, Ocala, Florida.

Ansata Gala Nile #0410753 April 4, 1988
Grey mare by Ansata Halim Shah. Transferred Ownership to Jean Kayser. Exported to Luxembourg.

Ansata Nile Rose

Ansata Stari Nile

Ansata Nile Wine Sparagowski photo.

Ansata Nile Charm with Ansata Nile Sand as a foal

Ansata Nile Rose

PRODUCE of Ansata Nile Charm continued

Ansata Nile Sand #0453431 March 28, 1990
Grey stallion by Ansata Halim Shah. Gelded
Transferred to Cindy Mize, Kadiz, Kentucky.

Ansata Nile Emir #0472012 March 31, 1991
Grey stallion by Ansata Halim Shah. Sold to Marcella
Ospina. Exported 1992 to Columbia, South America.

Bint BintSabbah/Fa-Habba/Falima/Ansata Nile Queen/Ansata Nile Charm/Ansata Nile Rose

Ansata Nile Rose #0259986
Chestnut Mare; March 21, 1982
Sire: Ansata Ibn Shah   Dam: Ansata Nile Charm
Sold to Jim Fleming, Regal Arabians, Ontario, Canada,
December 1990.

PRODUCE
Ansata Nile Magic #0356046 March 16, 1986
Grey mare by *Jamilll. Sold to Robert and
Merry Carole Falke, Fredericksburg, Virginia
December 15, 1990, in foal to Ansata Halim Shah.
Produced 1991 filly, **Nile Allure** for new owners.

Ansata Nile Emir

Ansata Nile Magic

Ansata Nile Silk

Halimas Rose

Nile Allure

Ansata Prima Nile

PRODUCE *of Ansata Nile Rose continued*

Shams El Nil #0394580 May 18, 1987
Chestnut stallion by Dorian Shah el Shams.
Transferred to Shauna Hansbrough, Mena, Arkansas
December 1987.

Ansata Nile Silk #0410764 June 7, 1988
Grey mare by Ansata Halim Shah. Sold to Karen Sptizer,
Temecula, California, December 1990.

Ansata Nile Beau #0453418 April 6, 1990
Grey stallion by Ansata Halim Shah. Gelded and
transferred to Cindy Mize, Cadiz, Kentucky, July 1991.

Halimas Rose #0474064 June 7, 1991
Grey mare by Ansata Halim Shah. Owned
By Jim Fleming, Regal Arabians, Canada.

**Bint BintSabbah/FaHabba/Falima/Ansata Nile Queen
/Ansata Nile Charm/Ansata Stari Nile*

Ansata Stari Nile #0302848
Bay Mare; April 12, 1984
Sire: Ansata Ibn Shah   Dam: Ansata Nile Charm
Sold to Peter and Vera Stoessel-Vocka, Veruska Arabians,
Switzerland and Spain. Exported to Switzerland,
September 1994.

PRODUCE

Ansata Nile Eagle #0419890 June 23, 1988
Grey stallion by Ansata Halim Shah. Gelded.
Transferred to Jeffrey R. Peal, Hernshaw, West Virginia.

Ansata Blue Nile #0453457 June 2, 1990
Grey mare by Prince Fa Moniet.

Ansata Nile Genie #0471986 September 10, 1991
Grey stallion by Prince Fa Moniet.
Gelded and sold to Nancy Gates, Rison, Arkansas.

Ansata Nile Starr #0483235 August 20, 1992
Grey mare by Prince Fa Moniet.

Ansata Prima Nile #0493683 August 13, 1993
Grey mare by Prince Fa Moniet.

1994 serviced by Prince Fa Moniet.
Exported in foal to Europe.

Ansata Blue Nile

*BintBintSabbah/Fa-Habba/Falima/Ansata Nile Queen/Ansata Nile Charm/Ansata Stari Nile/Ansata Blue Nile

■ Ansata Blue Nile #0453457

Grey Mare; June 2, 1990
Sire: Prince Fa Moniet   Dam: Ansata Stari Nile
Sold to Sheikh Abdulaziz Al Thani, Al Rayyan Farm, Qatar, December 1995.
Exported to Qatar, February 1996.

PRODUCE

Ansata Nile Blue #0518749, April 18, 1995
Grey stallion by Ibn Al Mareekh.
Sold to Tay Smith, Eureka Springs, Arkansas, November 1996. Gelded.

Exported in foal to Ansata Hejazi.
Produced filly in Qatar, **RN Hejaziah.**

RN Hejaziah

*BintBintSabbah/Fa-Habba/Falima/Ansata Nile Queen/Ansata Nile Charm/Ansata Stari Nile/Ansata Nile Starr

■ Ansata Nile Starr #0483235

Grey Mare; August 20, 1992
Sire: Prince Fa Moniet   Dam: Ansata Stari Nile

PRODUCE

Ansata Nile Gypsy #0529844 May 23, 1996.
Grey mare by Ansata Hejazi. Sold to Al Naif Stud, Sheikh Abdullah Bin Nasir Al Thani. Exported to Qatar February 1997.

Ansata Nile Jade #0543391 August 11, 1997
Grey mare by Ansata Hejazi. Sold to Prince Khalid S.K. Al Saud, Nejd Stud. Exported to Saudi Arabia, December 1997.

Ansata Nile Starr

Ansata Nile Jade

Ansata Nile Gypsy

Ansata Nile Comet

PRODUCE of Ansata Nile Starr continued

Ansata Asila Nile #0563808 March 15, 1999
Grey mare by Ansata Hejazi. Sold to Ray and Jamie Roberts, Nirvana Farms, Lees Summit, Missouri.

Ansata Nile Comet #0578067 May 18, 2000
Grey stallion by Ansata Sirius. Gelded. Sold to Dr. Carmen Arick, Little Rock, Arkansas.

Ansata Nile Spice #589167 September 16, 2001
Grey filly by Ansata Sirius.
Sold in 2002 to Sheikh Mishaal Al Thani, Doha, Qatar.

Ansata Nile Starlight #0596947 October 16, 2002
Grey mare by Ansata Osiron.

Above and below, Ansata Nile Spice

Ansata Asila Nile

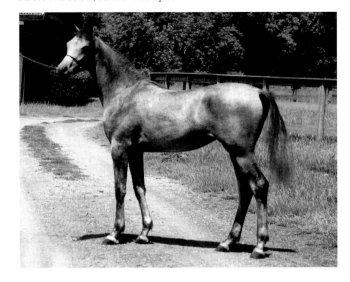

**BintBint Sabbah/Fa-Habba/Falima/Ansata Nile Queen/
Ansata Nile Charm/Ansata Stari Nile/Ansata Prima Nile*

▪ Ansata Prima Nile #0493683
Grey Mare; August 13, 1993
Sire: Prince Fa Moniet   Dam: Ansata Stari Nile
Sold to Sheikh Abdullah Bin Nasir Al Thani, Al Naif Stud, Doha, Qatar.

PRODUCE
Ansata Nile Cameo #0578070 March 9, 2000
Grey filly by Ansata Iemhotep.

Exported to Qatar.

*BintBintSabbah/FaHabba/Falima/Ansata Nile Queen/
Ansata Nile Dawn*

▪ Ansata Nile Dawn #0196864
Grey Mare; May 16, 1979
Sire: Nabiel   Dam: Ansata Nile Queen
Sold to Jan Van Ness, Flaxman Arabians, December 1990.
Exported to the Netherlands.

Ansata Nile Dawn

PRODUCE
Ansata Nile Bay #281129 August 29, 1983
Bay stallion by Ansata Shah Zaman. Sold to
Lyle and Virginia Bertsch, Fort Wayne, Indiana.
Purchased by Ansata in June 1992 at Dean Parker
Sale, Lexington, Kentucky. Sold 1993 to Tracy Davis, Ozark, Missouri.

Ansata Nile Gold #328033 June 8, 1985
Chestnut stallion by Ansata Shah Zaman.
Gelded and transferred to Marilyn Wright,
Summer, Texas, April 1987.

Ansata Vali Nile #0355856 June 9, 1986
Grey mare by *Jamilll.

Unnamed grey stallion by Ansata Haji Halim March 26, 1988. Transferred unregistered to Peggy Stanz, Marshall, Texas.

Ansata Nile Bay

Ansata Nile Dove #0430129 April 7, 1989
Grey mare by Ansata Halim Shah. Sold 1990 to
Guillermo Henin, Buenos Aires, Argentina, South America. Exported.

Ansata Nile Hawk #453443 April 20, 1990
Grey stallion by Ansata Halim Shah. Gelded.
Transferred to Michelle Mistretta-Hermann, Plantation, Florida.

Ansata Nile Dove

Ansata Vali Nile

Ansata Nile Jewel

*BintBintSabbah/Fa-Habba/Falima/Ansata Nile Queen/
Ansata Nile Dawn/Ansata Vali Nile

■ Ansata Vali Nile #0355856
Grey mare; June 9, 1986
Sire: *Jamilll   Dam: Ansata Nile Dawn
Sold to Gary and Gemma Barclay, Ontario, Canada,
1987 Pyramid Society Sale.

PRODUCE
Ansata Nile Bey #0459512 June 30, 1990.
Bay stallion by Imperial Imdal. Owned by Gary and Gemma
Barclay, Ontario, Canada.

*BintBintSabbah/Fa-Habba/Falima/Ansata Nile Jewel

■ Ansata Nile Jewel #61935
Grey Mare; April 20, 1970
Sire: *Morafic Dam: Falima
Sold to John Hacklander, Janesville, Wisconsin,
January 1976.

PRODUCE
Ansata Nile Gem #105819 April 10, 1974
Grey mare by *Ansata El Wazir. Sold to Jarrell
McCracken, Waco, Texas, November 1976.

Ansata Nile Moon #127431 April 17, 1985
Grey mare by Ansata Shah Zaman. Owned by John
Hacklander, Janesville, Wisconsin.

Ansata Nile Gift #151962 August 9, 1976.
Grey mare by Ansata Ibn Sudan. Owned by
John Hacklander, Janesville, Wisconsin.

Ansata Nile Gem

Ansata Nile Moon

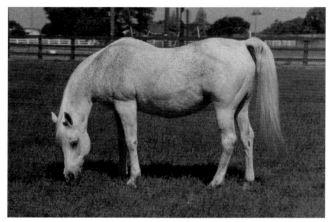

Ansata Nile Gift

BintBintSabbah/Fa-Habba/Falima/Ansata Nile Jewel/ AnsataNileGift

■ Ansata Nile Gift #0141962

Grey Mare; August 9, 1976
Sire: Ansata Ibn Sudan   Dam: Ansata Nile Jewel
Owned by John Hacklander, Janesville, Wisconsin
Sold to Lenor Romney, Somerset Farm. Leased by Ansata from Romney Estate. Sold to Dr. Chess Hudson, Lawrenceville, Georgia.

PRODUCE

Ansata White Nile #0459529 May 16, 1990.
Grey mare by Prince Fa Moniet.

Ansata White Nile

BintBintSabbah/Fa-Habba/Falima/Ansata Nile Jewel/ Ansata Nile Gift/Gift of the Nile

■ Gift of the Nile #0352745

Grey Mare; October 31, 1985
Sire: Ruminaja Bahjat   Dam: Ansata Nile Gift
Purchased 1999 from Dennis Key, Renaissance Arabians, Texas.

PRODUCE

Ansata Nile Splendour #0596952 October 1, 2001
Grey mare by Ansata Sirius.

Gift of the Nile

Ansata Nile Splendour

Ansata White Nile

Ansata Nile Echo   Filsinger photo.

*BintBintSabbah/Fa-Habba/Falima/Ansata Nile Jewel/
Ansata Nile Gift/Ansata White Nile*

Ansata White Nile #0459529
Grey Mare; May 16, 1990
Sire: Prince Fa Moniet   Dam: Ansata Nile Gift
Sold to Kuwait Equestrian Club (Kuwait Arabian Horse Center) May 1998. Exported to Kuwait.

PRODUCE
Ansata Nile Echo #0529845 June 4, 1996
Grey stallion by Ansata Hejazi. Sold to Sheikh Abdullah Bin Nasir Al Thani, Al Naif Stud, Doha, Qatar. Exported February 1997.

Ansata Nile Pearl #0543393 August 15, 1997
Grey mare by Ansata Hejazi. Sold to H.R.H. Prince Khalid S.K. Al Saud, Nejd Stud, Saudi Arabia, December 1997. Exported.

1997 - Serviced by Ansata Hejazi. Exported. Foaled filly in Kuwait.

Ansata Nile Pearl

BintBintSabbah/Fa-Habba/Falima/Ansata Nile Star

Ansata Nile Star #080323
Grey Mare; April 11, 1972-January 5, 1979
Sire: Ansata Shah Zaman   Dam: Falima

PRODUCE
Ansata Nile Joy #139699 March 21, 1976
Bay mare by *Ansata El Wazir. Sold to Robin and Joan Arheim, Parker, New South Wales, Australia. Exported.

Ansata Nile Pride #153297 May 9, 1977
Grey mare by Ansata Ibn Sudan. Died September 5, 1978.

Unnamed grey stallion July 5, 1978 by *Ansata Ibn Halima. Died July 31, 1978.

Ansata Nile Joy

BintBintSabbah/Fa-Habba/Falima/Ansata Nile Mist/Imperial Mistry/Imperial Impress

■ Imperial Impress #0224844
Chestnut Mare; April 15, 1981 – August 1998
Sire: Ansata Imperial   Dam: Imperial Mistry
Bred by Imperial Egyptian Stud
Purchased 1994 by Ansata from David and Martha Lucas Bishopville, South Carolina.

PRODUCE
Ansata Misty Nile #0529847 February 26, 1996
Grey mare by Ansata Hejazi.

BintBintSabbah/Fa-Habba/Falima/Ansata Nile Mist/Imperial Mistry/Imperial Impress/Ansata Misty Nile

■ Ansata Misty Nile #0529847
Grey Mare; February 26, 1996

PRODUCE
Ansata Nile Diva #578068 June 10, 2000
Grey mare by Ansata Sinan
Sold 2001 to Joseph and Kimberly Murgola, Rocky Top Ranch, Brenham, Texas.

Ansata Nile Nadir #589374 September 18, 2001
Grey colt by Ansata Sinan. Sold 2002 to Jeff and Judy Barth, Qadar Arabians, Weyauwega, Wisconsin.

Imperial Impress

Ansata Misty Nile

Ansata Nile Diva

Ansata Nile Nadir

MB Adallah

*BintBintSabbah/Fa-Habba/Falima/Ansata Nile Mist/Imperial Mistique/Imperial Imballora/Imperial Imballanah/MB Adallah

MB Adallah #0530044
Grey Mare; February 1, 1996
Sire: Imperial Madheen   Dam: Imperial Ballanah
Purchased by Ansata 1998 from IEB Texas Auction.

PRODUCE
Ansata Nile Satin #0579386 September 13, 2000
Grey mare by Ansata Sinan.

Ansata Nile Lace #589160 October 22, 2001
Grey mare by Ansata Sinan.

Ansata Nile Princess #0596945 October 10, 2002
Grey mare by Ansata Iemhotep.

Ansata Nile Satin

Ansata Nile Lace

Asal Sirabba with Raalima

Silima

Raalima

Halim Pasha

The Habba Family

BintBintSabbah/Habba/Asal Sirabba

Asal Sirabba #13379

Grey Mare; April 20, 1958
Sire: Sirecho   Dam: Habba
Bred by Mr. John Ekern Ott, Hinsdale, Illinois
Leased by Ansata from Ray and Jane Davis, Reeds Springs, Missouri. Sold 1970 to Robert Cowling, Houston, Texas.

PRODUCE

Silima #37769 May 12, 1966
Grey mare by *Ansata Ibn Halima. Sold to Ray and Jane Davis. Exported to Europe.

Raalima #43039 May 21, 1967
Grey mare by *Ansata Ibn Halima.
Sold to Robert Cowling, Houston, Texas.

Halim Pasha #47126 May 10, 1968
Grey stallion by *Ansata Ibn Halima. Returned to Ray and Jane Davis. Sold to Dan and Joan Appleby, Lewisville, Texas.

Sir Halim #65445 February 20, 1970
Grey stallion by *Ansata Ibn Halima.
Owned by Ray and Jane Davis.

The Fa Saana Family

**BintBintSabbah/Fa Saana/Aana*

Aana #3389
Grey Mare; April 2, 1945
Sire: Fay-El-Dine Dam: Fa Saana
Bred by Henry B. Babson, Dixon, Illinois
Leased by Ansata, Chickasha, Oklahoma.

PRODUCE
1963 serviced by *Ansata Ibn Halima.
1964 serviced by *Ansata Ibn Halima.
Returned to Babson in 1964.
Died without further production.

Above and below, Aana

The Fa-Abba Family

**Bint Bint Sabbah/Fa-Abba/Rose of Egypt*

Rose of Egypt #13500
Bay Mare; April 4, 1958 –1982
Sire: Fa-Serr   Dam: Fa-Abba
Bred by Henry B. Babson, Dixon, Illinois
Purchased from Walter Schimanski on October 7, 1979.

PRODUCE
1979 - 1981 serviced by *Ansata Ibn Halima. Barren.
1982 died.

Rose of Egypt

The Bint Sabah family through Kamar {Kamr}

Kamar was one of the most beautiful of the Nazeer daughters - extreme head with very refined foreface and muzzle. She was typical Dahma in overall type, but again, the exceptional head was married to a peaky short croup.

Kamar produced Hagir (by El Sareei), a pretty-headed, well-balanced grey mare, but somewhat light in croup. Hagir was the dam of Akhtal (by Amrulla – "Ziada"), a refined and elegant well-conformed stallion used as a sire at the E.A.O. when they went off on their ill-fated non-Nazeer program. Hagir also produced the tall and classic-headed mare *Hegrah (by Alaa El Din), who inherited a poor croup through both sides of her family.

AK Rishafa (x *Hegrah), was a rather plain mare sired by AK Shah Moniet. Ansata acquired her as a challenge to reclaim the beauty of the line through use of Ansata stallions. However, she was non-productive.

The Kamar family at Babolna:

The Babolna State Stud of Hungary laid claim to one of the special mares of this line: Tamria (x Kamar) by *Tuhotmos. She was an extremely elegant and beautiful-headed grey mare that had been imported from Egypt as one of the foundation mares of the Egyptian program. She foundered and remained lying down most of the time when we saw her during our visits to Hungary. She was hauntingly special. Bred to Farag she had produced *Pharrah.

*Pharrah was imported and used in the Imperial Egyptian Stud breeding program and later that of the Fortunas. An elegant grey mare, she was acclaimed for her beautiful head. She was somewhat long in foreface but exquisitely chiseled, having that "teacup" muzzle. Her daughter, Imperial Pharalima, and others such as Salon du Cheval World Champion, Imperial Phanilah, endeared this broodmare to the world.

Imperial Pharalima was a stellar daughter by *Ansata Ibn Halima, and she became an important mare of this lineage, producing Imperial Daeemah (by Imperial Daeem, full brother to Imperial Imdal), a rather compact and atttractive broodmare, very pretty-headed with big black eyes.

Sherifa Tamria (x Imperial Daeemah), sired by Royal Jalliel, doubles *Pharrah in tail female and *Ansata Ibn Halima on both sides of the pedigree. Sherifa Tamria, who was named in honor of Tamria, inherited the beautiful head of this family plus a stronger topline through the doubling of Ansata Imperial. She was acquired by Ansata as a two-year old, and her first foal, Ansata Aly Sherif, (by Ansata Iemhotep) indicates her promise as a valued broodmare.

In assessing the general merits and demerits of this Bint Sabah family over many generations, on the plus side one finds extreme type, refinement, and good balance , but on the minus side, somewhat short straight necks needing that "extra piece" and better mitbah. There were also offset knees (the latter more particularly through the Bukra line).

The shape of neck and mitbah was basically corrected in the Bukra line through the use of *Ansata Ibn Halima. Length of neck was improved through Ansata Ibn Sudan, Ansata Shah Zaman and Prince Fa Moniet, and they generally corrected the offset knees. *Jamilll get always had straight legs and clean knees, although he was not dominant in head-type.

In the *Bint Bint Sabbah line the incorporation of *Ansata Ibn Halima, Ansata Ibn Sudan and Ansata Shah Zaman improved the length and shape of necks; later use of *Jamilll and Prince Fa Moniet improved the legs.

Greys predominated in the Bukra and Kamar lines; the *Bint Bint Sabbah's were more colorful.

AK Rishafa

Orient Queen

The Kamar Family

El Dahma/Obeya/Bint Obeya/Sabah/
*Bint Sabah/Layla/Komeira/Kamar/Hagir/*Hegrah/AK Rishafa*

AK Rishafa #0305968
Grey Mare; March 11, 1983
Sire: AK Shah Moniet Dam: *Hegrah
Purchased by Ansata, June 1991, at Dean Parker Sale, Lexington, Kentucky with Alidaar filly, Orient Queen, at side. Sold at Addis Oklahoma City Auction to Donald or Eva Jenkins, April 1993.

PRODUCE

Orient Queen #0472052 May 2, 1991
Chestnut mare by Alidaar.
Sold to a Canadian Partnership, January 1994.

Ansata Simoom #0480192 May 17, 1992
Grey stallion by Ansata Halim Shah.
Gelded.

El Dahma/Obeya/Bint Obeya/Sabah/Bint Sabah/Layla/Komeira/
*Kamar/Tamria/*Pharrah/Imperial Pharalima/Imperial Daeemah/*
Sherifa Tamria

Sherifa Tamria #0563738
Grey Mare; March 18, 1997
Sire: Royal Jalliel Dam: Imperial Daeemah
Purchased by Ansata July 1999 in Canada at the Edwards Sale.

PRODUCE

Ansata Aly Sherif #0597632 March 27, 2002
Grey stallion by Ansata Iemhotep.

Sherifa Tamria

Ansata Aly Sherif

Ansata Majesta as a yearling in Arkansas. Sparagowski photo

The Farida Family

The Farida family, although tracing to the same root mare, El Dahma of Ali Pasha Sherif, took on some different nuances from that of the Bint Sabah family. The Farida family generally was shorter headed with shorter foreface, better shaped neck, more compact, stronger loins and topline, and flatter croups. The latter points, by comparison to the Bint Sabah's, probably related to the fact that she was by Kazmeen (Blunt breeding) while Farida was by Saklawi II (Ali Pasha Sherif stock).

When breeding a stallion of this tail female line to a mare of the Sabah tail female, or the reverse, the qualities of the Dahman strain become intensified, and Dahman type is relatively assured. Both lines produced good broodmares and good breeding stallions with considerable regularity.

Having purchased Ibn Halima, we later tried to purchase his sister, Nawarra. She was much like her brother, though not as extreme in the head. Our plans were shattered when she contracted a spinal disease, now known as EPM, and died in Egypt before we could buy her.

The first opportunity to acquire the tail female Halima line came through our association with Jean Kayser of Luxembourg. Along with other mares, we leased *JKB Masouda, the full sister to the well-known stallion, *Messaoud (Madkour x Maymoonah), who traced to the Marbach foundation mare, Moheba (x Halima).

The Farida family through Halima:

Halima (by Sheikh El Arab) was an attractive refined bay mare with a pleasing, but not special head, good eyes, well-shaped average neck, and good overall well-balanced structure and no glaring faults. She was not a super classic individual, as one would expect considering she was the dam of *Ansata Ibn Halima whose type and quality are legendary. Nevertheless, she was another prime example that teaches "blood will tell" and that one should assess all the hidden potentials when choosing foundation bloodstock. Ibn Halima's full brother, a bay, was not in a class with him, yet was still a handsome individual; he favored his mother, and Ibn Halima favored his grandsire, Sheikh El Arab.

Moheba (by Sid Abouhom), was purchased as a foundation mare by the Marbach State Stud of Germany during the time Von Szandtner was managing El Zahraa. She was in foal when exported and produced the bay mare, Malacha.

Malacha (by El Sareei), was rather ordinary, judging from her photos, but she was a good brood mare who went on to found an important line for breeders in Europe particularly through the full sisters, Mohebba II and the grand matriarch, Malikah.

Malikah (by Hadban Enzahi), inherited classic beauty through the Nazeer/Kamla line, thus picking up the Nazeer and doubling the Sheikh El Arab blood as well as adding the Bukra and Nazeer blood through her sire, Ghazal (Nazeer x Bukra). She founded a branch much improved over the initial import, Mohebba. We saw Malikah at the Filsinger farm in Germany. She was a very classic broodmare who made a major contribution to Egyptian breeding programs through her many descendants.

Maymoonah (by Hadban Enzahi) we did not see, but she produced a number of important get, including *JKB Masouda, who was imported on lease from Jean Kayser of Luxembourg.

*JKB Masouda (by Madkour), was a handsome grey mare, good head, dark eyes, well-shaped neck of average length, strong body and topline, good tail set, good legs and very good gaits. She produced the following first generation progeny.

Ansata Malika (by *Jamilll), an attractive grey broodmare typical of this line resembling, but prettier than, her dam with better length of neck, not as strong in the topline, but good overall balance, clean legs and good movement. An incredible broodmare, she produced American and international champions by Ansata Halim Shah that founded dynasties of their own: Ansata Majesta, Ansata Magnifica, Ansata Malaha, Ansata Majesty and Ansata Malaka - each ideal individuals representative of this family. Four became champions.

Ansata Meryta (by Ansata Halim Shah x *JKB Masouda), produced two excellent mares, Ansata Marha and Ansata Hamama. Marha, a stunning bay mare of wonderful proportions and excellent way of going, was sold to Sheikh Abdulaziz al Thani of Qatar where she won national and international championships. Hamama was sold to Europe where she has become a valued broodmare.

The influence of this family tracing to *JKB Masouda is appreciated worldwide and remains at Ansata in 2002 through Ansata Majesty, Ansata Malaka and her son and daughter, Ansata Malik Shah and Ansata Millennia.

Another Halima line to Mohebba was provided through the acquisition of Dal Macharia. Mahari combined Nazeer/Kamla lineage, thus doubling Mansour, while her daughter, Maharia, was a Gharib daughter. We had seen Gharib in Egypt as youngster and had been there when Dr. Wentzler purchased him to add pigment to the Marbach herd. He was an attractive black stallion, quite striking as a young horse, but as he aged, he got coarser and lacked classic type as an individual, nor was his sire, Antar, known for producing stallions. Nevertheless, the Halima female line shone through.

Dal Macharia (by Dalul), was typical of the bay Halima line; a well-proportioned foundation broodmare, good conformation, with a pleasing, but not exotic head. Bred to Ansata Halim Shah she produced champions. Her progeny by Ansata Hejazi were also of high quality, some of which were exported. Bred to Ansata Sokar she produced one of the most extraordinary fillies born at Ansata - very similar to Ansata Samantha and Ansata Selket in type, but she died soon after birth due to a defective intestine. Again, this was one of those times when the valleys seem very far below the mountain peaks.

The Farida Family through Futna

Futna was an old mare when we saw her. Unlike most Shahloul daughters, she was rather plain in the head, but she was a tall and good bodied broodmare. It seems the Sheikh El Arab blood on this line gave better heads; nevertheless, blood will tell, and when bred into the Sheikh El Arab/Mansour/Nazeer line she either produced beautiful individuals or those who would produce in the future.

Our first acquisition of this lineage was

*Ansata Bint Sameh "Amal" (by Sameh). She was a tall very well-balanced, good-bodied, basically correct mare, but plain-headed, as one would have expected from this combination. Nevertheless, she bred on perfectly with *Ansata Ibn Halima, which doubled the Farida line in tail female and brought back the type.

Ansata Jezebel (by *Ansata Ibn Halima), was a big very pretty-headed mare with excellent conformation. She was never shown, but her progeny were valuable individuals, especially Ansata Ibn Shah, who became a multi-champion stallion. We experimented at in-breeding Ansata Jezebel to her father, but the result was not satisfactory.

Ansata Jamila (by *Ansata Ibn Halima), was a wonderful mare with a superb body and an outstanding trot. She was pretty too, but not as beautiful in the head as her sister. We also bred her to her sire, but this also was not satisfactory. She was shown successfully and became a halter champion.

This family has carried on at Ansata and produced some exceptional individuals. The Futna line differs somewhat from the Halima tail female with descendants being taller, with more scope, and somewhat longer but very handsome heads.

Another line to Futna was provided through *Bint Dahma (x Dahma II), a well-bred mare imported by Richard Pritzlaff. Dahma II was an especially beautiful Nazeer daughter, and her line through the imported mare *Deenaa, whom we selected for H. J. Huebner, exceeded the Mossa line in influence, producing the famed Bint Deenaa (by *Ansata Ibn Halima). Mossa RSI was imported on lease from Jean Kayser, as was her daughter, *Dantilla.

Mossa RSI (by Faarad), was a dark bay somewhat plain-headed mare of average size and good though somewhat heavy body. She outproduced herself when bred to Ansata stallions.

***Dantilla** (by Saab), Mossa's chestnut daughter, was of average type and quality, well-balanced and strong-bodied, and, like her mother, she outproduced herself when bred to Ansata stallions. The line to these mares no longer exists at Ansata.

The Farida Family through Helwa
- (Antar/Abla daughters)

The most prolific family through Helwa was through her daughter, Abla, a very classic, pretty-headed and strong-bodied Nazeer daughter. The story of lovers Antar and Abla is famed throughout Arabia, and the Egyptians never deviated from breeding this beautiful white mare to the powerful red chestnut stallion, Antar. The result was a number of daughters who became influential in Egypt and America. We acquired three mares tracing to this line through three different daughters: Looza, Adaweya, and *Somaia. We saw all three daughters in Egypt.

Looza was a good solid foundation broodmare type with a rather straight profile but pleasing head. Her daughter, *Lamees (by Karoon) was imported by Lowe, and the subsequent generations combining the beautiful-headed *Farazdac with The Egyptian Prince blood produced two sisters whom Ansata acquired:

SH Alleya, a lovely kind mare with good head and expressive large dark eyes, long-neck, and good body and topline, and the full sister to SH Say Anna. Her only two fillies proved to be exceptionally classic in type, the first, Ansata Afifa, (by Ansata Hejazi). She died the day she gave birth to the second one, Ansata Semiramis (by Ansata Sinan).

SH Say Anna, was gifted with a remarkably beautiful head, long neck, good balance and good length of hip, but again, the wonderful head was married to a not so wonderful croup. She produced two excellent fillies by Ansata Hejazi who were much better conformed.

Adaweya was one of the most refined and prettiest headed of these Antar-Abla daughters, and she was an important foundation mare in Egypt. *Rasheeka imported by Lancer Arabians rivaled her in beauty, but she was rather heavy bodied and non-productive. Adaweya's daughter, Enayah, wasn't as beautiful in the head, but she was a good broodmare. Her son, Adl, was a useful stallion in Egypt, and his full sister, *Nageia, produced well in the U.S.

***Nageia** (x Enayah), was by the popular E.A.O. stallion, Ikhnatoon, a very refined individual with an exceptionally beautiful head. *Nageia was imported to America, and we took her on lease for a short time to acquire a line to Adaweya. Bred to Ansata Hejazi, she produced Ansata Najiba who in turn bred to Ansata Iemhotep and Ansata Sinan, produced fillies of good size with very pretty heads, long well-shaped necks, good bodies, clean legs and outstanding movement.

*Somaia (x Abla) was imported by Gleannloch Farms and became an influential broodmare with a strong line in America. She, too, was a good bodied mare typical of this family, with a decent, but not exotic, head. Numerous champions descend from her including Dorian Kahra-Lima (by Halim El Nefous).

Simurgh (x Dorian Kahra-Lima), by Dalul, was acquired by Ansata at an auction. She was a broodmare of average type, strong body and good size, and she was purchased as a replacement for Dal Macharia, who had been sold to Kuwait. Simurgh has bred on very well when crossed with Ansata bloodlines.

Mansoura El Halima (x Simurgh) is sired by Royal Jalliel, a beautiful Ansata Imperial son that we have always admired. She was acquired at the time we purchased her dam. A pretty-headed and strong-bodied young mare, very Dahma in type, she was shown to a championship win at halter as a two-year-old.

The Farida Family through Helwa/Nefisa

Nefisa became an influential mare of this family in Egypt, Europe, America, and subsequently, the Arab world. She was rather plain-headed (typical of her sire's get), but nicely balanced and good-bodied; she was by the stallion, Balance, thus doubling her line to Farida. Her daughter, *Bint Nefisaa, imported by Gleannloch, was an exquisite Nazeer daughter - small, beautiful head, large dark eyes, clean neck and refined like a fine piece of porcelain. *Bint Nefisaa produced El Hilal, by *Ansata Ibn Halima - again doubling the Farida line in tail female - and he became a leading living sire of champions. Her half-sister, *Bint Nefisa I, a bay El Sareei daughter imported by Pritzlaff, was heavy by comparison, plain-headed, but having big, dark wonderful eyes.

In Germany, the good grey Nefisa daughter, Nadia (Nadja), by Nazeer, became a valued broodmare at the Marbach State Stud. She was bred to Hadban Enzahi, thus doubling Nazeer, and produced Nabya, a mare of extreme head, but again the beautiful head married to the peaky croup inherited from her sire. She was the most exquisite of the full sisters from these matings. Noha, the plainer full sister to Nabya, produced one of the most exquisite mares of this era, the beautiful Qatar National Champion Mare, RN Farida (by Salaa El Dine).

Ansata acquired the line to Nabya through purchasing three mares:

*Nadima (by Gharib), a very beautiful, well-balanced, fleabitten mare with very typey head and an excellent body far better than her dam's. She was purchased when she was an old mare and did not produce at Ansata.

Alajneha Nahme (by Anaza El Farid), was a mature mare when Ansata obtained her with her dam, *Nadima. She produced Ansata Haisam (by Ansata Hejazi) exported to Qatar where he became a Qatar National Jr. Champion Stallion, and Ansata Alida, a pretty filly by Ansata Sirius.

Il Nadheena (by Imperial Madheen), was out of the *Nadima daughter, Il Naheefa (by Ruminaja Ali). A small fleabitten mare, she inherited the exquisite head of Nabya, but was better in topline and croup. She was purchased at auction in very thin condition. It was said she wouldn't eat much, but on examination it was found she had a twisted tooth. After removing it she ate well, gained weight, and looked like a different mare. Due to foaling complications she lost a foal by Ansata Iemhotep, and then became a problem mare. She died in 2002 before producing a live foal at Ansata.

In reviewing the Halima line and the Futna lines at Ansata, both were extremely consistent and produced some of the best balanced, as well as the most typey mares and stallions. Beautiful heads with good dark expressive eyes were a trademark. Necks were well-shaped and proportionate, but could have been slightly longer. Legs were set out at the corners with well-defined forearms. This line had a tendency to offset-knees that could be overcome by the use of certain stallions. The addition of the Helwa family came later, through acquisitions in America. They bred on very well when served by Ansata stallions; however, we did not have them long enough to assess their performance over many generations as we did in comparison to the Halima and Futna lines.

Above, Halima (Sheikh El Arab x Ragia)

Above left and right, Abla (Nazeeer x Helwa)

El Dahma through Farida

Nadra El Saghira

El Dahma
c. 1880 Dahmah Shahwania mare
from the stud of Ali Pasha Sherif

Nadra El Khebira 1891 x **Nader El Khebir**

Nadra El Saghira x **Samhan**

Farida x **Saklawi II**

Ragia x **Ibn Rabdan**

Halima x **Sheikh El Arab**

Moheba x **Sid Abouhom**

Malacha x **El Sareei**

Farida

Futna

Futna x **Shahloul**

Dahma II x **Nazeer**

*Bint Dahma x **El Sareei**

Soja RSI x ***Rashad Ibn Nazeer**

Mossa RSI x **Faarad**

*Dantilla x **Saab**

Dahma II

*Bint Dahma

Ragia

Malikah x **Ghazal**

Maymoonah x **Hadban Enzahi**

*JKB Masouda x **Madkour**

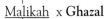

Mahari x **Hadban Enzahi**

Maharia x **Gharib**

Dal Macharia x **Dalul**

Soja

Halima

Mossa

Malacha

*JKB Masouda

Dal Macharia

*Dantilla

*Ansata Bint Sameh

Bint Farida

Helwa

Nefisa

Abla

Bint Farida x **Mansour**
Helwa x **Hamran II**
*Ansata Bint Sameh x **Sameh**
Nefisa x **Balance**
Abla x **Nazeer**

Adaweyah x **Antar**   Looza x **Antar**   *Somaia x **Antar**
Enayah x **Nawaf**   *Lamees x **Karoon** Bint Somaia x ***Morafic**
*Nageia x **Ikhnatoon** Bint Farazdac x ***Farazdac**

Adaweya

Enayah

Nadia x **Nazeer**
Nabya x **Hadban Enzahi**
*Nadima x **Gharib**

Looza

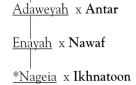

Bint Farazdac

Bint Somaia

*Nageia

Nadia/Nadja

Nabya

*Nadima

Full sisters, Ansata Majesty, left, and Ansata Malaka, right, (Ansata Halim Shah x Ansata Malika). Sparagowski photo.

FARIDA BRANCH of El Dahma

The Halima Family

El Dahma/Nadra El Kebira/Nadra El Saghira/Farida/Ragia/Halima/
Moheba/Malacha/Malikah/Maymoonah/*JKB Masouda

*JKB Masouda #245049
Grey Mare; May 23, 1980
Sire: Madkour   Dam: Maymoonah
Bred by Jean Kayser, Ettlebruck, Luxembourg.
Imported October 1981 on lease to Ansata Arabian Stud.

PRODUCE
Ansata Malika #328031 June 2, 1985
Grey mare by *Jamilll.

Ansata Majida #0396750 March 12, 1987
Grey mare by Ansata Halim Shah.
Name changed by Kayser to JKB Majida.
Exported to Luxembourg, July 1989.

Ansata Meryta #0419884 March 1, 1988
Grey mare by Ansata Halim Shah.

Ansata Menkaure #0430301 February 25, 1989
Grey stallion by Ansata Halim Shah.
Died February 10, 1991.

*JKB Masouda

Ansata Malika

Ansata Majida

Ansata Meryta

*JKB Masouda/Ansata Malika

Ansata Malika #0328031
Grey Mare; June 2, 1985 – August 24, 1994
Sire: *Jamilll   Dam: *JKB Masouda.

PRODUCE

Ansata Majesta #0430171 March 9, 1989
Grey mare by Ansata Halim Shah.
Sold to Sheikh Abdulaziz Al Thani, Al Rayyan Farm.
Exported to Qatar, 1992.

Unnamed grey stallion, April 30, 1990 by Ansata Halim Shah. Gelded.

Ansata Magnifica #0472000 April 12, 1991
Grey mare by Ansata Halim Shah.
Sold to Sheikh Abdulaziz Al Thani, Al Rayyan Farm.
Exported to Qatar, 1993.

Ansata Majesty #0479621 April 1, 1992
Grey mare by Ansata Halim Shah.

Ansata Malaha #0492640 March 24, 1993
Grey mare by Ansata Halim Shah. Sold to H.H. Sheikh Hamad bin Khalifa Al Thani,
Al Shaqab Farm. Exported to Qatar.

Ansata Malaka #0411000 August 15, 1994
Grey mare by Ansata Halim Shah.

Ansata Majesta. Photo in Qatar by Gigi.

Ansata Magnifica. Photo in Qatar by Filsinger.

Ansata Malaha

Ansata Malaka

Ansata Majesty

*JKB Masouda/Ansata Malika/Ansata Majesty

Ansata Majesty #0479621
Grey Mare; April 1, 1992
Sire: Ansata Halim Shah   Dam: Ansata Malika

PRODUCE
Ansata Majestic #0529851 February 24, 1996
Grey mare by Prince Fa Moniet. Sold to Usamah
Al Kazemi, May 1997. Exported to Kuwait.

1997 serviced by Ansata Iemhotep. Aborted filly at term.

Ansata Mital #568654 March 23, 1999
Grey stallion by Ansata Iemhotep. Sold to Judith Wich,
Orienta Arabians. Exported to Germany.

Unnamed grey stallion by Farres, March 20, 2002. Died
March 26, 2002.

*JKB Masouda/Ansata Malika/Ansata Malaka

Ansata Malaka #0511000
Grey Mare; August 15, 1994
Sire: Ansata Halim Shah   Dam: Ansata Malika

PRODUCE
Ansata Malik Shah #0566536 February 22, 1999
Grey stallion by Ansata Hejazi.

Ansata Millennia #0578191 March 2, 2000
Grey filly by Ansata Hejazi.

Ansata Mameluke Bey unregistered April 15, 2002
Grey stallion by Farres. Died December 2002.

Above left and right, Ansata Majestic

Ansata Mital

Ansata Malik Shah

Ansata Millennia

Ansata Mansoura

*JKB Masouda/Ansata Meryta

Ansata Meryta #0419884
Grey Mare; March 1, 1988
Sire: Ansata Halim Shah   Dam: *JKB Masouda
Sold to Kuwait Equestrian Club (Arabian Horse Center)
May 1998. Exported to Kuwait.

PRODUCE
Ansata Marduk #0481157 January 17, 1992
Grey stallion by Prince Fa Moniet. Gelded.
Sold to John Watkins, Mt. Ida, Arkansas.

Ansata Montu #0493708 January 26, 1993
Grey stallion by Prince Fa Moniet.
Sold at Auction. To be gelded.

Ansata Marha #0505509 March 11, 1994
Bay mare by Ansata Manasseh.
Sold to Sheikh Abdulaziz Al Thani, Al Rayyan Farm,
December 1995.
Exported to Qatar.

Ansata Maya #0518760 March 16, 1995
Grey mare by Arabest Kalid. Sold to H.R.H. Prince Khalid
S.K. Al Saud, Nejd Stud. Exported to Saudi Arabia, October
1996.

Ansata Munira #0529849 April 28, 1996
Grey mare by Ansata Manasseh. Sold to Jan and
Arja Lancee. Exported to the Netherlands, January 1997.

Ansata Mansoura #0542154 May 29, 1997
Grey mare by Ansata Iemhotep. Sold to H.R.H.
Prince Khalid S.K. Al Saud, Nejd Stud. Exported to Saudi
Arabia, November 1998.

1997 serviced by Ansata Hejazi. Exported in foal. Foaled colt
in Kuwait.

Ansata Marha

Ansata Maya

Ansata Munira

<u>Mohebba/Malacha/Mahari/*Maharia/Dal Macharia</u>

■ Dal Macharia #0414337

Bay Mare; April 18, 1988
Sire: Dalul   Dam: *Maharia
Bred by Andress Brothers, Tiverton, Ontario, Canada.
Purchased by Ansata in 1991 at the Addis Sale.
Sold to Khaled Ben Shokr, Kuwait, November 1999.

PRODUCE

Ansata Marjaneh #0494039 January 29, 1993
Grey mare by Ansata Halim Shah.

Ansata Maharani #0505513 February 9, 1994
Grey mare by Ansata Halim Shah
Sold to Jan and Arja Lancee.
Exported to the Netherlands, January 1997.

Ansata Mariam #0534381 May 25, 1996
Grey mare by Ansata Hejazi.
Sold to Sheikh Abdullah Bin Nasir Al Thani, Al Naif Stud.
Exported to Qatar, February 1997.

Ansata Majeeda #0542104 May 17, 1997
Grey mare by Ansata Hejazi.
Sold to H.R.H. Prince Khalid Bin Sultan Bin Abdulaziz
Al Saud, Saudi Arabia, December 5, 1997.
Exported to Saudi Arabia.

Unregistered chestnut mare June 5, 1998 by Ansata Sokar
Died.

Ansata Mahasen #0563812 May 23, 1999
Bay filly by Ansata Sokar. Sold to Jessie MacLean, New Brunswick, Canada.

Bred to Ansata Sokar 2000 and exported to Kuwait.
Foaled chestnut filly in Kuwait.

Above left and right, Dal Macharia

Ansata Marjaneh

Ansata Mariam

Ansata Maharani

Ansata Mahasen

Ansata Majeeda

Dal Macharia/Ansata Marjaneh

■ Ansata Marjaneh #0494039
Grey Mare; January 28, 1993
Sire: Ansata Halim Shah   Dam: Dal Macharia
Sold to Sheikh Abdullah Bin Nasir Al Thani, Al Naif Stud, Qatar. Exported 2001.

PRODUCE

Ansata Majdi #0545590 October 15, 1997
Grey stallion by Ansata Hejazi. Sold to Khaled Ben Shokr. Exported to Kuwait.

Ansata Mariha #0578066 March 18, 2000
Grey filly by Ansata Hejazi.

Exported in foal to Ansata Sinan.
Foaled filly in Qatar.

Ansata Majdi

Ansata Mariha

FARIDA branch of El Dahma

The Futna Family - *Ansata Bint Sameh

*El Dahma/Nadra El Kebira/Nadra El Saghira/Farida/Futna/ *Ansata Bint Sameh*

*Ansata Bint Sameh #69406
Grey Mare; April 13, 1962 – May 1, 1973
Sire: Sameh Dam: Futna
Bred by E.A.O., Cairo, Egypt
E.A.O. #307, Volume II
Registered as **Amal** in Egypt
Imported by Mr. and Mrs. Donald L. Forbis, Chickasha, Oklahoma, March 3, 1970.

PRODUCE

*Ansata El Wazir #70443 January 23, 1971
Grey stallion by *Tuhotmos. Sold to Bill and Betty Nottley, Newberg, Oregon, July 1977.

Ansata Jezebel #80321 March 27, 1972
Grey mare by *Ansata Ibn Halima.

Ansata Jamila #94539 March 26, 1973
Grey mare by *Ansata Ibn Halima.

Died at Ansata 1973 without further production.

*Ansata Bint Sameh

*Ansata El Wazir

Ansata Jamila

Ansata Jezebel

Ansata Ibn Shah

Ansata Jacinda

Ansata Joy Halima

<u>El Dahma/Nadra El Kebira/Nadra El Saghira/Farida//Futna/*Ansata Bint Sameh/Ansata Jezebel</u>

■ Ansata Jezebel #80321
Grey Mare; March 27, 1972 – July 4, 1987
Sire: *Ansata Ibn Halima   Dam: *Ansata Bint Sameh

<u>PRODUCE</u>

Unnamed colt April 28, 1976 by Ansata Ibn Sudan. Died May 26, 1976.

Ansata Ibn Shah #163143 October 17, 1977
Grey stallion by Ansata Shah Zaman.

Ansata Jacinda #196865 April 24, 1979
Bay mare by Ansata Ibn Sudan.

Ansata Aly Halim #237271 February 26, 1981
Grey stallion by *Ansata Ibn Halima. Sold to Gibson Arabians, Loomis, California January 1985.

Ansata Joy Halima #259987 May 27, 1982
Grey mare by *Ansata Ibn Halima.

1982 serviced by Ansata Shah Zaman. Foaled filly September 4, 1983. Died.

Ansata Ibn Zaman #327935 May 16, 1985
Bay stallion by Ansata Shah Zaman. Sold to Lee Romney, Santa Ynez, California, December 1985. Sold by Romney Estate to Count Federico Zichy-Thyssen. Exported to Argentina.

Ansata Jasmina #365207 October 6, 1986
Grey mare by Ansata Haji Halim. Sold to Larry and Roberta Kirberger, Blanchard, Oklahoma, February 1991.

Ansata Ibn Zaman

Ansata Bint Sameh/Ansata Jezebel/Ansata Jacinda

Ansata Jacinda #196856
Bay Mare; April 24, 1979
Sire: Ansata Ibn Sudan Dam: Ansata Jezebel
Sold to Jennifer Shaw, Callente, California,
December 1990.

PRODUCE
Ansata Amir Ali #0308079 February 24, 1984
Chestnut stallion by Ansata Shah Zaman
Gelded. Transferred to Fielding Lewis, Mena, Arkansas.

Ansata El Jeddi #0328034 February 26, 1985
Chestnut stallion by *Jamilll. Gelded.

Unnamed chestnut mare by *Jamilll
February 27, 1986. Died March 7, 1986.

Ansata Karim Shah #0392619 April 7, 1987
Grey stallion by Ansata Halim Shah. Died April 9, 1988.

Ansata Jahara #0410802 March 26, 1988
Grey mare by Ansata Halim Shah.

Ansata El Jaffar #0430146 March 25, 1989
Grey stallion by Ansata Halim Shah. Gelded.
Transferred to Jim and Diane Kruger, Sheridan, Arkansas.

Ansata Jahara

Ansata Bint Sameh/Ansata Jezebel/Ansata Jacinda/Ansata Jahara

Ansata Jahara #0410802
Grey Mare; March 26, 1988
Sire: Ansata Halim Shah Dam: Ansata Jacinda
Sold to Kuwait Equestrian Club (Arabian Horse Center) May 1998. Exported.

PRODUCE
Ansata Princessa #0481159 February 22, 1992
Grey mare by Prince Fa Moniet.
Sold to Jan and Arja Lancee, October 26, 1996.
Exported to the Netherlands.

Ansata Palmyra #0493648 March 19, 1993
Grey mare by Prince Fa Moniet.

(continued)

Ansata Princessa

Ansata Palmyra

Full sisters, Ansata Jezebel, above left
and Ansata Jamila, above right
(*Ansata Ibn Halima x *Ansata Bint Sameh)

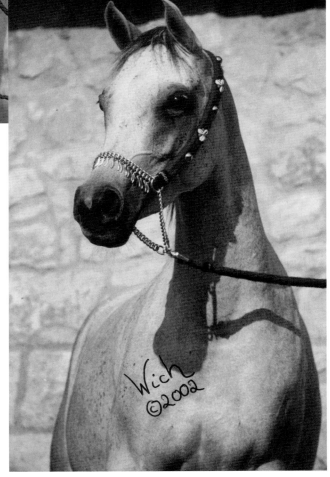

Ansata Jeylan in Germany
(Ansata Iemhotep x Ansata Jahara)
Judith Wich photo

Ansata Jubayl

Ansata Paloma

PRODUCE of Ansata Jahara continued

Ansata Paloma #0505507 March 26, 1994
Grey mare by Prince Fa Moniet.

Ansata Jubayl #0545589 April 22, 1997
Grey stallion by Ansata Hejazi.
Sold to Marty Garrett, Cairo, Egypt, June 1992. Exported.

Ansata Jeylan #0551193 June 12, 1998
Grey mare by Ansata Iemhotep. Sold to Judith Wich,
Orienta Arabians. Exported to Germany, May 1999.

*Ansata Bint Sameh/Ansata Jezebel/Ansata Jacinda/Ansata Jahara/
Ansata Palmyra*

Ansata Palmyra #0493648
Grey Mare; March 19, 1993
Sire: Prince Fa Moniet   Dam: Ansata Jahara

PRODUCE
Ansata Petra #589218 March 4, 2001
Grey mare by Ansata Sinan.

Ansata Phaidra #0596953 May 20, 2002
Grey mare by Ansata Sirius.

Ansata Petra

Ansata Bint Sameh/Ansata Jezebel/Ansata Jacinda/Ansata Jahara/Ansata Paloma

Ansata Paloma #0505507
Grey Mare; March 26, 1994
Sire: Prince Fa Moniet Dam: Ansata Jahara
Sold to Sheikh Abdullah Bin Nasir Al Thani, Al Naif Stud, Doha, Qatar. Exported.

PRODUCE

Ansata Julima #0551159 May 11, 1998.
Grey mare by Ansata Iemhotep. Sold
to Kuwait Arabian Horse Center, Kuwait, 1999. Exported.

Ansata Jasour #575127 August 20, 1999
Grey colt by Ansata Iemhotep. Gelded.
Sold to Roger Daniels, Atlanta, Georgia.

Ansata Jalala #579275 August 25, 2000
Grey mare by Ansata Iemhotep.

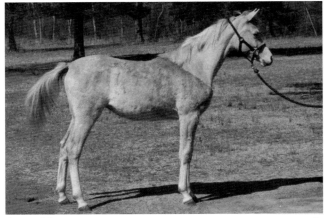
Ansata Julima

Ansata Bint Sameh/Ansata Jezebel/Ansata Jacinda/Ansata Joy Halima

Ansata Joy Halima #0259987
Grey Mare; May 27, 1982
Sire: *Ansata Ibn Halima   Dam: Ansata Jezebel
Sold to Guillermo Henin, Buenos Aires, Argentina, South America, December 1990. Exported.

PRODUCE

Ansata Joy Jamila #0355891 March 18, 1986
Grey mare by *Jamilll.
Sold to Richard Marshall, Ocala, Florida.

Ansata Joy Amira #0389340 June 18, 1987
Grey mare by Dorian Shah El Shams.

Ansata Yildiz #0418466 June 24, 1988
Grey stallion by Ansata Mourad Bey.
Gelded and transferred to Mr. and Mrs. Mark Gober, Wilmar, Arkansas, April 1989.

Ansata Joy Moniet #0431987 June 19, 1989
Grey mare by Prince Fa Moniet.
Sold to Dean Homstead, Pitsfield, Maine, December 1990.

Ansata Jalala

Ansata Joy Jamila

Ansata Jamila

Ansata Joy Halima produce continued

Ansata El Jalil #0456413 June 24, 1990
Grey stallion by Prince Fa Moniet. Gelded and transferred to Virgie Lawrence, Mena, Arkansas.

El Wajeeb #0479103 August 14, 1991
Grey stallion by Ansata Cairo Bey. Owned by Guillermo Henin. Sold to Abdullah Al Sweilam. Exported to Saudi Arabia.

Ansata Bint Sameh/Ansata Jezebel/Ansata Jacinda/Ansata Joy Halima/Ansata Joy Amira

Ansata Joy Amira #0389403
Grey Mare; June 18, 1987
Sire: Dorian Shah El Shams   Dam: Ansata Joy Halima
Sold to Jim Dent, Enid, Oklahoma, December 1990.

PRODUCE
Andara Joy Amira #0473664 April 20, 1991
Grey mare by Prince Fa Moniet. Owned by Jim Dent, Enid, Oklahoma.

Ansata Joy Amira

Ansata Bint Sameh/Ansata Jamila

Ansata Jamila #094539
Grey Mare; March 25, 1973 – October 2, 1983
Sire: *Ansata Ibn Halima   Dam: *Ansata Bint Sameh

PRODUCE
Ansata Judea #179148 July 29, 1978
Grey mare by Ansata Abu Sudan.

Ansata El Halim #219548 October 7, 1980
Grey stallion by *Ansata Ibn Halima. Exchanged with Gibson Arabians, Loomis, California, for Ansata Aly Halim, February 1987.

Ansata Shah Zahir #237273 September 20, 1981
Grey stallion by Ansata Shah Zaman. Sold to Joan Skeels, Hope Farm, Mendham, New Jersey, November 1983.

Ansata Ibn Jamila #259989 September 4, 1982
Grey stallion by Ansata Ibn Shah. Sold to Frank Wyer, Jenks, Oklahoma, February 1985.

Ansata Jamila

Produce of Ansata Jamila

Ansata Judea

Ansata El Halim

Ansata Shah Zahir

Ansata Ibn Jamila

Ansata Jumana

*Ansata Bint Sameh/Ansata Jamila/Ansata Judea

Ansata Judea #179148
Grey Mare; July 28, 1978 –1986
Sire: Ansata Abu Sudan   Dam: Ansata Jamila

PRODUCE

Ansata Jessica #260507 March 27, 1982
Grey mare by Ansata Ibn Sudan.

Ansata Jumana #278159 March 27, 1983
Grey mare by Ansata Halima Son.

Ansata Justina #308071 March 20, 1984
Bay mare by *Jamilll. Sold to
Ansata Justina Partnership, Waco, Texas,
January 1989.

Ansata El Jadib #0327857 March 10, 1985
Grey stallion by *Jamilll. Gelded and
transferred to Yvonne Head, Mena, Arkansas.

Ansata Julnara #0355854 February 18, 1986
Bay mare by *Jamilll. Donated to
Society for the Preservation of the Egyptian Arabian Horse,
Lexington, Kentucky.

*Ansata Bint Sameh/Ansata Jamila/Ansata Judea/Ansata Jessica

Ansata Jessica #0260507
Grey Mare; March 27, 1982 – 1990
Sire: Ansata Ibn Sudan   Dam: Ansata Judea
Leased to Light of the Mountain Partnership.
Died while on lease.

PRODUCE

Ansata Jalisa #0355855 March 17, 1986
Grey mare by *Jamilll

Unnamed grey colt by Ansata Halim Shah April 1, 1988.
Transferred unregistered to Ray Goodner July 1988.

LOM Karima #0462076 May 23, 1990
Grey mare by DAJ Abbas Pasha.
Owned by Light of the Mountain Partnership.

Ansata Justina

*Ansata Bint Sameh/Ansata Jamila/Ansata Judea/Ansata Jessica/
Ansata Jalisa

Ansata Jalisa #0355855
Grey Mare. March 17, 1986
Sire: *Jamilll   Dam: Ansata Jessica
Leased by Light of the Mountain Arabians 1989-1992.
Sold to Dr. Chess Hudson,
Lawrenceville, Georgia, February 1994.

PRODUCE
LOM Jafar #0461971 June 30, 1990
Chestnut stallion by DAJ Abbas Pasha.
Owned by Light of the Mountain Partnership.

LOM Tajalli #0500551 April 20, 1992
Grey mare by Ansata Ibn Tulun.
Owned by Light of the Mountain Partnership.

*Ansata Bint Sameh/Ansata Jamila/Ansata Judea/Ansata Jumana

Ansata Jumana #278159
Grey Mare; March 21, 1983
Sire: Ansata Halima Son   Dam: Ansata Judea
Sold in 2000 to Dr. Linda Ford, Mountain View, Arkansas.

PRODUCE
Ansata Janina #0390050 April 5, 1987
Bay mare by Dorian Shah El Shams.
Sold to Gary Shores, Houston, Texas, December 1990.

Ansata Rami #0410745 May 27, 1988
Grey stallion by Ansata Halim Shah.
Gelded.

Ansata Vizier #0440739 September 26, 1989
Grey stallion by Prince Fa Moniet.
Gelded and transferred to Fred Creekmore, Mena, Arkansas.

Ansata Jumana Leased to Barbara Cole,
Hammond, Louisiana, from 1990 -1994.

Asjahs Dominion #0580169 June 14, 1991
Grey stallion by Asjah Ibn Faleh.
Owned by Barbara Cole, Hammond, Louisiana.

Ansata Bint Asjah #0485693 June 7, 1992
Bay mare by Asjah Ibn Faleh. Owned by Ansata.

Asjahs Black Jewl #516969 June 18, 1993
Bay mare by Asjah Ibn Faleh.
Owned by Barbara Cole, Hammond, Louisiana.

Ansata Jumana

Ansata Bint Asjah

Ansata Exemplar

Ansata Exemplar

Ansata Julnar

Ansata Empress

PRODUCE *of Ansata Jumana continued*

Ansata Exemplar #0510991 June 12, 1994
Black stallion by Asjah Ibn Faleh.
Owned by Ansata. Sold to Heinrich and Margaret Kretschmann. Exported to Germany, June 1996.

1995 foaled stillborn filly by Ansata Mourad Bey.

Ansata Julnar #05349191 September 10, 1996
Grey mare by Prince Fa Moniet.
Sold in 1999 to Rocky Top Ranch, Brenham, Texas.

Ansata Jemlah #0551160 April 8, 1998
Grey mare by Ansata Iemhotep.
Sold in 2001 to Mohammed A. Al-Tobaishi, Saudi Arabia.

1998 serviced by Ansata Sokar. Foaled colt April 3, 1999. Died.

Ansata Jessamine #580178 September 15, 2000
Grey mare by Ansata Sirius.

Serviced 2000 by Ansata Sokar. Foaled filly 2001 owned by Dr. L. Ford.

*Ansata Bint Sameh/Ansata Jamila/Ansata Judea/Ansata Jumana/ Ansata Bint Asjah

Ansata Bint Asjah #0485693
Bay Mare; June 7, 1992
Sire: Asjah Ibn Faleh   Dam: Ansata Jumana

PRODUCE

Ansata Nibal #pending May 18, 1998
Grey colt by Ansata Iemhotep. Gelded and transferred to Roger Shuler, Fayetteville, Arkansas.

Ansata Empress #056381 May 24, 1999
Black mare by Ansata Sokar. Sold to H.R.H. Prince Khalid bin Fahd Al Saud. Exported to Saudi Arabia.

Ansata Shukri #pending September 22, 2000
Chestnut colt by Ansata Sokar. Sold at auction.

Sold to Khalid Al- Haddad, Jeddah, Kingdom of Saudi Arabia.

Ansata Bint Sameh/Ansata Jamila/Ansata Judea/Ansata Justina

Ansata Justina #308071

Bay Mare; March 20, 1984
Sire: *Jamilll   Dam: Ansata Judea
Sold to Ansata Justina Partnership, Austin, Texas,
January 1986

PRODUCE

Ali Jamila #0382614 March 16, 1987
Black mare by Ruminaja Ali.
Owned by Ansata Justina Partnership, Austin, Texas.

Jubillee #0427077 February 5, 1989
Bay mare by The Minstril.
Owned by Ansata Justina Partnership, Austin, Texas.

The Futna Family - *Bint Dahma

*El Dahma/Nadra El Kebira/Nadra El Saghira/Farida//Futna/Dahma II/*Bint Dahma-Kismat/Soja RSI*

Mossa RSI #050207

Bay Mare; April 16, 1968
Sire: Faarad   Dam: Soja RSI
Bred by Richard G. Pritzlaff, Sapello, New Mexico.
Owned by Jean Kayser, Luxembourg.
Leased to Ansata, Mena, Arkansas.

PRODUCE

Ansata Diana #0287112 August 23, 1983
Bay mare by Ansata Ibn Sudan. Sold to
Mike Albertini and Tom Patterson, Mena, Arkansas.

Ansata Justina

Ali Jamila

Mossa RSI

Ansata Diana

*Dantilla

*Bint Dahma-Kismat/Soja RSI/Mossa RSI/*Dantilla

*Dantilla #245048
Chestnut Mare; April 1, 1979
Sire: Saab AHS   Dam: Mossa RSI AHS
Bred by Mr. and Mrs. R. J. Hutchings, Dover, Kent, England
Owned by Mr. Jean Kayser, Ettlebruck, Luxembourg October 30, 1981.
Imported and leased to Ansata, Mena, Arkansas.

PRODUCE

Ansata Dafina #284228 April 2, 1983
Chestnut mare by Ansata Ibn Sudan.
Sold to Gary Jones, Kingwood, Texas.

Ansata Cleopatra #0298426 March 27, 1984
Chestnut mare by Ansata Ibn Sudan.
Sold to Roy L. Fuller, Baytown, Texas.

Al Shahab #0327827 April 10, 1985
Chestnut stallion by Ansata Ibn Sudan.
Bred and owned by Jean Kayser.

Ansata El Waseem #0372542 August 20, 1986
Grey stallion by Ansata Halim Shah.
Ownership transferred to Jean Kayser May 1990.
Sold by Kayser to Cindy Mize, Cadiz, Kentucky.

Ansata El Farid #0396389 September 8, 1987
Grey stallion by Ansata Halim Shah.
Transferred to Philippe and Yvonne de Bourbon, Haras Jinnah Al Tayr. Exported to Uruguay, South America.

FARIDA Branch of El Dahma

The Helwa Family

El Dahma/Nadra El Kebira/Nadra El Saghira/Farida/Helwa/Abla/Looza/*Lamees/Bint Farazdac/SH Alleya

SH Alleya #0367708
Grey Mare; April 24, 1986 - March 17, 2001
Sire: MFA Saygah   Dam: Bint Farazdac
Purchased by Ansata in 1999 from C. Scott Crawford.

PRODUCE

Ansata Afifa #0578069 March 18, 2000
Grey mare by Ansata Hejazi.

SH Alleya

Ansata Afifa

Ansata Semiramis

SH Say Anna

PRODUCE of SH Alleya continued

Ansata Semiramis #587268 March 17, 2001
Grey mare by Ansata Sinan.
Sold to Sheikha Sarah Al Sabah, Kuwait, 2002. Exported.

Lamees/Bint Farazdac/SH Say Anna

SH Say Anna #0417644
Grey Mare; January 28, 1988
Sire: MFA Saygah   Dam: Bint Farazdac
Bred by Harry and Sue Belinski and W.J. Pat Trapp,
Dousman, Wisconsin.
Purchased by Ansata at The Egyptian Elegance Sale,
Lexington, Kentucky, June 1991.

Ansata Aniq

PRODUCE

Ansata Aniq #0505492 February 1, 1994
Grey mare by Ansata Manasseh.
Sold to Paul and Anne Walker of Canada in 1995.

Ansata Amira #0519646 February 21, 1995
Grey mare by Arabest Kalid.
Sold to Debby Wood, Eagle, Colorado, October 1995.

Ansata Aldara #0529846 May 27, 1996
Grey mare by Ansata Hejazi. Sold to
Claudia Quentin. Exported to Argentina.

Ansata Anna Maria #0563810 February 21, 1999
Grey mare by Ansata Hejazi. Sold to Sheikh Abdullah Bin
Nasir Al Thani, Al Naif Stud, Doha, Qatar. Exported.

Ansata Aldara

*Beautiful descendants of the E.A.O.'s mating of Antar and Abla,
in keeping with the tradition of the beloved Arabic poem about these two lovers.
Below left, Ansata Nahida, descending from the Antar/Abla daughter, Looza.
Below right, SH Say Anna, descending from the Antar/Abla daughter, Adaweya.*

Ansata Amira

Ansata Anna Maria

<u>El Dahma/Nadra El Kebira/
Nadra El Saghira/Farida/Helwa/Abla/Adaweyah/Enayah/*Nageia</u>

*Nageia #0380688
Grey Mare; August 25, 1984
Sire: Ikhnatoon   Dam: Enayah
Bred by the E.A.O., Cairo, Egypt.
Imported from Egypt by George El Cassabgui and leased by Ansata, Mena, Arkansas.

PRODUCE
Ansata Najiba #0529832 February 25, 1996
Grey mare by Ansata Hejazi.

*Nageia

Ansata Najiba

Nageia/Ansata Najiba

Ansata Najiba #0529832
Grey Mare; February 25, 1996
Sire: Ansata Hejazi   Dam: *Nageia

PRODUCE
Ansata Nahida #0578071 June 4, 2000
Grey mare by Ansata Iemhotep.

Ansata Nasrina #589163 September 28, 2001
Grey mare by Ansata Sinan.
Sold 2002 to Steve and Erica Parsons. Exported to England.

El Dahma/Nadra El Kebira/Nadra El Saghira/Farida/Helwa/Abla/
*Somaia/Bint Somaia/Dorian Shahramana/Dorian Kahra-Lima/
Simurgh

Simurgh #0440433
Bay Mare; June 6, 1989
Sire: Dalul   Dam: Dorian Kahra-Lima
Purchased by Ansata in Canada at the Edwards Sale, July 1999.

PRODUCE
Ansata Shakilah #589164 Octoer 27, 2001
Grey mare by Ansata Sirius.

*Somaia/Bint Somaia/Dorian Shahramana/Dorian Kahra-Lima/
Simurgh/Mansoura El Halima

Mansoura El Halima #563739
Grey Mare; August 27, 1997
Sire: Royal Jalliel   Dam: Simurgh
Purchased in Canada by Ansata at the Edwards Sale, July 1999.

Ansata Nahida

Ansata Nasrina

Simurgh

Mansoura El Halima

FARIDA Branch of El Dahma

The Nefisa Line

El Dahma/Nadra El Kebira/
Nadra El Saghira/Farida/Helwa/Nefisa/Nadia/Nabya/*Nadima

▪ *Nadima #235513
Grey Mare; February 1, 1980
Sire: Gharib S.B.W.M. 172 Dam: Nabya S.B.W.M. 199
Bred by Wurttemburg Landgestut Marbach, Germany.
Imported by Gleannloch Farms.
Purchased in 1998 by Ansata from Wingspread Farm, Virginia.

PRODUCE
Mare was non-productive at Ansata.

*Nadima at age 22

*Nadima/Alajneha Nahma

▪ Alajneha Nahme #0526963
Bay Mare; May 15, 1995
Sire: Anaza El Farid   Dam: *Nadima
Leased by Ansata from Wingspread Farm in 1998 and purchased in July 1999.

PRODUCE
Ansata Haisam #566462 May 14, 1999.
Grey stallion by Ansata Hejazi.
Sold to Sheikh Abdullah Bin Nasir Al Thani, Al Naif Stud.
Exported to Qatar, February 2000.

Foaled pre-mature filly by Ansata Hejazi on May 13, 2000. Died.

Ansata Alida #589166 August 29, 2001
Grey mare by Ansata Sirius.

Ansata Al Nadim #0596949 August 6, 2002
Grey stallion by Ansata Iemhotep.

Alajneha Nahme

Ansata Alida   Ansata Haisam

*Nadima/Il Naheefa/Il Nadheena

▪ Il Nadheena #0426913
Grey Mare; March 3, 1989   Died 2002
Sire: Imperial Madheen Dam: Il Naheefa
Purchased by Ansata, Mena, Arkansas, from Montebello Farms, Canada, September 1997.

PRODUCE
1998 serviced by Ansata Iemhotep. 1999 colt died during foaling.

Il Nadheena

Bint El Bahreyn
of Lady Anne Blunt

This Dahman family stemmed from a different root mare than the Bint Sabah and Farida line. Bint El Bahreyn came directly from the horses of the Al Khalifahs of Bahreyn, who were preservers of this strain. Lady Anne Blunt described her as a bright bay with four white feet and saying that "...this mare had one defect, ears like bat's ears, but they need not be transmitted; her filly did not have them and otherwise she is very fine." Among her foals were Dalal, by Jamil (Blunt), "A very beautiful filly with gazelle-like head," who was influential in the Inshass herd.

The Bint El Bahreyn/Durra/Zareefa descendants were also structurally different from those of the Ali Pasha Sherif El Dahma line, being taller and bigger-bodied in general, but also of good balance with strong toplines. Some exceptionally beautiful individuals stemmed from this line in Egypt through Zareefa, particularly her progeny by Shahloul: the mare Assila (non-productive), her sister, Maisa, and the bay stallion, El Sareei, one of the most beautiful stallions we have ever seen within Egyptian bloodlines. El Sareei distinguished himself as a sire of daughters, but his only son of some significance, though not a dominant breeding stallion, was *Tuhotmos, who was the result of a mating we had requested Dr. Marsafi to make in order to double the Shahloul blood (sire of sire is sire of dam).

Two daughters of Zareefa are found in many pedigrees: Elwya and Maisa. Elwya (by Sid Abouhom) was the dam of Seef, a Mashhour son, used by the E.A.O., good-bodied, but having had a rather large, plain though pleasant head, as did his sire.

Maisa, the daughter of Zareefa, was an extremely handsome fleabitten mare, one of the outstanding Shahloul daughters we saw at the time we first visited the E.A.O. The importation of *Bint Maisa El Saghira by Gleannloch solidified this strain in America, and from her descended important breeding stallions such as the exquisite-headed Amaal (unshown due to injuries), Sheikh El Badi, a champion and sire of renown, while daughters such as Dahma Shahwania, Dahma El Ashekwar, Rihahna, Radia, and others contributed to the broodmare band. Another male line to Maisa went to Germany through the exportation of Madkour (*Morafic x Maisa), who became influential in Europe.

For further reference about this family, consult *Authentic Arabian Bloodstock, Vol. I.*

Elwya Branch

We did not see Elwya at the E.A.O., nor do we have any pictures of her. The Ansata connection to the Elwya family was acquired by importation of her daughter, Bint Elwya.

*Ansata Bint Elwya (by Antar). An attractive mare with nice, but somewhat plain head, rather long ears, well-shaped sufficiently long neck and strong body and topline. She produced well with *Ansata Ibn Halima, her colt, Ansata El Hakim, being exported to Australia, her daughter, Ansata Haliwa, was an attractive bay mare. However, *Ansata Bint Elwya was later sold and this line was carried on by other breeders.

Maisa Branch

*Bint Maisa El Saghira (by Nazeer), imported by Gleannloch, was an ideal representative of this line, winning championships at halter and performance along with her stablemate, *Ansata Ibn Halima, when he was leased to and campaigned by Gleannloch. She was an elegant, tall, bold mare with good length of neck and pleasing head, though not exotic. She endeared herself to breeders everywhere because of her demeanor and style, and as a broodmare she was very prolific. One of her most beautiful daughters was Dahma Shahwania, sired by *Ansata Ibn Halima when he was at Gleannloch. The combination of these two families has been extremely successful. We acquired three mares tracing to *Bint Maisa el Saghira to develop this combination in the Ansata herd.

AK Princess Maisa (by The Egyptian Prince), was an attractive chestnut mare with pleasing head, good neck and strong body who produced well, but was later sold. She was a daughter of Rannana x Rahmaa x the beautiful Dahma Il Ashekwar, an extremely classic daughter of *Bint Maisa El Saghira.

Rayya (by Al Fattah), was a very pretty mare, big bodied, tall, and well-balanced. She was out of the wonderful white mare, Rihahna, a *Morafic daughter out of *Bint Maisa El Saghira and full sister to Shaikh Al Badi and Amaal. Ansata purchased Rayya at the Legacy Sale, but after she produced several foals, she was sold.

Ibriiah (by Ruminaja Ali), was shown successfully and was a typical representative of this family; a tall grey mare with attractive head, good length of neck, strong body, and a good mover. She was out of Il Maddah, a *Soufian daughter out of Radia, a *Morafic daughter who was out of *Bint Maisa El Saghira. Ansata acquired Ibriiah on lease, then purchased her. She produced several foals and was later sold to Khalid Al Haddad, who had earlier bought and exported her Ansata Hejazi daughter, Ansata Al Halima, to Saudi Arabia.

In summary of this family at Ansata, we did not breed it long enough to make a clear judgment about it's contribution to our particular program. There were many beautiful individuals from this line, and it gave both stallions and mares that bred on well in other studs. Some stallions were rather high on the leg in comparison to the length of their bodies (Madkour, Amaal, Ezz el Arab), and there appears to be more white markings and white in the eyes than we experienced in the Farida or Sabah family. While we have always admired this family and use it within the program, particularly through grand parents and great-grandparents, we found the Farida and Sabah family were more predictable in our program.

*Ansata Bint Elwya

*Bint Maisa El Saghira

Radia
(*Morafic x
*Bint Maisa El Saghira)

Bint El Bahreyn Through Maisa and Elwya

(Dahma Shahwanieh Strain. Bay mare foaled 1898. Bred by Aissa Ibn Khalifeh, Sheikh of Bahreyn, and given by him to Khedive Abbas Pasha II in 1903. She was bought from the Khedive by Lady Anne Blunt on December 26, 1907 and sold by her in 1912 to an unnamed purchaser.)

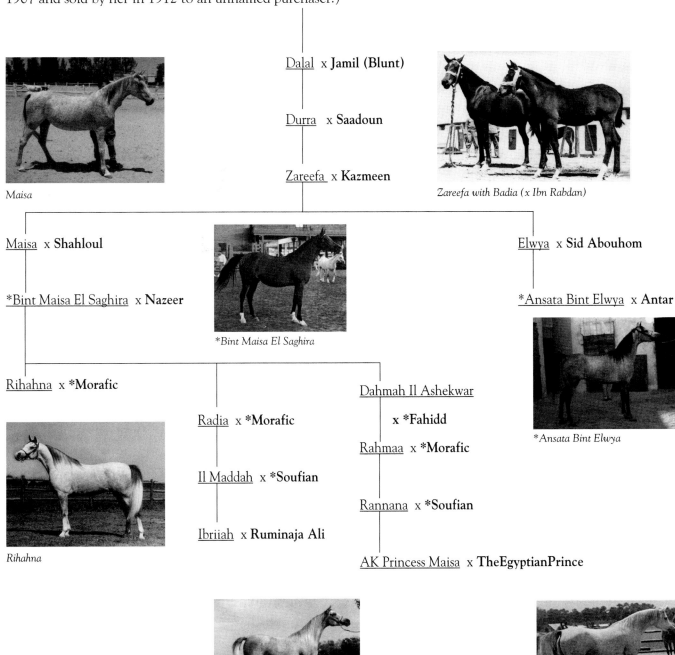

Dalal x **Jamil (Blunt)**

Durra x **Saadoun**

Zareefa x **Kazmeen**

Maisa

Zareefa with Badia (x Ibn Rabdan)

Maisa x **Shahloul**

*Bint Maisa El Saghira x **Nazeer**

Bint Maisa El Saghira

Rihahna x ***Morafic**

Radia x ***Morafic**

Il Maddah x ***Soufian**

Ibriiah x **Ruminaja Ali**

Dahmah Il Ashekwar x ***Fahidd**

Rahmaa x ***Morafic**

Rannana x ***Soufian**

AK Princess Maisa x **TheEgyptianPrince**

Elwya x **Sid Abouhom**

*Ansata Bint Elwya x **Antar**

Ansata Bint Elwya

Rihahna

Ibriiah

Rahmaa

245

BINT EL BAHREYN (Bahrein)
of Lady Anne Blunt

The Elwya Family

*Bint El Bahreyn/Dalal/Durra/Zareefa/Elwya/*Ansata Bint Elwya*

■ *Ansata Bint Elwya #60569
Dahma Shahwania
Grey Mare; February 28, 1961 – April 17, 1984
Sire: Anter   Dam: Elwya
Bred by E.A.O.
E.A.O. #301, Volume II
Registered as Bint Elwya in Egypt
Imported by Mr. and Mrs. Donald L. Forbis,
Chickasha, Oklahoma.
Sold in foal to *Ansata Ibn Halima to Trudy Mangels
and Carol Thompson, Standfordville, New York.

PRODUCE

Ansata Al Caliph #64612   May 5, 1970
Bay stallion by Ibn Hafiza. Gelded and
sold to Susan Morstad, Rochester, Minnesota.

Ansata El Hakim #80320   February 17, 1972
Bay stallion by *Ansata Ibn Halima.
Sold to Mr. P. S. James, Australia. Exported.

Ansata Haliwa #91008   February 15, 1973
Bay mare by *Ansata Ibn Halima.
Owned by Dr. Carol Thompson.

Mona El Ajzaa #0113089   March 20, 1974
Grey mare by *Sanaad. Owned by
Trudy Mangels.

*Ansata Bint Elwya

Ansata Al Caliph

Ansata Haliwa

Ansata El Hakim

The Maisa Family

*Bint El Bahreyn/Dalal/Durra/Zareefa/Maisa/*Bint Maisa El Saghira/
Dahma El Ashekwar/Rahma/Rannana/AKPrincess Maisa*

■ AK Princess Maisa #0313005
Chestnut Mare; May 12, 1984
Sire: TheEgyptianPrince   Dam: Rannana
Bred by Bentwood Farms, Waco, Texas.
Purchased with AK Bint Maisa at side from the Bentwood
Sale, July 1988.
Sold July 1996 to Joan Skeels and Sue Burnham, Hope Farm,
Cove, Arkansas.

PRODUCE
AK Bint Maisa #0404013   March 27, 1988
Grey mare by Prince Fa Moniet.
Sold to Jan Van Ness, December 15, 1990.
Exported to Holland.

El Mais #0441692   August 12, 1989
Grey stallion by Ansata Haji Jamil.
Gelded.

AK Princess Maisa was on lease to Barbara Cole, Hammond,
Louisiana from 1990 to 1994.

Anazeha #0487628   March 13, 1992
Chestnut mare by Asjah Ibn Faleh.
Owned by Barbara Cole, Hammond, Louisiana.

Ali Asjah #0540432   April 3, 1994
Chestnut stallion by Asjah Ibn Faleh.
Owned by Barbara Cole, Hammond, Louisiana.

Ansata Astra #0520273   August 6, 1995
Grey mare by Ansata Orion.
Sold July 29, 1996 to Joan Skeels and Sue Burnham,
Cove, Arkansas.

Ansata Astoria #0538764   September 20, 1996
Grey mare by Ansata Hejazi.
Owned by Joan Skeels and Sue Burnham,
Cove, Arkansas.

*Above and at right,
AK Princess Maisa*

AK Bint Maisa

Ansata Astra

Rayya

Bint El Shaikh

Ansata El Kaswa

*Bint El Bahreyn/Dalal/Durra/Zareefa/Maisa/*Bint Maisa El Saghira/ Rihahna/Rayya*

Rayya #278110

Grey Mare; May 18, 1983
Sire: Al Fattah   Dam: Rihahna
Bred by Mr. and Mrs. Douglas Marshall, Gleannloch Farms, Barksdale, Texas U.S.A. Purchased by Ansata in August 1986 at Legacy Sale, in foal to Ibn Shaikh.
Sold at Bittersweet Sale to Robert and Merry Carole Falke, Fredericksburg, Virginia.

PRODUCE

Bint El Shaikh #0389334   January 31, 1987
Bay mare by Ibn Shaikh.

Ansata Ibn Rayya #0412843   February 28, 1988.
Grey stallion by Ansata Halim Shah. Gelded and transferred to Cindy and Rick Moody,
Natural Dam, Arkansas.

Ansata Bint Rayya #0433438   March 2, 1989
Grey mare by Ansata Halim Shah. Sold to Jef and Joop Smarius, Smaroel Arabians,
Tilburg, Holland. Exported to the Netherlands.

Unnamed grey mare by Ansata Halim Shah, March 22, 1990. Died March 28, 1990.

Prince Fa Amir #0470354   March 26, 1991
Grey stallion by Prince Fa Moniet. Gelded. Owned by Robert and Merry Carole Falke,
Fredericksburg, Virginia.

*Bint El Bahreyn/Dalal/Durra/Zareefa/Maisa/*Bint Maisa El Saghira/ Rihahna/Rayya/Bint El Sheikh*

Bint El Sheikh #0389334

Bay mare; January 31, 1987
Sire: Ibn Shaikh   Dam: Rayya
Bred by Gleannloch Farms. Foaled at Ansata.

PRODUCE

Ansata Ibn Asjah #0488930   April 20, 1992
Chestnut stallion by Asjah Ibn Faleh.

Ansata El Kaswa #0520274   August 2, 1995
Bay mare by Ansata Manasseh.

*Bint El Bahreyn/Dalal/Durra/Zareefa/Maisa/*Bint Maisa El Saghira/
Rihahna/Rayya/Bint El Sheikh/Ansata El Kaswa*

■ Ansata El Kaswa   #0520274
Bay mare; August 2, 1995
Sire: Ansata Manasseh.   Dam: Bint El Sheikh

PRODUCE
Sold in foal to Ansata Sokar to H.R.H. Prince Khalid Bin Fahad Al Saud. Exported to Riyadh, Saudi Arabia.

*Bint El Bahreyn/Dalal/Durra/Zareefa/Maisa/*Bint Maisa El Saghira/
Radia/Il Maddah/Ibriiah*

■ Ibriiah #0323276
Grey Mare;   April 1, 1985
Sire: Ruminaja Ali   Dam: Il Maddah
Leased in 1994 and later purchased by Ansata from David and Martha Murdoch, Rocaille Ranch, Texas. Sold 2001 to Khalid Al-Haddad. Exported to Jeddah, Saudi Arabia.

Ibriiah when young

PRODUCE
Ansata Al Halima #0526779 March 6, 1996
Grey mare by Ansata Hejazi.
Sold to Khalid Al-Haddad. Exported to Jeddah, Saudi Arabia, October 1996.

Ansata Khanjar April 19, 1999
Grey stallion by Ansata Iemhotep. Gelded.

Ansata Khaleed unregistered April 8, 2000
Grey stallion. Transferred to Sheila Theriot. Died.

2001 serviced by Ansata Sirius and exported in foal to Saudi Arabia.

Ansata Al Halima

Ibriiah at maturity

Al Saqlawiyat

Omar, 19th century lithograph by Alfred DeDreux

The Saqlawi Jedran Strain

The Saklawi strain was much venerated by Abbas Pasha I; collecting and breeding it became his passion. He spared no effort or cost to secure the finest specimens in the desert, eventually acquiring the largest and best collection ever assembled under one roof. When his emissaries visited the tribes in Arabia to record the history of this strain and its descendants, they compiled a seemingly endless dialogue from their Bedouin hosts about how the strain was passed from breeder to breeder. This intense scrutiny is noteworthy. The Saklawi strain appears third in preference in *The Abbas Pasha Manuscript* and begins with a history of Al Samniyat as follows:

The assembly took place in the presence of Johayl ibn Jedran, a whitehaired man who is the eldest of the present descendants of al Jedran, and also Shara'an ibn Hadhud and Hamad of al Jedran and all the children and relatives of the family of al Jedran, the Ra'i of the stud..

They were asked," O descendants of Jedrani, tell us about your horses and from whom did they originate, and how they passed to you, and from whom they passed to you, and to whom they passed from you, and how they passed from you to the outsiders, and whether they were beautiful, famous and swift at the time of your grandfather, al Jedrani. Tell us about the cherished ones that belonged to him descending from al Jedraniya herself. Tell us the positive truth on your honor and good fortune, and religion, and faith. Tell us the true history that you have heard from the grandfathers of your grandfathers."

The discourses take place from tribe to tribe with the revelation that she was Kuhayla Ajuz from among Al Khamsa, from the time of Al Sahaba, the followers of the Holy Prophet, Mohammed. In the presence of the majless of the sheiks of Shammar, and in the presence of Nasr al-Sohaymi, the Sheikh of 'Onayzah of the people of Al Qasim, it was learned how Al Jedraniyah passed to Ibn Jedran in the old times, and that:

"....the Kuhaylah was 'awdah,' a saqla mare. And the strain was named after her, a saqla Kuhaylah mare. And Saqlawi is a name."

Saqla means a kick, and *awdah* means old or aged broodmare; therefore, the strain was named Saqlawi after the old broodmare who was a kicker! It is spelled different ways in English transliteration; sometimes with a K, sometimes with a Q, but originally it was written with a *qaf* (Q).

At this point in time, there is little reference to any particular visual characteristics other than markings, colors, acts of courage, distinguishing battles, or odd traits, that may have related to an appended strain name, such as Saklawi of Ibn Jedran of Ibn Sudan. In time, however, this strain became associated with certain characteristics. Dr. Skorkowski, described the Saklawi horses within the Polish programs as being a heavier, masculine type, whereas within the Egyptian programs it was exactly the opposite. The Egyptian Saklawis have been known for feminine refinement and elegance.

H.H. Prince Mohamed Aly said he found the Saklawi Jedran were the most courageous, as have we, and he commented:

"*A Saklawi will fight for his master, and in charging, nothing will frighten him; he will charge a gun, a lion, and even a locomotive if put to it.*"

With this nucleus of Saklawi mares and stallions, the royal family of Egypt developed lines that were generally prized for their refinement and elegance. Chestnuts and greys predominate in Egypt; in America a bay line also developed through use of the more substantial and less typey bay Kuhaylan Rodan stallion, *Rashad Ibn Nazeer, on the elegant chestnut, *Bint Moniet El Nefous, regrettably almost wiping out the refinement and type the best greys and chestnuts of this line were noted for. The black Saklawi stallion, Fa-Serr, also sired color, but he was of heavier body type and did not contribute refined feminine elegance to the line bred at Babson's. However, his elegant refined grey brother, Fay El Dine, did.

Branches of this strain are extant today in Egyptian bloodlines through Roda (Roga) El Beida and Ghazieh of Abbas Pasha I.

Roda/Roga El Beida Family
1882 Grey mare from Ali Pasha Sherif

Moniet El Nefous

Moniet El Nefous Family

This branch has been known since the 1960's for its beloved matriarch, Moniet El Nefous, who was General Von Szandtner's favorite mare. She was a chestnut with what is likened to the "camel head," a long foreface with very tapered muzzle that could drink from a teacup. She was of average size with a most refined feminine head, good length and shape of neck, good overall balance, although light in the croup and hindquarters, typical of that family, and she also had good legs. Her personality set her apart from the others; she was a queen; she was spoiled; and she insisted on receiving favors, particularly sugar, from visitors. Her line carries on in Egypt and around the world through families founded by her daughters.

Moniet's chestnut daughter, *Bint Moniet El Nefous (by Nazeer), was the finest of the Pritzlaff imports, but unfortunately, she was bred most of the time to his bay stallion, *Rashad Ibn Nazeer, who was completely out of type to her, and she produced mostly bays who were unlike her, being taller, heavier boned and plainer-headed by comparison. The chestnuts she produced by him were more her type, and the grey progeny by another stallion., who was not straight Egyptian, were the most beautiful. The grey mare, Maya, a full sister to *Bint Moniet El Nefous, remained in Egypt, a stunning individual whom we nearly purchased, but who later suffered an accident and died.

Two other chestnuts, Mona and Mabrouka by Sid Abouhom, were most important contributors to this family; Mona having the spectacular head, even more extreme than her dam, but married to the weak croup. Mabrouka was not as beautiful as her sister, but was better conformed with better neck, shoulder, and body structure. Other chestnuts out of Moniet by Alaa El Din and Galal were pleasing individuals, but not usally as pretty-headed as her Nazeer and Sid Abouhom progeny. However, several of them founded important families, as will be noted in Chapter VI.

Mabrouka Family

*Ansata Bint Mabrouka was an exceptionally classic and refined individual - beautiful broad forehead with huge black eyes, long neck, well-shaped and well-set on to well laid back shoulders, with a good body and topline and clean legs. Her dam produced no other fillies. *Ansata Bint Mabrouka also produced only one filly, who died young, thus dashing our hopes of founding a family based on her tail female line. Three sons, however, were outstanding individuals, excellent sires, and were also significant champions in the show ring: Ansata Abbas Pasha, Res. European Champion and other Class A championships in America; Ansata Ibn Sudan, U.S. National Champion Stallion; and Ansata Shah Zaman, many times a champion. All were sires of champions. Of the three, Ansata Shah Zaman made the greatest impact as a sire within the Ansata program.

*Morafic, full brother to *Ansata Bint Mabrouka, was imported by Gleannloch, and he went on to win at halter and park, and, of course, is legendary as a sire. The Mabrouka branch of the Moniet family did not carry on in tail female, but her male progeny contributed immeasurably to the breed. The line to Mona, however, was prolific. *Bint Mona (Mouna), whom we had wanted to purchase in Egypt, was eventually imported by Gleannloch Farms. An exquisite full sister in blood to *Ansata Bint Mabrouka, she was similar in type though finer in head and muzzle, but she did not have the huge eyes or as broad a forehead; nor was she as good in body. She went on to produce excellent foals and champions for Gleannloch when bred to her full brother in blood, *Morafic, including the important sire, The Egyptian Prince.

Mona (Mouna) Family

*Ansata Adeeba (by Antar x Ibtsam x Mona), was acquired to represent this Mona family. She was an exquisite filly with the wonderful though shorter head and better body than Mona and typier than her dam, Ibtsam (by Nazeer). She was in foal to *Ansata Ibn Halima when she choked on hairy oats, telescoped her esophagus, and died in Chickasha, Oklahoma, before foaling. It was an irreplaceable loss and a set-back at that time to our breeding program.

*JKB Mamdouhah (x Moneera), provided another opportunity to obtain this Mona line. She was acquired on lease from Jean Kayser of Luxembourg. A long-bodied mare, she lacked the extreme quality in head type of this line, exhibiting more the type of her dam. She produced a good filly by *Ansata Ibn Halima, but neither mother, nor daughter were typical of the much sought after Moniet/Mona head type.

G. Fantazia (x *Fantasia), was acquired from the Gucci dispersal. Her dam, Fantasia, was one of the most beautiful and elegant mares that sold at the Gleannloch dispersal, and we were pleased to purchase her daughter, who was much like her in refinement and type. She had sold with her dam to Gucci, and when Gleannloch repossessed most of the horses he had purchased, G. Fantazia was among the group sent from Gucci's farm in New York to a nearby farm for boarding until they could be resold. Marshall gave the horses to the Pyramid Society. The people where they were boarded said G. Fantazia, among others, was in poor condition; we agreed to take her, along with the mare, Hasbah, and with careful attention, she returned to good health and became a beautiful broodmare, producing several good foals before she was exported to Europe.

*JKB Bint Nehaya (x Nehaya), was another mare of this line leased from Kayser. She was a pleasing, but not very typey, mare with neither her dam, nor granddam, being as elegant as Mona, from whom they descended. *Hoyeda (Nehaya's dam), was a rather plain chestnut, along with her daughter, the bay Nehaya (by *Ansata Ibn Halima), and then *JKB Bint Nehaya (by *Ibn Moniet El Nefous), none of which were very classic type examples of this line. Nevertheless, *JKB Bint Nehaya produced well by Ansata Ibn Sudan and, particularly, by Ansata Halim Shah.

AK Faressa (x AK Namessa), also provided a line to Mona. Faressa was a tall very beautiful grey mare with excellent head and neck and strong body. She produced two colts, including the very classic, Ansata Shaamis, who was sold to Italy. AK Faressa was exported in foal to Australia where she produced the pretty filly, Ansata Bint Faressa by Ansata Hejazi.

Talmona (x Falmona), a white mare very much resembling Bint Mona in type and beauty, was acquired at a dispersal sale. She was in foal to the very classic and refined stallion, Farres. An exceptionally beautiful filly, Ansata Mouna, resulted from that mating. She became a champion in U.S. as a two year old before being sold to Sheikh Abdullah Bin Nasir Al Thani of Qatar who showed her to championships in Europe. Talmona promised to be an important addition to the program, but, unfortunately, she died from a foaling problem the year after Ansata Mouna was born.

MB Moneena (x Talmona) was acquired when we purchased Talmona. We bred her to Farres, and her first foal resulted in the very classic grey colt, Ansata Qasim, an Egyptian Event Top Ten Futurity and World Class winner, with quality similar to Ansata Mouna. Therefore Moneena should be capable of carrying on the excellent type of this family. This line to Mouna (Mona) was acquired to try and maintain the type that we had early-on admired in Egypt, but that we had not seen being perpetuated in like-kind recently.

*Bint Moniet El Nefous Family

MB Sateenha (by Safeen, a handsome stallion tracing to the Serra line in tail female) brought Ansata a line to *Bint Moniet El Nefous. A tall, elegant and bold moving chestnut mare with the Moniet style head, she was acquired at a dispersal sale in Canada. Sateenha's dam was by Moniet El Sharaf, a bay stallion inbred to Moniet El Nefous. He had an exquisite front end - extreme head and neck - but his rear end repeated the combination of beautiful front, weak rear, that his get tended to inherit. Nevertheless, Sharaf daughters are much sought after as broodmares. Through careful selection of stallions, the opportunity exists through Sateenha to recapture the type both in chestnut and in grey that distinguished this family in the late 50's and 60's. Her first foal, Ansata Bint Sokar, an exquisite chestnut, was an Egyptian Top Ten World Class filly at the 7 months of age against 32 top competitors.

El Bataa Family

The El Bataa line was closely related to the Moniet line, but they differed in type and quality from the

Moniet family. El Bataa, a bay, was by Sheikh El Arab x Medallela; Moniet el Nefous, a chestnut, was by Shahloul out of Wanisa x Medallela. El Bataa was a longer-lined mare than her full sister, Wanisa, with longer, straighter neck and attractive, but straighter, head. She closely resembled Halima (also by Sheikh El Arab), in type, except she was not as short coupled and had a longer, straighter neck. While both Moniet and El Bataa tended to lightness in the hindquarters, the El Bataa line was not as strong. El Bataa produced the beautiful-headed black mare, *Bint El Bataa (Amal), imported by Pritzlaff, and also three grey daughters by Nazeer, who were imported to America. The mare, Korima, by *Tuhotmos (who was double Shahloul), incorporated Shahloul into this line, as did Wanisa in producing Moniet El Nefous, but the El Bataa branch generally produced a different type and hindquarter structure from the Moniet line.

*Ansata Bint Nazeer (*Fulla*) by Nazeer, was a very pretty mare of this family. She had been sold to the Cairo Police, and when we and the Marshalls were purchasing additional stock, we visited the Police stables where they maintained a breeding program based on horses purchased from the E.A.O. *Ansata Bint Nazeer was much like her mother in type and body structure with a noble head, long neck, somewhat long back, deep hip and good tail set. She had straight legs, which she passed on consistently to all her foals, but she, like other sisters, was not strong enough in the hindquarters. Her sons were not significant in the stud.

Ansata Salome (by *Ansata Ibn Halima), was a grey filly, smaller than her dam, not as pretty, more compact, and with similar conformation.

Ansata Samira (by *Ansata Ibn Halima), also grey, was very pretty and well-proportioned and a better individual than her full sister.

Ansata Wanisa (by Ansata Ibn Sudan), a chestnut, was much like the Moniet family in head, length and shape of neck, and overall type. She was one of the prettiest and refined young mares in the early phases of the Ansata program.

Ansata Nazeera (by Ansata Ibn Sudan), a grey, much resembled Ansata Bint Mabrouka, an absolutely exquisite filly; she died young of pneumonia.

Ansata Saklawia (by Ansata Ibn Sudan), a chestnut, was similar to Ansata Wanisa in type, but her foreface was not as refined.

Ansata Vanessa (by Ansata Abu Sudan), a grey, was much like her dam in overall type and conformation.

Ansata Nazira (by Ansata Ibn Sudan), a chestnut, again was more like Wanisa and Saklawia in type; not as pretty in the head as the former, but better than the latter.

Korima Family

Another line to El Bataa came through the mare, JKB Blue Belkies, a broodmare bred by Jean Kayser. She was out of Korima, who was by El Sareei thus adding the Shahloul blood to this line which generally enhanced and strengthened it structurally.

*Ansata Karima (by Kaisoon), a grey mare, was a rather plain-headed average quality broodmare, the combination of Galal (her granddam's sire) and Kaisoon not strong enough to provide the classic type needed.

The El Bataa line was interesting and challenging line to work with, providing Ansata with several exceptionally beautiful individuals, although we much preferred the Moniet line over time. The tail female El Bataa line has carried on successfully in other breeding programs around the globe.

Inshass Ghazalah Family

The Inshass family to Om Dalal was refined and elegant; and the mare Baheera had one of the most beautiful heads in Egypt during her time. A noble individual of good size, long neck, good body and topline and good legs, she was a fine broodmare.

*Bint Baheera (by Emad), was acquired after she had been imported by Lowe and Heber. She was a very elegant young mare, and we liked the fact she was out of the beautiful Baheera and that she harked back to El Arabi through her sire, Emad, although he was a very plain individual and did not have any of El Arabi's redeeming outward attributes. *Bint Baheera produced mostly colts that were of good quality, while her filly, Ansata Orienta, by Ansata Halim Shah, inherited many of the best qualities of the line.

Bint Arrieta (by Hadaya El Tareef), represented another line to this family through the bay mare, *Subayha, imported by Tom and Rhita McNair. A big bold-moving grey mare of pleasing proportions, but not as refined as *Bint Baheera, Bint Arrieta was exported to Saudi Arabia in foal to Ansata Hejazi, where she produced a good foundation-quality filly.

Saklawi Jedran Family Through Roda (Roga El Beida) Ancestral Chart

Roda - Saqlawiah Jedraniya of Ibn Sudan

Bint Roda x **Jamil El Ahmar**

Om Dalal x **Sabbah**

Dalal x **Rabdan**

Khafifa x **Ibn Samhan**

Medallela x **Awad**

Khafifa with Medallela as a foal

Wanisa with foal

Dalal

Wanisa x **Sheikh El Arab**

Moniet El Nefous x **Shahloul**

*Bint Moniet El Nefous

Moniet El Nefous (l) with Mabrouka (r)

Mona (Mouna)

*Bint Mona

*Bint Moniet El Nefous x **Nazeer**

Muniet Nefous RSI x **Umi**

Fatiha x ***Ibn Moniet El Nefous**

AK Bint Fatiha x **Moniet El Sharaf**

MB Sateenha x **Safeen**

Mabrouka x **Sid Abouhom**

*Ansata Bint Mabrouka x **Nazeer**

*Ansata Bint Mabrouka

Mona (Mouna) x **Sid Abouhom**

*Bint Mona x **Nazeer**

Norra x ***Morafic**   Falmona x ***Faleh**

Nashwa x ***Sakr**   Talmona x ***Talal**

AK Nameessa   MB Moneena x **Safeen**
  x **NaIbn Moniet**

AK Faressa
  x **Prince Fa Moniet**

Ibtisam x **Nazeer**

*Ansata Adeeba x **Antar**

Ibtisam

*Hoyeda x ***Morafic**

Nehaya
  x ***Ansata Ibn Halima**

JKB Bint Nehaya
  x ***Ibn Moniet El Nefous**

Moneera x **Alaa El Din**

*JKB Mamdouhah x **Kaisoon**   Nafteta x **Kaisoon**

*Fantasia

*Fantasia x **Farag**

G. Fantasia
  x **Maar Ibn Amaal**

Medallela with El Bataa as a foal

Rooda

El Bataa

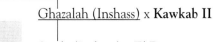

Ghazalah (Inshass) x **Kawkab II**

Saada (Inshass) x **El Deree**

Ragaa x **Rasheed**

Ilham

Rateeba x **El Moez**

Rayana x **Ezzat**

Raghda x ***Morafic**

Baheera x **Shaarawi**

*Bint Baheera x **Emad**

Ragha x **Adham**

Rooda x **Hamdan**

Ilham x ***Morafic**

*Subhaya x ***Tuhotmos**

Sakani x **Ibn Morafic**

Talkani x ***Talal**

Arrieta x ***Kanas**

Bint Arrieta x **Hadaya El Tareef**

El Bataa x **Sheikh El Arab**

*Binte El Bataa x **Nazeer**   *Ansata Bint Nazeer x **Nazeer**

Korima x **El Sareei**

*JKB Blue Belkies x **Galal**

*Ansata Karima x **Kaisoon**

Korima

*Binte El Bataa

Baheera

*Subhaya

*JKB Blue Belkies

*Bint Baheera

Sakani

257

*Ansata Bint Mabrouka.   Sparogowski photos.

His mother's son, Ansata Ibn Sudan, 1971 U.S. National Champion Stallion.

Saklawi Jedran Strain
RODA/ROGA EL BEIDA FAMILY

Moniet El Nefous Family

*Roda-Roga/Bint Roda-Roga/Om Dalal/
Dalal/Khafifa/Medallela/Wanisa/Moniet El Nefous/
Mabrouka/*Ansata Bint Mabrouka*

▪ *Ansata Bint Mabrouka #15898
Grey Mare; September 9, 1958-June 12, 1974
Sire: Nazeer   Dam: Mabrouka
Bred by E.A.O., Cairo, Egypt
E.A.O. #37, Volume II
Registered as Bint Mabrouka in Egypt
Imported by Mr. and Mrs. Donald L. Forbis, Chickasha, Oklahoma.

*Ansata Bint Mabrouka as a yearling in Egypt. Forbis photo.

Above, below, and at right, *Ansata Bint Mabrouka

PRODUCE of *Ansata Bint Mabrouka

Ansata Abbas Pasha #28122
June 17, 1964.
Grey stallion by *Ansata Ibn Halima. Transferred to Henry Babson, Dixon, Illinois, October 20, 1966, in exchange for Sabrah. Later sold to Bentwood Farms, Waco, Texas. Repurchased from Bentwood by Ansata, January 27, 1988. Died October 7, 1989.

Ansata Ibn Sudan #32342 July 1, 1965
Grey stallion by *Ansata Ibn Halima.
Died March 9, 1987.

Foaled grey mare by *Ansata Ibn Halima, October 3, 1966. Died.

Ansata Shah Zaman #47967 January 28, 1968
Grey stallion by *Morafic.
Sold half interest to Count Federico Zichy-Thyssen. Died January 23, 1985.

Serviced 1968 by *Ansata Ibn Halima. Aborted filly.

1969-1973 serviced by *Ansata Ibn Halima.
Died without further progeny.

Ansata Abbas Pasha

Above and below, Ansata Ibn Sudan

Below, Ansata Shah Zaman

Mona Family

Roda-Roga/Bint Roda-Roga/Om Dalal/Dalal/Khafifa/Medallela/Wanisa/ Moniet El Nefous/Mona-Mouna/Ibtisam/*Ansata Adeeba

■ *Ansata Adeeba #50735
Chestnut Mare; September 18, 1966—October 11, 1969
Sire: Anter #10HB28   Dam: Ibtisam #45
Bred by E.A.O., Cairo, Egypt
E.A.O. # 543, Volume III
Registered as Adeeba in Egypt
Imported by Mr. and Mrs. Donald L. Forbis, Chickasha, Oklahoma, November 1, 1968.

PRODUCE
1969 serviced by *Ansata Ibn Halima.
Died in foal with unborn filly.

Above and below, *Ansata Adeeba. Forbis photos in Egypt.

Roda-Roga/Bint Roda-Roga/Om Dalal/Dalal/Khafifa/Medallela/Wanisa/ Moniet El Nefous/Mona-Mouna/Moneera/*JKB Mamdouhah

■ *JKB Mamdouhah #186547
Chestnut Mare; April 24, 1973
Sire: Kaisoon   Dam: Moneera
Owned by Jean Kayser, Ettlebruck, Luxembourg
Imported by Kayser and leased to Ansata, Lufkin, Texas

PRODUCE
Ansata Monahalima #0240698 April 28, 1981.
Bay mare by *Ansata Ibn Halima. Owned by Jean Kayser.

1982 Chestnut filly by Ansata Ibn Sudan. Died after birth.

1984 serviced by *Jamilll. Delivered stillborn foal.

*JKB Mamdouhah

Ansata Monahalima

*Roda-Roga/Bint Roda-Roga/Om Dalal/Dalal/Khafifa/Medallela/
Wanisa/Moniet El Nefous/Mona-Mouna/Moneera/*JKB
Mamdouhah/Ansata Monahalima*

Ansata Monahalima #240698
Bay Mare; April 18, 1981
Sire: *Ansata Ibn Halima   Dam: *JKB Mamdouhah
Transferred to Jean Kayser, Luxembourg.

PRODUCE

Ansata Ahmed Bey #0412844 May 17, 1988
Bay stallion by Ansata Mourad Bey.
Sold to Ruth Gold, Yale, Michigan.

*Roda-Roga/Bint Roda-Roga/Om Dalal/Dalal/Khafifa/Medallela/
Wanisa/Moniet El Nefous/Mona-Mouna/Moneera/Nafteta/
Fantasia/G Fantazia

G Fantazia

G Fantazia #0510991
Grey Mare; February 23, 1992
Sire: Maar Ibn Amaal   Dam: *Fantasia
Bred by Gleannloch Farms, Barksdale, Texas
Donated by Gleannloch to The Pyramid Society.
Purchased from The Pyramid Society by Ansata, Mena,
Arkansas, May 1994. Sold to Azienda Agricola San Pietro Di
Damilano Paolo and exported to Italy.

PRODUCE

Ansata Hakima #0534520 August 30, 1996
Grey mare by Ansata Hejazi.
Sold to Claudia Quentin. Exported to Argentina, South
America.

Ansata Fantazi #0543392 August 26, 1997
Grey mare by Ansata Hejazi.
Sold to Jessie MacLean, New Brunswick, Canada, in foal to
Ansata Sirius.

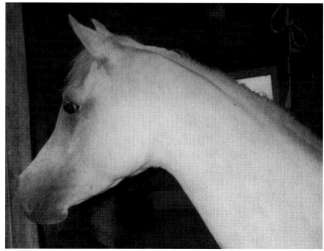
Ansata Fantazi

Ansata Fantastiq #0556197 August 12, 1998
Chestnut mare by Ansata Sokar. Sold to HRH Prince Khalid
bin Fahad Al Saud. Exported to Saudi Arabia.

Ansata Fayza #0566535 August 17, 1999
Bay mare by Ansata Sokar. Sold to Kathleen Horn,
Little Rock, Arkansas.

Serviced 1999 to Ansata Sokar. Exported to Europe in foal.

Ansata Fantastiq

*Roda-Roga/Bint Roda-Roga/Om Dalal/Dalal/Khafifa/Medallela/
Wanisa/Moniet El Nefous/Mona-Mouna/*Hoyeda/Nehaya/
JKB Bint Nehaya

■ *JKB Bint Nehaya #137650

Grey Mare; April 10, 1976
Sire: *Ibn Moniet El Nefous   Dam: Nehaya
Bred by Jarrell McCracken, Waco, Texas.
Owned by Jean Kayser, Ettelbruck, Luxembourg.
Imported on lease to Ansata, Lufkin, Texas.

PRODUCE

Ansata Zaki Halim #242582 April 2, 1981
Grey stallion by *Ansata Ibn Halima.
Sold to Ansata Zaki Halim Syndicate,
Gadsden, South Carolina.

Ansata Suhaya #0280820 February 13, 1983
Grey mare by Ansata Ibn Sudan. Sold to Count
Federico Zichy-Thyssen at The Pyramid Society Sale.
Exported to Argentina, South America.

Ansata Narjisa #311260 September 9, 1984
Grey mare by Ansata Ibn Sudan.

Ansata Taya #0372543 August 23, 1986.
Grey mare by Ansata Ibn Sudan. Sold to Paolo Gucci,
Millfield Stables, New York, November 1988.

Ansata Ibn Tulun #396751 September 8, 1987
Grey stallion by Ansata Halim Shah. Owned by Kayser.

*JKB Bint Nehaya

Ansata Zaki Halim

Ansata Suhaya

Ansata Taya

*Roda-Roga/Bint Roda-Roga/Om Dalal/Dalal/Khafifa/Medallela/
Wanisa/Moniet El Nefous/Mona-Mouna/*Hoyeda/Nehaya/
JKB Bint Nehaya/Ansata Narjisa

■ Ansata Narjisa #0311260

Grey Mare; September 9, 1984 –1994
Sire: Ansata Ibn Sudan   Dam: JKB Bint Nehaya #137650
Sold to Al Shaqab Farm, HH Sheikh Hamad Bin Khalifa Al
Thani, Doha, Qatar, 1994.

PRODUCE

Ansata Nabiha #0410744 August 4, 1988
Grey mare by Ansata Halim Shah. Sold to Mark and
Candace Otto, Unadilla, Georgia, September 1989.
Mare leased to Dr. Chess Hudson, Lawrenceville, Georgia In
1989.

Zandai Alexandria #0477506 April 27, 1991.
Grey mare by Ansata Ibn Aziza.

Ansata Azza #0487410 April 21, 1992.
Grey mare by Ansata Ibn Aziza.

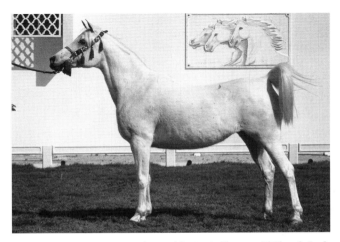
Ansata Narjisa in Qatar at Al Shaqab Stud.

PRODUCE of Ansata Narjisa continued

Zandai Tabitha #0502681 May 17, 1993.
Grey mare by Ansata Ibn Aziz.
Owned by Dr. Chess Hudson, Lawrenceville, Georgia.

Exported to Qatar in foal to Ansata Omar Halim, January 1994.

Ansata Bint Faressa

Ansata Shaamis

*Roda-Roga/Bint Roda-Roga/Om Dalal/Dalal/Khafifa/Medallela/ Wanisa/Moniet El Nefous/Mona-Mouna/*Bint Mona/Norra/ Nashwa/AK Nameesa/AK Faressa*

AK Faressa #0405122
Grey Mare; May 2, 1988
Sire: Prince Fa Moniet   Dam: AK Nameesa
Bred by Bentwood Farms, Waco, Texas.
Leased by Ansata from John Stevenson, New York, New York. Sold to Peter and Jenny Pond, Forest Hills Arabian Stud. Exported to Australia.

PRODUCE

Omar Bey #0493686 April 9, 1993.
Grey stallion by Ansata Omar Halim. Gelded

Ansata Shaamis #0507420 April 8, 1994.
Grey stallion by Ansata Halim Shah. Sold to Giuseppe Fontanella, Fontanella Magic Arabians. Exported to Italy, January 1998.

Ansata Bint Faressa (registered in Australia)
Grey mare by Ansata Hejazi. Born in Australia. Owned by Forest Hill Arabian Stud.

*Roda-Roga/Bint Roda-Roga/Om Dalal/Dalal/Khafifa/Medallela/ Wanisa/Moniet El Nefous/Mona-Mouna/*Bint Mona/Falmona/ Talmona*

Talmona #0554803
Grey Mare; May 13, 1986 - 2000
Sire: Talmaal   Dam: Falmona
Bred by James M. Kline, Kline Arabians, Whittier, California
Purchased by Ansata at the IEB Texas auction in 1998 in foal to Farres.

PRODUCE

Ansata Mouna #563813 March 30, 1999
Grey mare by Farres. Sold to Sheikh Abdullah Bin Nasir Al Thani, Al Naif Stud, Doha, Qatar. Exported 2001.

Serviced 1999 to Ansata Sokar. Died 2000 after giving birth to dead colt.

AK Faressa

Talmona

Ansata Mouna in Qatar at Al Naif Stud.

*Roda-Roga/Bint Roda-Roga/Om Dalal/Dalal/Khafifa/Medallela/
Wanisa/Moniet El Nefous/Mona-Mouna/*Bint Mona/Falmona/
Talmona/MB Moneena*

■ MB Moneena # 0554803

Grey Mare; June 16, 1997
Sire: Safeen   Dam: Talmona
Purchased by Ansata at the 1998 IEB Texas auction.

PRODUCE

Ansata Qasim #0589144 April 24, 2001
Grey stallion by Farres.

Ansata El Jedrani #0596951 September 14, 2002
Grey stallion by Ansata Osiron.

MB Moneena

*Bint Moniet El Nefous Family

*Roda-Roga/Bint Roda-Roga/Om Dalal/Dalal/Khafifa/Medallela/
Wanisa/Moniet El Nefous/*Bint Moniet El Nefous/Muniet Nefous
RSI/Fathia/AK Bint Fathia/MBSateenha*

■ MB Sateenha #0563740

Chestnut Mare; March 20, 1995
Sire: Safeen   Dam: AK Bint Fatiha
Purchased July 1999 in Canada by Ansata at the Edwards Sale.

PRODUCE

Ansata Bint Sokar #589162 October 5, 2001
Chestnut mare by Ansata Sokar.

Ansata Qasim

MB Sateenha

Ansata Bint Sokar

El Baaa Family

Roda-Roga/Bint Roda-Roga/Om Dalal/Dalal/Khafifa/Medallela/ El Bataa/*Ansata Bint Nazeer

▪ *Ansata Bint Nazeer #69407
Grey Mare; January 27, 1960 – December 19, 1984
Sire: Nazeer   Dam: El Bataa
Bred by E.A.O., Cairo , Egypt
E.A.O. # 41, Volume II
Registered as **Fulla** in Egypt
Imported by Mr. and Mrs. Donald L. Forbis, Chickasha, Oklahoma.

*Ansata Bint Nazeer. Polly Knoll photos

*Ansata Bint Nazeer

PRODUCE of *Ansata Bint Nazeer

Unnamed bay stallion by *Ibn Hafiza, May 1, 1971
Died August 16, 1971.

Ansata Salome #81382 May 14, 1972
Grey mare by *Ansata Ibn Halima.
Sold to Bill and Janet Lowe, Beaver Dam,
Wisconsin, November 1977.

Ansata Samira #94538 May 8, 1973
Grey mare by *Ansata Ibn Halima.
Sold to Barbara Hall Edwards, Las Vegas, Nevada.

Ansata Ibn Meheyd #105816 April 30, 1974
Grey stallion by *Ansata Ibn Halima. Gelded.

Ansata Shah Zeer #122313 May 9, 1975
Grey stallion by Ansata Shah Zaman.
Died September 15, 1982.

Ansata Wanisa #137120 July 30, 1976.
Chestnut mare by Ansata Ibn Sudan.
Sold to Lyle and Virginia Bertsch, Fort Wayne, Indiana.

Ansata Nazeera #153299 July 19, 1977.
Grey mare by Ansata Ibn Sudan.
Died September 16, 1977.

Ansata Saklawia #196859 February 6, 1979.
Chestnut mare by Ansata Ibn Sudan.

Ansata Vanessa #219544 February 1, 1980
Grey mare by Ansata Abu Sudan.

Ansata El Rakkad #237270 March 17, 1981
Grey stallion by Ansata Halima Son.
Gelded and transferred to Candy Foster, Mena, Arkansas.

Ansata Nazira #279329 June 6, 1983
Chestnut mare by Ansata Ibn Sudan.

Ansata Nazira

Ansata Samira

Ansata Shah Zeer

Ansata Wanisa

Ansata Wanisa

Roda-Roga/Bint Roda-Roga/Om Dalal/Dalal/Khafifa/Medallela/ El Bataa/*Ansata Bint Nazeer/Ansata Salome

Ansata Salome #81382
Grey Mare; May 14, 1972
Sire: *Ansata Ibn Halima Dam: *Ansata Bint Nazeer
Sold to Bill and Janet Lowe, Beaver Dam, Wisconsin, November 1977

PRODUCE

Phalom #144713 June 23, 1976
Grey stallion by *Farazdac. Bred and owned by Bill and Janet Lowe.

Ansata Dalala #170402 June 14, 1977
Grey mare by *Tuhotmos. Sold to Bill and Susan Malloy, Days Creek, Oregon.

Roda-Roga/Bint Roda-Roga/Om Dalal/Dalal/Khafifa/Medallela/ El Bataa/*Ansata Bint Nazeer/Ansata Samira

Ansata Samira #94538
Grey Mare; May 8, 1973
Sire: *Ansata Ibn Halima Dam: *Ansata Bint Nazeer
Sold to Barbara Hall Edwards, Las Vegas, Nevada.

PRODUCE

Ansata Saroya #0153296 April 26, 1977
Grey mare by Nabiel.

Asmara Bintsamira #0187271 February 22, 1979
Chestnut mare by Ansata Ibn Sudan. Owned by Barbara Hall Edwards, Las Vegas, Nevada.

Roda-Roga/Bint Roda-Roga/Om Dalal/Dalal/Khafifa/Medallela/ El Bataa/*Ansata Bint Nazeer/Ansata Wanisa

Ansata Wanisa #137120
Chestnut Mare; July 30, 1976
Sire: Ansata Ibn Sudan Dam: *Ansata Bint Nazeer
Sold to Lyle and Virginia Bertsch, Zahara Arabian Stud, Ft. Wayne, Indiana

PRODUCE

Ansata Amir Zarif #227619 January 15, 1981
Bay stallion by Ansata Halima Son. Owned by Zahara Arabian Stud, Ft. Wayne, Indiana.

Nile Splendor #0252655 May 5, 1982
Bay mare by AK Shah Moniet.
Owned by Zahara Arabian Stud, Ft. Wayne, Indiana.

Ansata Saroya

Roda-Roga/Bint Roda-Roga/Om Dalal/Dalal/Khafifa/Medallela/ El Bataa/*Ansata Bint Nazeer/Ansata Saklawia

Ansata Saklawia #196859
Chestnut Mare; February 6, 1979
Sire: Ansata Ibn Sudan Dam: *Ansata Bint Nazeer
Leased to Betty Jones, Texarkana, Arkansas
February 1989

PRODUCE

Ansata Sakkara #0278160 March 23, 1983
Chestnut mare by Ansata Halima Son. Transferred to J. Parks, Pilot Point, Texas.

Ansata Sahhara #0303388 May 13, 1984
Bay mare by Ansata Halima Son. Transferred to Ronald J. Hutchings. Exported to England.

Ansata Amir Zarif

Ansata Sakkara

Ansata Sahhara

PRODUCE of Ansata Saklawia continued

Ansata El Khalil #0389992 February 16, 1987
Grey stallion by Ansata Halim Shah. Gelded and transferred to Elizabeth Smith, Russellville, Arkansas March 1988.

Roda-Roga/Bint Roda-Roga/Om Dalal/Dalal/Khafifa/Medallela/ El Bataa/*Ansata Bint Nazeer/Ansata Vanessa

▪ Ansata Vanessa #0219544
Grey Mare; February 1, 1980
Sire: Ansata Abu Sudan   Dam: *Ansata Bint Nazeer

PRODUCE

Ansata Caressa #0302781 June 4, 1984.
Grey mare by Ansata Halima Son. Transferred to Dick and Lucille Edgerton, Mena, Arkansas.

Ansata El Harik #0340687 August 17, 1985.
Grey stallion by Ansata Shah Zaman. Gelded and transferred to Betty Jones, Texarkana, Arkansas.

Ansata Valeria #0365242 October 1, 1986.
Grey mare by Ansata Ibn Shah.

PRODUCE of Ansata Vanessa continued

Unnamed stallion by Dorian Shah El Shams
February 6, 1988. Donated to 4-H Program.

My Amir #0442547 August 5, 1989.
Grey stallion by Ansata Amir Zaman.
Transferred to Betty Jones, Texarkana, Arkansas.

Roda-Roga/Bint Roda-Roga/Om Dalal/Dalal/Khafifa/Medallela/ El Bataa/*Ansata Bint Nazeer/Ansata Nazira

▪ Ansata Nazira #279329
Chestnut Mare; June 6, 1983
Sire: Ansata Ibn Sudan   Dam: *Ansata Bint Nazeer
Transferred to Betty Jones, Texarkana, Arkansas

PRODUCE

Ansata El Shaqra #0392669 June 20, 1987.
Chestnut mare by Dorian Shah El Shams
Transferred to Betty Jones, Texarkana, Arkansas.

Roda-Roga/Bint Roda-Roga/Om Dalal/Dalal/Khafifa/Medallela/ El Bataa/*Ansata Bint Nazeer/Ansata Salome/Ansata Dalala

▪ Ansata Dalala #170402
Grey Mare; June 15, 1977
Sire: *Tuhotmos   Dam: Ansata Salome
Sold to Bill and Susan Malloy, Days Creek, Oregon.

PRODUCE

Shahliah #9233930 August 12, 1981
Bay mare by Ansata Shah Zaman.
Owned by Bill and Susan Malloy.

Roda-Roga/Bint Roda-Roga/Om Dalal/Dalal/Khafifa/Medallela/ El Bataa/*Binte El Bataa/Korima/Belkies/*JKB Blue Belkies/ *Ansata Karima

▪ *Ansata Karima #185324
Grey Mare; September 24, 1977
Sire: *Kaisoon GASB   Dam: JK Blue Belkies GASB
Bred by Jean Kayser, Luxembourg
Imported by Mr. and Mrs. Donald L. Forbis, Lufkin, Texas, April 1979.
Sold to Charles and Joanne Gibson, Gibson Arabians, Loomis, California, October 1979.

PRODUCE

GA Magic Melody #0233907 April 23, 1971
Grey mare by Ansata Majid Shah. Owned by Charles and Joanne Gibson, Loomis, California.

GA Fire Magic #0270086 June 6, 1986
Grey gelding by Ansata Majid Shah. Owned by Charles and Joanne Gibson, Loomis, California.

Ghazalah (Inshass) Family

*Roda-Roga/Bint Roda-Roga/Om Dalal/Dalal/Ghazalah/Saada/Ragaa/
Rateeba/Rayana/Raghda/Baheera/*Bint Baheera*

*Bint Baheera #193411
Grey Mare; May 9, 1978 – August 16, 1990
Sire: Emad (E.A.O.) Dam: Baheera (E.A.O.)
Bred by E.A.O., Cairo, Egypt
E.A.O. # 1040, Volume V
Imported by Rick Heber and Bill Lowe.
Purchased by Mr. and Mrs. Donald L. Forbis, Mena, Arkansas, October 1982.

PRODUCE

Unnamed bay filly by *Ansata Ibn Halima, March 23, 1982. Died.

Ansata El Baheer #282331 March 16, 1983
Grey stallion by Ansata Ibn Sudan. Gelded and transferred to Bill and Karen Barnes, Mountain Harbor, Arkansas.

Ansata Ali Khan #302801 March 31, 1984
Grey stallion by Ansata Halim Shah. Sold to Nagib Audi, December 1984. Exported to Sao Paulo, Brazil.

PRODUCE of *Bint Baheera continued

Ansata Araby Bey #0327964 March 18, 1985
Bay stallion by *Jamilll. Gelded and transferred to Carrie Hatcher, Palestine, Texas.

Ansata Mehemet Aly #0355921 February 19, 1987
Grey stallion by *Jamilll. Gelded and transferred to Mark and Viki Varley, Mena, Arkansas.

Ansata Al Haroun #0396925 February 19, 1987
Grey stallion by Ansata Halim Shah. Transferred to Dorian Farm, Goshen, New York, April 1988.

Ansata Sulayman #0410772 March 8, 1988.
Grey stallion by Ansata Halim Shah. Gelded and transferred to Ben F. Thomson, Jr., Sheridan, Arkansas, June 1991.

Ansata Orienta #0453398 February 13, 1990.
Grey mare by Ansata Halim Shah. Sold to Wolfgang Effkemann. Exported to Germany. Owned by Judith Wich, Orienta Arabians, Germany.

*Bint Baheera

Ansata Ali Khan

Ansata Araby Bey

Ansata Sulayman

Ansata Orienta

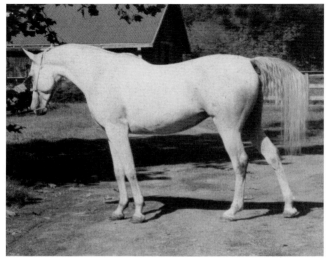

Bint Arrieta

<u>Roda-Roga/Bint Roda-Roga/OmDalal/Dalal/Ghazalah/Saada/Ragaa/ Ragha/Rooda/Ilham/*Subhaya/Sakani/Talkani/Arrieta/Bint Arrieta</u>

■ Bint Arrieta #0504863
Grey Mare; June 6, 1994
Sire: Hadaya El Tareef   Dam: Arrieta
Bred by Willis Flick, acquired by Ansata in 1998.
Sold 1999 in foal to Ansata Hejazi to HRH Prince Khalid Bin Fahad Al Saud, Saudi Arabia.   Exported.

Ghazieh Family

c. 1850s Grey mare from Abbas Pasha I

Ghazieh Family through Horra

The Ghazieh line, which is also the Saklawi Jedran strain, stems from a different root mare, and we observed it differed in structure from the Dalal's. It was predominantly grey, and handsome stallions (Mesaoud, Shahloul, Hamdan) and mares (*Ghazala el Beida, Radia, Serra, Samira) came through this line. Samira was considered one of the most beautiful mares at the Royal Agricultural Society, winning a prize for beauty, and from her descended the main line that we perpetuated.

*Ansata Bint Zaafarana (by Nazeer), was a tall, well-balanced mare with classic head, good dark eyes, and a beautifully shaped neck, that could have been longer, attached to very deep shoulders. She was much taller and more compact than her full brother, *Talal, who was longer in head, neck, and body, and shorter in stature. She was strong in body and coupling, deep of hip, and extremely powerful when moving, an asset inherited from her dam who produced top race horses consistently, and whose sire, Balance, was the most celebrated Egyptian race horse of his era. *Ansata Bint Zaafarana was much valued as a broodmare, but she also won a major Class A championship when she was 15 years old. She was bred continually to *Ansata Ibn Halima, and when tried with Ansata Ibn Sudan, did not carry. Her progeny were very uniform in type, of course, some being typier than others, but nevertheless there was high degree of consistency in their overall structure and how they produced. Her sons, Ansata Ali Pasha and Ansata El Nisr, were important show champions; Ali Pasha died young, but Nisr became a significant sire, as did Ansata Abu Nazeer. The daughters were also beautiful, and Ansata Aziza and Ansata Divina were show winners.

Ansata Fatima (by *Ansata Ibn Halima) was a very elegant, beautiful and refined filly with a lovely head. Unfortunately, she went down in the back while still a foal. After producing two daughters she became non-productive.

Ansata Aziza (by *Ansata Ibn Halima), was a very pretty and strong-bodied champion mare who went on to found a significant family for the Gibson Stud.

Ansata Divina (by *Ansata Ibn Halima), was perhaps the most beautiful and well-balanced of all the daughters. She became a show champion and had a great future ahead of her. However, she came down with EPM before it was easily diagnosed or treated, and died.

Ansata Zareifa (by *Ansata Ibn Halima), was more like her sire, very compact, extremely pretty short head and short-coupled.

The *Ansata Bint Zaafarana line was very consistent in producing classic type, good-moving individuals, and it was very much appreciated within the Egyptian community.

*Bint Serra Line

This family was relatively scarce in the Babson herd in comparison to the *Maaroufa's and *Bint Bint Sabbah's. It does breed on in tail female today, but it is not prolific.

Allah Ateyyah (x Khamsa) was an attractive and noble grey mare with a refined head, long foreface, somewhat long-lined, quite fine-boned and feminine. She had crosses to the Serra Saklawi Jedran line (same tail female as Zaafarana) that we wanted to add to the Ansata program. While owned by Jean Jennings she had been bred to *Ansata Ibn Halima and produced the good stallion, Raadin Inshallah. We leased her when she was an older mare to breed to *Ansata Ibn Halima again. She proved to be non-productive and was returned to her owner.

Ghazieh Ancestral Chart

Ghazieh - Saqlawiah Jedraniya of Ibn Sudan from Abbas Pasha
|
Horra I x **Zobeyni**
|
Helwa x **Shueyman**
|
Bint Helwa x **Aziz**
|
*Ghazala x **Ibn Sherara**
|
Radia x **Feysul**
|
Bint Radia x **Mabrouk**
|
Samira x **Ibn Rabdan**
|
Zaafarana x **Balance**
|
*Ansata Bint Zaafarana x **Nazeer**

Bint Helwa

Radia

Bint Radia

Samira

Ghazala

Ansata Bint Zaafarana

Zaafarana

Serra (Sara)

*Bint Serra

Fa Deene

Khedena

Khamsa

Allah Ateyyah

Jemla x **Jamil I**

Serra (Sara) x **Sahab**

*Bint Serra x **Sotamm**

Fa-Deene x ***Fadl**

Khedena x **Khebir**

Khamsa x **Fay El Dine**

Allah Ateyyah x **Ar'Raad**

Saklawi Jedran Strain
GHAZIEH (of Abbas Pasha I) FAMILY

Zaafarana Family

Ghazieh/Horra/Helwa/Bint Helwa/Ghazala/Radia/(Ghadia)/Bint Radia/Samira/Zaafarana

■ ***Ansata Bint Zaafarana #15896**
Grey Mare; September 12, 1958- March 6, 1976
Sire: Nazeer   Dam: Zaafarana
Bred by E.A.O., Cairo, Egypt
E.A.O. # 38, Volume II
Registered as Bint Zaafarana in Egypt.
Imported by Mr. and Mrs. Donald L. Forbis, Chickasha, Oklahoma, October, 1959.

*Ansata Bint Zaafarana

*Ansata Bint Zaafarana as a yearling in Egypt

*Ansata Bint Zaafarana

PRODUCE

Ansata Ali Pasha #28121 May 27, 1964
Grey stallion by *Ansata Ibn Halima.
Sold to Willis Flick, Miami, Florida, November 1964.

Ansata Fatima #32341 September 2, 1965
Grey mare by *Ansata Ibn Halima.

Ansata El Nisr #41581 March 1, 1967
Grey stallion by *Ansata Ibn Halima.
Sold to Mrs. Elaine P. Alexander, Scott, Arkansas.

Ansata El Mamluke #47965 March 21, 1968
Grey stallion by *Ansata Ibn Halima.
Sold to T.D. Bogart, Hidden Hills, California.

Ansata Aziza #56302 March 26, 1969
Grey mare by *Ansata Ibn Halima.

Ansata El Khedive #61936 May 10, 1970
Grey stallion by *Ansata Ibn Halima.
Sold to Jarrell McCracken, Waco, Texas. Died.

Ansata El Alim #70166 April 25, 1971
Grey stallion by *Ansata Ibn Halima.
Sold to Pam and Mike McCauley, Santa Fe, New Mexico.

Ansata Abu Nazeer #105814 January 22, 1974
Grey stallion by *Ansata Ibn Halima.
Sold to Mike and Kiki Case, Glorieta Ranch, Lufkin, Texas.

Ansata Divina #122317 January 18, 1975
Grey mare by *Ansata Ibn Halima.
Died March 23, 1978.

Ansata Zareifa #137115 February 25, 1976
Grey mare by *Ansata Ibn Halima.

Ansata Ali Pasha

Ansata Fatima

Ansata El Nisr

Ansata El Mamluke

Ansata Aziza

Ansata Aziza

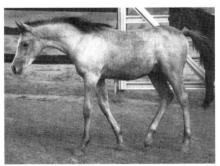
Ansata El Khedive

Ansata El Alim

Ansata Divina

Ansata Zareifa

Ansata Abu Nazeer

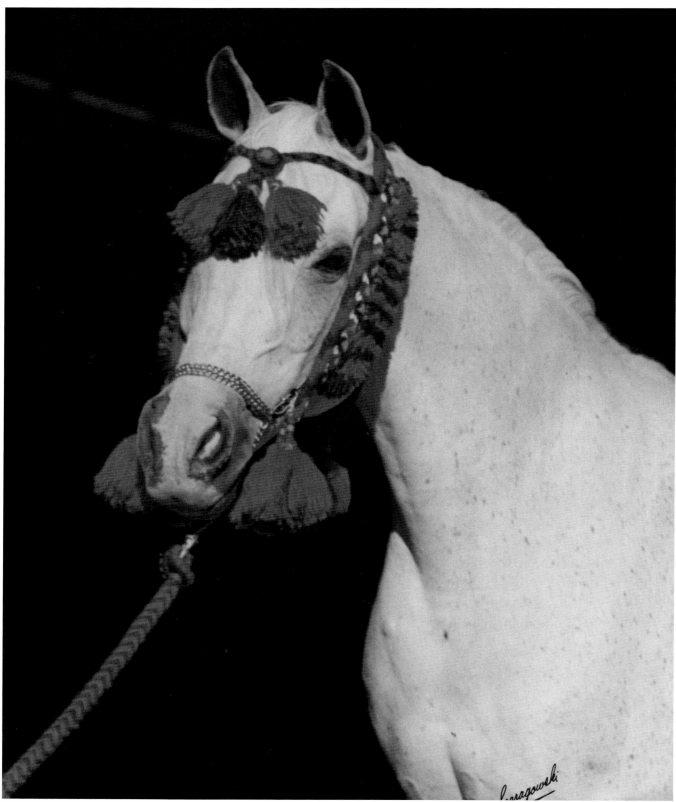

*Ansata Bint Zaafarana (Nazeer x Zaafarana) Sparagowski photo.

Ghazieh/Horra/Helwa/Bint Helwa/Ghazala/Radia/(Ghadia)/Bint Radia/Samira/Zaafarana/*Ansata Bint Zaafarana/Ansata Fatima

▒ Ansata Fatima #32341
Grey Mare; September 2, 1965 – August 12, 1976
Sire: *Ansata Ibn Halima   Dam: *Ansata Bint Zaafarana
Donated to Colorado State University in 1975.

PRODUCE
Unregistered grey filly April 6, 1969, by Ansata Ibn Sudan. Died 1969.

Bint Moftakhar #61937 May 20, 1970.
Grey mare by *Moftakhar.
Sold to Douglas Marshall, Gleannloch Farms, March 1971.

Bint Moftakhar

Ansata Jehan Shah

Ghazieh/Horra/Helwa/Bint Helwa/Ghazala/Radia/(Ghadia)/Bint Radia/Samira/Zaafarana/Ansata Bint Zaafarana/Ansata Aziza

▒ Ansata Aziza #056302
Grey Mare; March 26, 1969
Sire: *Ansata Ibn Halima   Dam: *Ansata Bint Zaafarana
Sold to Charles and Joanne Gibson, Gibson Arabians, Loomis, California, January 1981.

PRODUCE
Ansata Jehan Shah #0129019 September 14, 1975
Grey stallion by Ansata Shah Zaman. Sold to
Judy Lazenby, Toronto, Canada, August 1977.

Ansata Ra Harakti #0153294 March 16, 1977
Grey stallion by Nabiel. Died June 16, 1979.

Ansata El Aziz #196856 October 9, 1979
Bay stallion by *Zaghloul. Sold to Harold Hines,
Troy, Illinois.

Ansata Ibn Aziza #237268 February 19, 1981
Grey stallion by Ansata Halima Son. Sold to Fred Bennett,
Eastman, Georgia, December 1, 1983.

GA Alexis #0322517 July 27, 1984
Grey mare by Imperial Imperor
Owned by Gibson Arabians, Loomis, California.

GA Moon Tajmahal #0358652 September 5, 1985
Grey mare by Imperial Imperor.
Owned by Gibson Arabians, Loomis, California

Ansata Ra Harakti

Ansata El Aziz

Ansata Ibn Aziza

GA Moon Taj Mahal

Ansata Zareifa

Ansata Azalia

Ansata Zaafina

Ansata Zaahira

Ansata Zahra

Ghazieh/Horra/Helwa/Bint Helwa/Ghazala/Radia/(Ghadia)/Bint Radia/Samira/Zaafarana/Ansata Bint Zaafarana/Ansata Zareifa

■ Ansata Zareifa #137115
Grey Mare; February 25, 1976 - May 8, 1989
Sire: *Ansata Ibn Halima   Dam: *Ansata Bint Zaafarana

PRODUCE

Ansata Azalia #219934 May 1, 1980.
Chestnut mare by Maard.

Unnamed grey stallion by Ansata Ibn Shah, June 21, 1982. Died.

Unnamed grey stallion by Ansata Mourad Bey, July 29, 1983. Died.

Ansata El Zareef #356020 February 23, 1986
Grey stallion by *Jamilll. Died June 15, 1988.

Ansata Al Zafar #0392642 June 4, 1987
Grey stallion by Ansata Halim Shah. Gelded. Transferred to Rev. Randy Shankle, Marshall, Texas, December 1987.

*Ghazieh/Horra/Helwa/Bint Helwa/Ghazala/Radia/(Ghadia)/Bint Radia/Samira/Zaafarana/*Ansata Bint Zaafarana/Ansata Zareifa/Ansata Azalia*

■ Ansata Azalia #219934
Chestnut Mare; May 1, 1980
Sire: Maard   Dam: Ansata Zareifa
Sold to Jay Gormley, Mena, Arkansas, April 1990.

PRODUCE

Ansata Zaafina #0412845 February 6, 1988
Chestnut mare by Ansata Mourad Bey. Sold to Dr. Chess Hudson, Lawrenceville, Georgia, December 1990.

Ansata Zaahira #0430128 February 12, 1989
Grey mare by Ansata Mourad Bey.

Ansata Zahra #0453348 February 20, 1990
Grey mare by Ansata Halim Shah.

*Ghazieh/Horra/Helwa/Bint Helwa/Ghazala/Radia/(Ghadia)/Bint Radia/Samira/Zaafarana/*Ansata Bint Zaafarana/Ansata Zareifa/Ansata Azalia/Ansata Zaahira*

■ Ansata Zaahira #0430128
Grey Mare; February 12, 1989
Sire: Ansata Mourad Bey   Dam: Ansata Azalia
Sold 1995 to Paul and Anne Walker, Canada.

PRODUCE

Ansata Omar Zahir #0494030 May 10, 1993
Grey stallion by Ansata Omar Halim.
Transferred to Terry Nancy, Jackson, Tennessee

PRODUCE of Ansata Zaahira continued

Ansata Zia #0505498 May 23, 1994
Grey mare by Ansata Halim Shah. Died December 1994.

Ansata Zaahy #pending May 22, 1995.
Grey stallion by Ansata Orion. Gelded.

Ghazieh/Horra/Helwa/Bint Helwa/Ghazala/Radia/(Ghadia)/Bint Radia/Samira/Zaafarana/*Ansata Bint Zaafarana/Ansata Zareifa/ Ansata Azalia/Ansata Zahra

Ansata Zahra #0453348
Grey Mare; February 20, 1990
Sire: Ansata Halim Shah   Dam: Ansata Azalia
Sold to HRH Prince Khalid S.K. Al Saud, Nejd Stud.
Saudi Arabia, December 1997.

PRODUCE
Ansata Zaafara #0551162 March 16, 1998.
Grey mare by Ansata Iemhotep.
Owned by HRH Prince Khalid S.K. Al Saud, Nejd Stud.
Exported to Saudi Arabia.

Ansata Zahra

Serra Family

Ghazieh/Horra/Helwa/Bint Helwa/Ghazala/Jemla/Serra/ *Bint Serra/Fa Deene/Khedena/Khamsa/Allah Ateyyah

Allah Ateyyah #27711
Grey Mare; April 10, 1964
Sire: Ar-Ra'ad   Dam: Khamsa
Bred by Dr. Clark W. or Jean P. Jennings,
Newton, North Carolina. Leased from Jean Jennings.

PRODUCE
1977 serviced by *Ansata Ibn Halima. Open.
Returned to Jean Jennings.

Allah Ateyyah

Al Hadb

Arabe du desert. 19th century lithograph by A.J. Gros.

The Hadban Enzahi Strain

The Hadban of El Nazhi is a valued and important strain having produced sires and broodmares who have had a profound influence on Egyptian breeding programs. It is a great "blending" family and is placed fourth in *The Abbas Pasha Manuscript*. The Hadban strain took its name as follows:

Shabat al Mani' of al Suwayt, and he is advanced in age, and Ali Mani', the son of the brother of Shabat, who attended the gathering, were questioned in the presence of Sultan ibn Suwayt and a large number of people.

"Tell us about their origin and what their strain was and from whom they came."

The above-mentioned replied, "Mani' was at the time of Beni Lam, who is from our grandfathers...and she is Kuhayla om Maarif, and the reason for calling her Hadba at Mani's was because he had a mare with a profusely long mane which covered her forehead completely {hadba salifa}, and for that reason she was called Hadba. And she passed from Mani' to Nazhi of al Fudul the day they forced them at the hillside of Massel.

And she was blessed at al Nazhi's, and she became Hadban Nazhi from al Nazhi. From al Nazhi her blood spread through the tribes, and we can recognize [claim] them from the time of our fathers to this day.

And the people who were connected with our fathers and our first white-haired men have told us and assured us that they are Hadb."

Following is a typical history about how a specific "family" within the Hadban strain, the Hadba Munseriqah, got its name.

About al Hadba Munseriqah of al Mani'

She was so named at al Mani', and this is the reason. He went on a razzia with the tribe, and the tribe was prevented [from winning] because some of his horsemen who were very greedy stole from the tribes who were near the razzia. And [these tribes who were plundered] alerted the Arabs that they were about to be attacked, which enabled them to prepare for battle. Al Mani' presented the Hadba to those who had been stolen from unfairly as a compensation. And she was very swift. {It was a law that those not participating in a razzia should not be harmed. In this case, those people near the raid were wronged, and, because his men had unfairly robbed them, al Mani' gave the Hadba as compensation.}

And she was for that reason called at al Mani', Hadba al Munseriqah [the stolen]. And we did not know any present strain from her until the time of al Tuwayni al Sa'doun who marched with cannons against Ibn Saud. Then we recognized her from Ibn Sa'doun of Muteer."

Mani so loved this mare that he composed a poem about her virtues and the courageous qualities that inspired Arab poets to wax lyrical about horses both in pre- and post Islamic eras. Particularly noteworthy is Mani's specific reference to women riding mares (see poem introducing the Hadban section of Chapter VI).

The Hadban strain in Egypt descends from the mare Venus "Shekra Zefra." She was bred by the Shammar and imported to Egypt in 1895 by the Agha of Hassan abu Amin. Her daughter, Hadba, produced the two R.A.S. foundation mares, Bint Hadba El Saghira and Gamila. The former carried on in female line; the latter produced the stallions, Ibn Rabdan and Baiyyad, but her tail female line died out.

Hadban Enzahi:
Bint Samiha and Bint Rustem Families

Two significant lines descend from Bint Hadba El Saghira through Samiha (by Samhan) and Bint Rustem (by Rustem), each having certain distinguishing characteristics that breed on to the present. Ansata acquired both lines - the Bint Samiha line early on and the Bint Rustem line later. Noble stallions and mares descend from each line. Both lines are distinguished by somewhat longer backs (as compared to the Dahmans, for example), and a slight weakness of loin, plus a shortness and peakiness of croup which often persists. When one examines the early stallions and mares with poor hindquarters (e.g. *Astraled, Rustem, Sotamm, Kasida, Kazmeen, Hind) among others that were used in the early Egyptian programs, the persistence of this trait is transmitted by linebreeding to certain individuals who carry forward these characters. Furthermore, it is more difficult to get rid of it in some families than others.

Bint Samiha Family

The most famous and influential sire in modern Egyptian pedigrees was Nazeer, Bint Samiha's son. Had she done no more than produce him, she would have endured throughout history as a mare having made an immeasurable contribution to the breed. However, she also produced excellent females.

The closest to the root mares of this line that we remember in Egypt were mares, Kamla and Hemmat. Hemmat (x Maysouna) was a nice broodmare, rather plain of face with some pigmentation loss. Her grey daughter, Neamat (by Morafic), was an especially beautiful tall mare with a pretty head, long neck, elegant lines - one of my favorite mares. She was a full sister to Afifa, imported to Germany, who was of less type and quality and a perfect example of absolute opposites in full siblings. From Neamat came *Nihal (by Antar), one of the finest Gleannloch imports, an exceptionally beautiful chestnut mare that was shown to many championships in both halter and performance. From her descended the mare, Hasbah, whom we eventually acquired. Afifa went to Germany, imported by Karl Heinz Domken, where she produced Ghazalah, a beautiful daughter of the Bukra son, Ghazal (by Nazeer), imported to Germany, and *Assiedeh (by Mahomed) an elegant pretty-headed very refined stallion with light hindquarters.

Kamla Family

Kamla, a Sheikh El Arab daughter, was a very noble, refined mare with an average length of neck and a most classic head married to the weak croup and hindquarter, which unfortunately she inherited and passed on as a result of her being a double Kazmeen granddaughter. She tended to lose pigment, as did her son, Hadban Enzahi, (by Nazeer), who became head sire of the Marbach Stud. Her daughter, Bint Kamla (by El Sareei), was also an exquisite mare, slightly stronger in the hindquarters than her dam, perhaps due to her sire's contribution. Bint Kamla produced the exceptionally beautiful and refined grey mare, Lutfia {Lotfeia} (by Alaa El Din), who was imported to Babolna State Stud and founded the marvelous chestnut line of Ibn Galal I-7 mares. Lutfia eventually went to Dr. Nagel's stud in Germany, where she produced the exquisite Nashua, among others. Lutfia was an excellent representation of Alaa El Din bred to El Sareei daughters, thus doubling the Shahloul blood. She, nevertheless, resembled the Kamla family, i.e., a beautiful head married to a weak rear.

Maysouna Family

*Ansata Bintmisuna (Bint Maysouna), (by Nazeer), was acquired by Ansata from the Cairo Police in 1970 along with *Ansata Bint Nazeer (Fulla). She was a very elegant fine-boned mare with a very pretty head despite a piece of one nostril having been cut off. A part of her tail bone had also been cut, so that she never grew a full tail. She possessed a long well-shaped neck was somewhat long-bodied, rather short in the croup and good-legged. Her son, Ansata Abu Sudan (by Ansata Ibn Sudan), an extremely beautiful stallion, was an excellent sire for Ansata and an outstanding producer of daughters that could "move, " as did he. Her daughter, Ansata Selene (by *Ansata Ibn Halima), unfortunately died as a young mare. Bint Misuna foundered when Selene was weaned, and a long story about her recovery is found in *Authentic Arabian Bloodstock* (pp. 349-352). She was the kind of mare that would have inspired El Mani to write a poem about her.

*Assiedeh (by Mahomed, a Hadban Enzahi son) was imported to the U.S. and spent much time at Ansata where she produced four daughters by Ansata Halim Shah (one died after foaling in Colorado). She was a nice broodmare, much improved in type over her dam through the introduction of the classic head qualities of her sire, though needing longer neck, better shoulder, and stronger topline. Her three daughters, Aissha, PWA Asherah, and PWA Asifa, by Ansata Halim Shah, were much improved both in type and conformation over their dam. They traced in many lines to Bint Samiha through the intense Nazeer breeding on both sides of their pedigree. These three mares were acquired by Ansata when Pyramid West Arabians dispersed. We had observed them throughout their life and were glad to obtain this tail female line since their pedigrees contained the elements of classic horses we admired.

Aissha (by Ansata Halim Shah), resembles her sire, but is longer in body and shorter on leg than her full sister,

PWA Asherah (by Ansata Halim Shah), was shorter-coupled than Aissha and had the most extreme head of the three sisters as well as the best conformation. She was shown successfully as a young mare.

PWA Asifa (by Ansata Halim Shah), an attractive broodmare, was improved in conformation over her dam and reflected the essence of her sire; she was shorter in head than Aissha, but not as extreme as Asherah.

All three sisters have produced well to Ansata stallions.

Another line to Maysouna was also acquired through a Gleannloch mare.

Hasbah (by *Messaoud) harked back to Neamat and *Nihal through her dam Hadibah (by Al Metrabbi) and granddam, Haddiyah (by Amaal), whom we had seen when Tish Hewitt owned her. She was a very pretty broodmare, well-conformed but somewhat short on leg, and a valuable representative of this line. Hasbah was acquired from the Gucci herd when it was repossessed by Gleannloch and given to the Pyramid Society. Originally we wanted to purchase her at the Gleannloch dispersal sale, but Gucci bought her. Now she came to Ansata. She was in quite poor condition, but she rapidly gained weight and soon made herself at home in the herd. She was bred to Ansata Iemhotep producing the very beautiful colt, Ansata Chiron, who became a show champion before his untimely demise, and his full sister, the beautiful Ansata Haalah, who won Extended Futurity Res. Champion Mare at the Egyptian Event.

Kamla Family

The line to Kamla came to Ansata through the exquisite chestnut mare, Ibn Galal I-7, a daughter of Lutfia and Ibn Galal I (Galal x Hanan).

* Ibn Galal I-7 ("Gala"). We were visiting the Babolna Stud in Hungary with Dr. Nagel in 1982 and saw this incredible young mare. One instinctively knows when they are in the presence of greatness. She was of that caliber. Tall, elegant, beautiful proportions, wonderful head with expressive eyes, good length and shape of neck, good topline that barely hinted at the marriage of beautiful head and weak croup in her background. Her disposition was sweet and her demeanor graceful. We arranged a lease to take her to America, show her as representing the Babolna breeding program, and then share in her offspring sired by Ansata Halim Shah. She won Class A championships as easily as she won the hearts of spectators. A strong contender for U.S. National Champion, she was beaten by the pretty chestnut mare, VP Kaluha, and the big, not very classic dark bay, Love Potion, who took second. "Gala" placed a very close third, but those who saw her that year will never forget her. Many there were who thought she should have won, and she was Champion to me!

Unfortunately, Gala did not relate in bravery to the Hadba mare in the poem of El Mani. She was very sensitive to any kind of pain, and we were always attentive to this trait. When it came time for her to foal, our veterinarian, Dr. Karen Hayes, brought her well-equipped mobile lab van into the foaling barn and parked it beside the stall, so she could watch her all night. Everything appeared in good order, but when she began to deliver, a serious problem arose. A team of surgeons from Sapulpa, Oklahoma, was called, and they quickly hired a plane and flew to Ansata. Another equine vet who had attended Ansata horses was also brought in. The Sapulpa team operated and straightened a torsion, but Gala did not survive the shock. Her foal, a filly by Halim Shah, could not be saved either.

This was one of the low moments of the low, the kind that makes one want to sell everything and "get out."

Ibn Galal I-7, a full sister, was also available in Hungary, and arrangements were made to import her. She was not as beautiful as Gala, but she was very comparable. She was flown in via Los Angeles, and after being released from quarantine, she was transported to Ansata by Dick Edgerton, who hauled all our horses and made a special trip just to pick her up. There had been problems with rhino in Babolna, and whether this had something to do with the mare getting sick, we did not know. However, within days of her arrival, she became terribly sick and died.

Fortunately, this family from Lutfia carries on in Europe and Kuwait through her descendents. For Ansata, it obviously was not meant to be, although the pleasure of having Gala for the time apportioned was a highlight in the Ansata story.

Bint Rustem Branch

Galila Family

This line descends from Hind, a daughter of Ibn Rabdan, who was not one of his most beautiful daughters and appeared to be weak behind, a trait that continued to breed on in this family. Galila was a rather plain-headed broodmare having beautiful big dark eyes, good body, but rather light in the hindquarters. She was the dam of *Ghalii (by Nazeer), given by the E.A.O. to the U.S. Secretary of Agriculture, Ezra Taft Benson, who in turn gave him to the Agricultural Division of Michigan State University to use as a sire. Her daughter, Aziza, was a refined mare of good proportions, pretty head, also light behind, but better than her dam. While this line has produced well, it has not had as strong an impact to date as other Hadban lines.

AK Bint Dalia II (by *Ibn Moniet el Nefous), was on lease to Ansata. She was a tall and attractive bay mare who performed well under saddle. She was not extreme in type, but she was a good broodmare and produced very typey progeny when bred to Ansata Halim Shah.

Yosreia Family

The Yosreia line, also descending from Hind, took on an entity of its own and one often hears reference to "the Yosreia line." She was one of the very attractive Sheikh El Arab daughters, pretty head, rather short croup and a bit peaky behind. Her daughter, Farasha by Sid Abouhom, was also a very attractive mare, similar in type and hindquarters. *Nabilahh, by Antar was stronger bodied, also very pretty, with a better body structure. Her daughter, Lohelia, by *Morafic, was elegant but not as beautiful in the head as her dam. Bred to the handsome, big-bodied AK Al Zahra Moniet, she produced RXR Lia Moniet, whom we acquired to pick up this Yosreia line.

RXR Lia Moniet (by AK Al Zahra Moniet) was a big, strong-bodied chestnut mare of good quality, but not exotic type. When bred to Ansata stallions, the resulting progeny were more refined and better quality.

The Hadban strain has played an important part in the Ansata program, both mares and stallions. It did not breed on in quantity as well as the other lines, nor did we have a sufficient number of Hadbans until later years to develop what we envisioned. Some bad luck along the way did not help, but then, that's the way the horse world turns. The peaks are high, the valleys are low. You win some, and you lose some individuals that cannot be replaced.

Hadban Enzahi Strain

Venus d.b. Hadba Enzahiyah

Hadba (1894) x **Saklawi I**

Bint Hadba El Saghira x **El Halabi**

Samiha x **Samhan**

Bint Samiha x **Kazmeen**

Shams x **Mashaan**

Maysouna x **Kheir**

*Ansata Bintmisuna x **Nazeer**

Hemmat x **Sid Abouhom**

Neamat x *****Morafic**

*Nihal x **Antar**

Haddiyah x **Amaal**

Hadibah x **Al Metrabbi**

Hasbah x *****Messaoud**

Afifa x *****Morafic**

*Assiedeh x **Mahomed**

Aissha
PWA Asherah
PWA Asifa
all x **Ansata Halim Shah**

Samha x **Baiyyad**

Kamla x **Sheikh El Arab**

Bint Kamla x **El Sarie**

Lutfia x **Alaa El Din**

*Ibn Galal I-7 x **Ibn Galal I**
(full sisters)
*Ibn Galal I-7 x **Ibn Galal I**

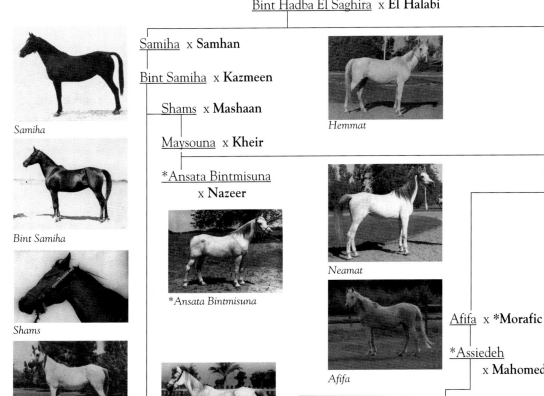

Samiha

Bint Samiha

Shams

Maysouna

Ansata Bintmisuna

Neamat

Afifa

Samha

Hemmat

Nihal

Kamla

Bint Kamla

Aissha

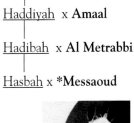
Assiedeh

Haddiyah

Lutfia

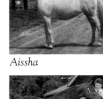
**Ibn Galal I-7*

Hadibah

Bint Rustem x **Rustem**

Hind x **Ibn Rabdan**

Fasiha x **Awad**

Rouda x **Sheikh el Arab**

Galila x **Sid Abouhom**

Aziza x **Alaa El Din**

*AK Dalia x ***Ibn Hafiza**

AK Bint Dalia II
    x ***Ibn Moniet El Nefous**

*AK Dalia

AK Bint Dalia II

Bint Rustem

Hind

Fasiha

Rouda

Galila

Aziza

Yosreia x **Sheikh El Arab**

Farasha x **Sid Abouhom**

*Nabilahh x **Antar**

Lohelia x ***Morafic**

RXR Lia Moniet x **AK El Zahra Moniet**

Yosreia

Farasha

*Nabilahh

RXR Lia Moniet

Ansata Bintmisuna resting in the warm Texas sun. JEF photo

Hadban Enzahi Strain
SAMIHA BRANCH

The Maysouna Family

Bint Samiha/Shams/Maysouna/*Ansata Bintmisuna

■ ***Ansata Bintmisuna #69408**
Grey Mare;   April 3, 1957 – January 11, 1977
Sire: Nazeer   Dam: Maysouna
Bred by E.A.O.
E.A.O. # 142, Volume II
Registered as Bint Maysouna in Egypt
Imported by Mr. and Mrs. Donald L. Forbis, Chickasha, Oklahoma, March 3, 1970.

PRODUCE

Ansata El Masri #76886 May 21, 1971
Bay stallion by *Ibn Hafiza
Sold to Bill Lowe, Beaver Dam, Wisconsin.
Exported by Lowe to Australia.

Ansata Abu Sudan #105813 January 20, 1974
Grey stallion by Ansata Ibn Sudan.

Ansata Selene #122316 January 24, 1975
Grey mare by *Ansata Ibn Halima.
Died August 2, 1976.

1977 foaled premature filly by
 *Ansata Ibn Halima that died after birth.

*Ansata Bintmisuna

*Ansata Bintmisuna

*Ansata Bintmisuna

*Some produce of *Ansata Bintmisuna*

Ansata El Masri

At left and right, Ansata Abu Sudan

Aissha

<u>Bint Samiha/Shams/Maysouna/Hemmat/Afifa/*Assiadeh</u>

▪ Aissha #0482510

Grey Mare; April 21, 1991
Sire: Ansata Halim Shah   Dam: *Assiadeh
Leased from Pyramid West Arabians, Colorado, 1997-1999
Purchased by Ansata in August 1999.

PRODUCE

Ansata Ashayet #0552503, May 30, 1998
Grey mare by Ansata Iemhotep. Died.

Ansata Anisah #0566461, August 14, 1999
Grey mare by Ansata Hejazi.

Ansata Bint Sinan #0579303 October 14, 2000
Grey mare by Ansata Sinan. Sold to Sheikh Abdullah Bin Nasir Al Thani, Al Naif Stud, Qatar. Exported.

Unnamed 2001 grey filly by Ansata Sinan. Died.

Ansata Azeem #0596946 October 10, 2002
Grey stallion by Ansata Iemhotep.

Ansata Anisah

Ansata Ashraf

Ansata Najdi

PWA Asifa

Ansata Ariana

<u>Bint Samiha/Shams/Maysouna/Hemmat/Afifa/*Assiadeh</u>

■ PWA Asherah #0493496
Grey Mare; April 26, 1993 - 2001
Sire: Ansata Halim Shah   Dam: *Assiadeh
Bred by Pyramid West Arabians, Colorado.
Purchased by Ansata from Pyramid West Arabians.

PRODUCE
Ansata Najdi #0560366 May 21, 1998.
Grey stallion by Ansata Iemhotep. Owned by Serene Egyptian Stud, Anna and Stefano Galber, Affi, Italy.

Ansata Ashraf #0566464 June 1,1999
Grey mare by Ansata Hejazi.

Mare died without further production.

<u>Bint Samiha/Shams/Maysouna/Hemmat/Afifa/*Assiadeh</u>

■ PWA Asifa #0505686
Grey Mare; April 14, 1994
Sire: Ansata Halim Shah   Dam: *Assiadeh
Bred by Pyramid West Arabians, Colorado.
Purchased by Ansata from Pyramid West Arabians.

PRODUCE
Prince Hal May 12, 1998
Grey stallion by Ansata Iemhotep. Transferred to Sheri Metcalf of Texas prior to registration and gelded.

Ansata Ariana #589219 March 7, 2001
Grey mare by Ansata Sinan.

Ansata Al Nitak #0597213 February 8, 2002
Grey stallion by Ansata Osiron

Ansata Chiron

Ansata Haalah

Ansata Hamama

Bint Samiha/Shams/Maysouna/Hemmat/Neamat/*Nihal/Haddiyah/ Hadibah

Hasbah #0468674

Grey Mare; June 17, 1991
Sire: * Messaoud   Dam: Hadibah
Bred by Douglas and Margaret Marshall, Gleannloch Farms, Barksdale, Texas. Purchased by Ansata, Mena, Arkansas, from The Pyramid Society, June 1994.

PRODUCE

Ansata Hamama #0520272 August 18, 1995
Grey mare by Prince Fa Moniet.
Sold to Jan and Arja Lancee. Exported to the Netherlands, March 1997.

Ansata Chiron #0537018 August 30, 1996
Grey stallion by Ansata Iemhotep. Died 2000.

Unnamed colt by Ansata Iemhotep September 14, 1997
Died September 15, 1997.

Ansata Haalah #0563809 February 27, 1999
Grey mare by Ansata Iemhotep.
Sold September 2001 to Christie and Henry Metz, Silver Maple Farm, Florida.

The Kamla Family

Venus/Hadba/Bint Hadba El Saghira/Samiha/Bint Samiha/Kamla/ Bint Kamla/Lutfia

*Ibn Galal I-7 #268277

Chestnut Mare; September 3, 1979 – July 6, 1986
Sire: Ibn Galal I (G.A.S.B.)   Dam: 7 Lutfia (E.A.O.)
Bred by Agraria Babolna, Hungary
Imported on lease to Ansata, February 1, 1983.
Serviced to Ansata Halim Shah 1985. Foaled dead filly.
Mare died foaling.

Venus/Hadba/Bint Hadba El Saghira/Samiha/Bint Samiha/Kamla/ Bint Kamla/Lutfia

*Ibn Galal I-7 (unregistered)

{Full sister to *Ibn Galal 1-7}
Chestnut Mare; April 14, 1983
Sire: Ibn Galal I G.A.S.B.   Dam: 7 Lutfia E.A.O.
Bred by Agraria Babolna, Hungary
Imported on lease to Ansata, February, 1987.
Died on arrival from quarantine.

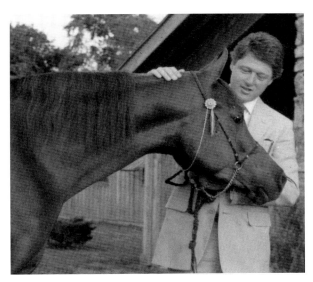

All pictures this page, *Ibn Galal I-7 - "Gala". Above left, with Arkansas Governor Bill Clinton, above right, winning champion mare, Richard Sanders handling. Below at an Ansata seminar.

BINT RUSTEM BRANCH

The Galila Family

Venus/Hadba/Bint Hadba El Saghira/Bint Rustem/Hind/Rouda/
Galila/Aziza/*AK Dalia

■ AK Bint Dalia II #018829
Bay Mare; June 22, 1979
Sire: *Ibn Moniet El Nefous   Dam: *AK Dalia
Bred by Jarrell McCracken, Waco, Texas.
Owned by Magness Arabians, Englewood, Colorado.
Leased by Ansata, Mena, Arkansas

PRODUCE

Dalia Halim #0470974   March 11, 1991
Grey stallion by Ansata Halim Shah. Owned
by Bob and Sharon Magness, Englewood, Colorado.
Sold by Ansata to Erwin and Annette Escher.
Exported to Germany.

Ansata Shahlia #0480677, March 4, 1992
Grey mare by Ansata Halim Shah. Owned by
Ansata. Sold to Erwin and Annette Escher, February
1995. Exported to Germany.

Dalima Shah #0491532, February 26, 1993
Grey mare by Ansata Halim Shah. Owned by
Bob and Sharon Magness, Englewood, Colorado.
Sold by Ansata to Erwin and Annette
Escher. Exported to Germany.

AK Bint Dalia II

Dalia Halim

Ansata Shahlia

Dalima Shah

The Yosreia Family

<u>Venus/Hadba/Bint Hadba El Saghira/Bint Rustem/Hind/Yosreia/
Farasha/*Nabilahh/Lohelia</u>

RXR Lia Moniet #0412429
Chestnut Mare; June 8, 1988
Sire: AK Elzahra Moniet   Dam: Lohelia
Bred by R.H. Niemeyer and G.P. Stillman and E.M. Isabelle,
Lancaster, Pennsylvania
Owned by Rocaille Ranch, David and Martha Murdoch,
Cleburne, Texas.
Transferred to Ansata, July 1994.
Sold to H.R.H. Prince Khalid Bin Fahad Al Saud. Exported
to Saudi Arabia.

PRODUCE

Ansata Aladdin #0505475 February 24, 1994
Grey stallion by Ansata Halim Shah. Sold to the Jaffar Stud,
H.R.H. Princess Zeyn Bint Hussein and Majdi Al Saleh.
Exported to Jordan, October 1995.

Unnamed colt by Ansata Hejazi May 26, 1996.

Ansata Laila #0542106 June 6, 1997
Grey mare by Ansata Hejazi. Sold to Sheikh Abdullah Bin
Nasir Al Thani, Al Naif Stud, Doha, Qatar. Exported.

Ansata Shams # Pending May 31, 1998
Chestnut stallion. Gelded and transferred to
Roger Shuler, Fayetteville, Arkansas.

1999 exported in foal to Ansata Hejazi.
Foaled filly in Saudi Arabia.

RXR Lia Moniet

Ansata Laila

Ansata Aladdin

Ansata Shams

Al 'Abeyyat

19th century lithograph by Victor Adam, Bobliothéque Nationale, Paris.

The 'Abeyyan Strain

The 'Abeyyan strain has always been valued by the Bedouins in desert Arabia, and Abbas Pasha placed it sixth in *The Abbas Pasha Manuscript*. Its colorful testimonials take up 43 pages, and it is one of the most interesting chapters due to tribal disputes as to whether certain horses were or were not to be mated. The history begins with questioning at a tribal gathering as follows:

Ahmed al Sherifi was questioned by Sheikh Faysal Sha'lan about the 'Abeyya of his stud, and from where she came to him and from what was her origin.

Ahmed al Sherifi replied, "The 'Abeyya of my stud, I, O Ahmed al Sherifi, originates from al Sherrak. And from al Sherrak she passed to Bani' al Sba'ani, my partner. And from Bani' al Sba'ani she passed to my stud, I, O Ahmed al Sherifi. And I testify by Allah on my grandfather's testimony and the grandfathers of my grandfathers, that the 'Abeyya of my stud is 'Abeyya Sherrakiya. And she is to be mated."

A particularly prized mare of this strain belonged to Hussein Ibn Safeer, and he composed a poem about her that appears at the beginning of Chapter VI.

There are several interesting histories about the horses of this strain within *The Abbas Pasha Manuscript*, and there are also controversial stories that indicate the lengths the Arabs went to in checking the truth of each oral history that was handed down from tribe to tribe, and how they uncovered lies and cover-ups about certain horses. A typical inquiry is noted on page 432 wherein Mustafa Bey, an Emissary of Abbas Pasha, is recording the testimony about the history of the Abeyya mare of Ibn Zabdan:

At the time Noqaz ibn Zabdan testified about al 'Abeyya and her strain, he sent a copy of this testimony to Talal ibn Rashid so that he could inquire about the truth of the origin of these histories: that Shabat al Mousa of Otaybah captured a shaqra mare from Noqaz ibn Zabdan two years ago when Noqaz was on a ghazu [a raid] against Otaybah. And the sword was thrown from the hand of Noqaz ibn Zabdan, and he went on his shaqra mare – and she was strong and fast – to recover his sword. But Shabat al Mousa came forth and hit him with an arrow and wounded him, and Shabat took the two year old mare. And Khuzam bought her from him.

Later on another person at the gathering was asked to verify the origin of the horse in question, but he is unable to verify it:

Shehata al Honaydis was asked about 'Abeyya ibn Zabdan after he had listened to the histories at the gathering.

"Is this 'Abeyya from your horses or not? Tell us about her on your honor and good fortune."

He replied: "I do not know about her, and she is not from my horses, and I do not recognize her."

Reference is also made to the 'Abeyyan Abu Jurays strain as follows:

Jada'an, the son of the brother of Saleh al Tayyar, the Sheikh of Waled Ali, was asked about the 'Abeyya al Lomaylimi, he replied: ...And we, O Waled Ali, mate the 'Abeyyan Abu Jurays. (That is, the strain is considered pure for breeding purposes, ed.)

The most famous breeder of the 'Abeyyan strain was Al Sherrak. The strain was popular among Al Sabaa, Al Dafeer, Al Bishr, Al Otaybah, and Al Dawasser tribes, among others. It was particularly appreciated by the Al Saud and Ibn Jaluwi families in the Arabian Peninsula and is maintained there to this day through the original Saudi stock.

The name originated when, during a raid, a Bedouin threw off his 'abeyya (cloak) to lighten his mare's burden, and when he finally outdistanced his pursuers, he looked back and found his 'abeyya had been carried on the high-borne tail of the mare. Thereafter, the strain was named 'Abeyya, for the mare who carried the 'abeyya on her tail.

The horses of this strain that we observed throughout Syria and other Arabian environs tended to be prettier headed and typier than many other desert-bred strains; they also showed some predisposition to a slightly soft back. It is interesting to note that one of the mares Faysal Ibn Turki took to Egypt with him when he was a prisoner of Mohamed Ali was a *"hamra fazrah"* (bay saddle-backed mare), who was born that way.

No tail female lines to the 'Abeyyan strain are found in the Royal Agricultural studbook; rather it descends fortuitously through the Inshass Herdbook of Kings Fouad and Farouk. The foundation mare was El Shahbaa, a white/grey mare foaled in 1925, whose dam was of the 'Abeyyan Om Jurays strain. El Shahbaa was purchased for the Inshass Stud in 1931, where she founded a line of much significance through her daughter, El Mahroussa. El Mahroussa's daughters Mahfouza and Mahdia carried on in tail female line; Mahfouza through Maysa and *Magidaa, Mahdia through Hanan and Montaha.

When the Inshass herd was culled by Von Szandtner, this family remained and was transferred to the El Zahraa Stud where it was incorporated within the EAO breeding program. It is usually characterized by good height, classic heads with rather long foreface, good length of neck, somewhat long soft backs, light hindquarters, fine bones, and good legs. Two branches are well known: the Inshass Mona line, the most prolific of this family being Hanan, while the Mahfouza line relates to Hafiza, Montaha, and *Magidaa. Both mares were bay, and one finds a number of bays of this strain mentioned in *The Abbas Pasha Manuscript*.

The mare, Hafiza, whom we saw in Egypt, was the result of inbreeding Hamdan to his daughter, Mahfouza. Hafiza was a tall bay broodmare of no special quality, rather plain of head, good length of neck, long of body and light in the hindquarters, the kind of mare many, who do not understand "blood will tell," would bypass for something more physically beautiful. Yet when bred to *Morafic, she produced the incredible bay stallion, El Arabi, one of the most noble and magnificent stallions we've ever seen; a horse worth a king's ransom - a one of a kind. Tall, elegant, very fine-boned, beautiful head with expressive eyes, long neck, and a good body, he was spirited and graceful, carrying himself with pride. One morning during our visit, the grooms turned him loose with Alaa El Din in a paddock beside the stallion barn. We could not believe the stablemen would free two breeding stallions together, but both horses ran and played, El Arabi trotting circles around Alaa El Din. We will never forget El Arabi's brilliant movement, free and ground-covering compared to Alaa El Din's who, although he was a successful race horse, was not a graceful mover. Unfortunately, El Arabi got loose from his handler one day, ran down El Zahraa's palm-fringed entry road directly into a bus and was killed. A gate that was erected thereafter still remains and is manned at the entrance to this day. One cannot help wondering what influence this marvelous stallion might have had if he had lived. For some reason he was not given the best mares during his introduction at stud, and he left only a few daughters to carry on and the son, Emad, who resembled him not an iota in type. El Arabi's daughters were better quality than Emad, although Emad sired a son who resembled El Arabi, but was sterile.

The stallion, *Ibn Hafiza (by Sameh), was a handsome individual, pleasing head, heavier in body than his half-brother, El Arabi. Sameh stallions, in general, proved not to be Class A sires. Of the Sameh sons he proved to be the best and is found in pedigrees around the world.

Alaa El Din nicked well with this 'Abeyyan strain, producing the two noteworthy mares, Hanan and Magidaa. In America the *Magidaa line (to Mahfouza) produced U.S. Res. National Champion, Ruminaja Ali, and his beautiful brothers, Alidaar, Ruminaja Bahjat, Ruminaja Fayez, among others, who have had immeasurable influence worldwide. The Inshass Mona branch (to Mahdia) has had more impact in Europe through the importation of Hanan to Germany. Dr. Nagel has essentially produced the "Nagel Family" of horses relating to this strain through linebreeding and inbreeding to Hanan. He also incorporated the Bukra and Lutfia lines that have added important classic traits and changed most of the herd from bay to a more classic grey.

In America the *Magidaa line has succeeded in tail female, her grey daughter, Bint Magidaa, was a very beautiful mare, as is her granddaughter, Bint Bint Magidaa. It is interesting to note the grey sons of Bint Magidaa (Ruminaja Ali, Ruminaja Bahjat, Ruminaja Fayez, Alidaar) were more dominant in the stud than the chestnut, Ruminaja Majed , a handsome horse who, nevertheless, was different in type. Close linebreeding and inbreeding to Ruminaja Ali through The Minstril and Thee Desperado have provided an exceptional preponderance of this blood in America,(particularly in bays and chestnuts).

We have found horses of this family that have been at Ansata, and in other observations, have attitudes and learning ability, especially as young horses, quite different from those of the more linebred Ansata horses. When we have crossed this line with Ansata bloodlines, or observed these combinations in some other instances, it appeared more docility, trainabilty, better movement and higher tail carriage often resulted while still retaining other traits of this valued 'Abeyyan strain.

El 'Abeyya Om Jurays - Inshass Mona Family

Hanan Family

We observed the Inshass Mona (not to be confused with Mouna/Mona of the Moniet El Nefous Saklawi Jedran strain) while she was among the group Von Szandtner had selected to keep when Inshass was disbursed. She was a refined well-balanced bay mare, fine bones, good size, pleasing head and nice length of neck, with a relatively long back with short croup. Although she had more quality than some of the Inshass mares, she still had that desert-bred look that set the Inshass horses apart from the old linebred R.A.S./E.A.O. stock until they were "bred up" to more classic Egyptian type through incorporation of the Nazeer blood. Of her descendants, the mare Hanan is particularly well known through her importation to Germany by Dr. Nagel. She was one of the most influential Alaa El Din daughters, and her intriguing story is related in the Nagel book *Hanan, The Story of An Arabian Mare and the Arabian Breed*. An elegant refined bay mare, she could be referred to as "an incubator mare." Most of her progeny resembled the stallion she was bred to, e.g. Ghazal, Madkour I, Ibn Galal, Malik, Mohafez, and Ansata Halim Shah. Her worldwide influence is now legendary.

The Hanan grandson, *Jamilll, was Ansata's first import of this family, and he went on to become a multiple champion and U.S. Top Ten Stallion. He sired many beautiful daughters that became broodmares, and handsome show-winning sons, but the latter did not measure up to the mares as breeding stock. Hanan's last foal, Salaa El Dine by Ansata Halim Shah, has become one of the most sought after stallions in Europe, siring good producing sons and daughters.

Another link to this Inshass Mona line was Montaha (*Serenity Montaha) by Galal, who was imported to Canada. She also provided descendants to this Mona family in America.

*Bint Amal (by Ansata Halim Shah x Amal x Hanan) was imported by Ansata on lease from Nagel, to be shown as a representative of her sire. She was sold to Robert and Erika Brunson, who let Ansata campaign her to many championships. A beautiful mare, she was one of the best show-quality Ansata Halim Shah daughters bred by Nagel while the stallion was on lease to him in Germany. She was eventually sold to Sheikh Abdulaziz Bin Khaled Al Thani who bred her to Ruminaja Ali before she was exported to his Al Rayyan Farm. She conceived and produced the excellent RN Ajeeba in Qatar, who became a show winning filly.

RAS Najma (by Thee Desperado x RAS Amala x *Bint Amal), was acquired from the Brunsons when they dispersed their breeding stock. She was an extremely refined and pretty-headed grey mare, well-balanced, but needing more strength in the hindquarters. She died before producing any foals at Ansata.

*Afifaa (x Ghazalah x Hanan), by *Jamilll, was an exquisite, well-balanced representative of this family with beautiful head and big black expressive eyes. She and her elegant bay full brother, *Aziz, were imported on lease with great hopes for showing and breeding them in America. Unfortunately, both were sterile. It was one of those inbred combinations that didn't work. *Aziz was gelded, and *Afifaa was returned to Germany.

Montaha Family

Serenity BtMontaha (by *Khofo), was productive in America and her daughter, **AK Malouma,** was a handsome, strong-bodied, liver-chestnut mare who lacked the refinement and pretty head of the Hanan/Mona line as an individual, but she produced the very beautiful foundation mare, RXR Farah by Al Metrabbi for Rocaille Ranch.

Mariyah (x AK Malouma), was on lease to Ansata for a year, but she was returned to Rocaille after producing a colt that was gelded.

This Malouma line has bred on with other breeders and provides a viable alternative to the inbred *Magidaa lines in America.

Magidaa Family

Although admirers of the *Magidaa line, particularly through her daughter Bint Magidaa, whose handsome sire, *Khofo, was obviously a true "nick" with *Magidaa, Ansata did not acquire any of this particular family line in tail female. However, daughters of the brothers, Ruminaja Ali, Alidaar, and Ruminaja Bahjat, were purchased, and some of their direct, but mostly indirect, progeny were incorporated into the Ansata herd.

The line to Hanan is much sought after in the Arab world, and the line to *Magidaa, originally much celebrated in America, is now appreciated in Europe and the Middle East, especially through the importation of Alidaar as a herdsire to Al Shaqab Farm in Qatar.

El Shahbaa ('Abeyyan Om Jurays) Family

1925 Grey mare from El Haj Mohamed Ibrahim

El Shahbaa Abayyah Om Jurays

El Mahrousa x **El Zafir**

Mahfouza x **Hamdan**   Mahmouda x **Beshier**   Mahdia x **Hamdan**

Maysa x **Antar**   Hafiza x **Hamdan**   Mona x **Badr**

Maysa   *Hafiza*   *Mona (Inshass)*

*Magidaa x **Alaa El Din**   Hanan x **Alaa El Din**   *Serenity Montaha x **Galal**

*Magidaa   *Hanan*   **Serenity Montaha*

Bint Magidaa x ***Khofo**   Amal x **Mohafez**   Ghazala x **Ghazal**   Serenity BtMontaha x ***Khofo**

*Bint Amal x **Ansata Halim Shah**   *Afifaa x ***Jamilll**   AK Malouma x **TheEgyptianPrince**

Mariyah x **Imperial Imperor**

Amal

*Bint Amal

Ghazala

AK Malouma

The 'Abeyyan Strain
EL SHAHBAA FAMILY

The Hanan Family

El Shahbaa/El Mahroussa/Mahdia/Mona (Inshass)/
Hanan/Amal/*Bint Amal

*Bint Amal #0352612
Grey Mare; March 4, 1985
Sire: Ansata Halim Shah   Dam: Amal G.A.S.B.
Bred by Dr. Hans Nagel, Bremen, Germany
Imported February 2, 1986 by Mr. and Mrs. Donald L. Forbis, Mena, Arkansas
Sold 1986 to Robert M. or Erika Brunson, Beverly Hills, California. Sold by Brunson to Sheikh Abdul Aziz Al Thani, Al Rayyan Farm, June 1994. Exported to Qatar

*Bint Amal

PRODUCE
RAS Prince Fareed #483473, March 27, 1992
Grey stallion by Prince Fa Moniet. Bred and owned by Robert M. or Erika Brunson, Beverly Hills, California. Donated November 1993 to Fern Leigh Foundation

RAS Amala #0494091, April 20, 1993.
Grey mare by Prince Fa Moniet. Bred and owned by Robert M. or Erika Brunson, Beverly Hills, California. Sold to Ray and Jamie Roberts, Nirvana Farms, Lees Summit, Missouri.

El Shahbaa/El Mahroussa/Mahdia/Mona (Inshass)/
Hanan/Amal/Bint Amal/RAS Amala/RAS Najma

RAS Najma #541295
Grey Mare; March 22, 1997
Sire: Thee Desperado   Dam: RAS Amala
Bred by Robert M. or Erika Brunson
Transferred to Donald and Judith Forbis, Mena, Arkansas.

Died in 2001 in foal to Ansata Sinan

RAS Amala

RAS Najma

El Shahbaa/El Mahroussa/Mahdia/Mona (Inshass)/
Hanan/*Jamilll

*Jamilll #0268275
(Jamil G.A.S.B.)
Grey Stallion; 1975
Sire: Madkour I   Dam: Hanan
Bred by Dr. Hans Nagel, Bremen, Germany
Imported on lease 1983 by Ansata, Mena, Arkansas.
Returned to Germany in 1985.

*Jamilll

El Shahbaa/El Mahroussa/Mahdia/Mona (Inshass)/
Hanan/Ghazala/*Afifaa

*Afifaa #0268276
Grey Mare; 1980
Sire: *Jamilll   Dam: Ghazala G.A.S.B.
Bred by Dr. Hans Nagel, Bremen, Germany
Imported 1983 by Ansata, Mena, Arkansas.
Returned to Germany.

*Afifaa

El Shahbaa/El Mahroussa/Mahdia/Mona (Inshass)/
Hanan/Ghazala/*Afifaa

*Aziz #0316918
Bay Stallion; 1983
Sire: *Jamilll   Dam: Ghazala G.A.S.B.
Bred by Dr. Hans Nagel, Bremen, Germany
Imported on lease by Ansata, Mena, Arkansas.
Full brother to *Afifaa. Gelded and used for riding.
Brother and sister were non-productive.
Beautiful individuals but a breeding that did not work.

*Aziz

The Montaha Family

El Shahbaa/El Mahroussa/Mahdia/Mona (Inshass)/
**Serenity Montaha/Serenity BtMontaha/AK Malouma*

■ AK Malouma #0302636
Chestnut Mare; May 23, 1982
Sire: TheEgyptianPrince   Dam: Serenity BtMontaha
Bred by Bentwood Farms, Inc.
Leased from Rocaille Ranch by Ansata Arabian Stud, Mena, Arkansas.

Non-productive. Returned to Rocaille.

Above and below, AK Malouma

El Shahbaa/El Mahroussa/Mahdia/Mona (Inshass)/
**Serenity Montaha/Serenity BtMontaha/AK Malouma/*
Mariyah

■ Mariyah #0479218
Chestnut Mare; February 28, 1992
Sire: Imperial Imperor #0204381   Dam: AK Malouma
Bred by David and Martha Murdoch, Cleburne, Texas.
Leased to Ansata Arabian Stud.

PRODUCE
Ansata Nassri #0542287 February 19, 1997.
Grey stallion by Ansata Hejazi. Gelded.
Transferred to Sheri Metcalf of Hawley, Texas, August 4, 1997.

Mariyah

The 'Abeyyan Strain
KARIMA/KAREEMA FAMILY
'Abeyyan (?)

(Note: The strain of the root mare, Obeya ('Abeyya) is not given in the Inshass Herdbook. Her granddaughter, Bint Karima, is listed in the 'Abeyyan section in the E.A.O. Studbook Vol. III, p. 214, and great-granddaughter, Kawthar, precedes her in the E.A.O. Studbook, Vol. II, p. 327.)

This family was not as prolific, or as significant in production, as the 'Abeyyan Om Jurays strain previously noted. Through the mare, Ithad, three daughters were produced: Hania by *Tuhotmos, Hodhoda by Alaa El Din, with representatives in Europe, and *Higran (by Seef), who was imported to the U.S.

We remember Hodhoda as standing out among this family in Egypt; she was a very refined grey mare with a beautiful head, but she had a somewhat peaky sloping croup. The mare, *Moshira (by Ibn Akhtal), tracing to this line through Hodhoda's daughter, Domia, was imported by the El Casabgui family and was sent to Ansata for breeding to Ansata stallions. She was an average-sized, dark bay mare, somewhat plain, and not as refined in comparison to others of her family noted above.

El Yasmine (x *Moshira) was a daughter of Ansata Halim Shah that came to Ansata on lease when the Cassabgui family dispersed its herd. She was a pleasing grey mare, more refined than her dam. She produced one colt at Ansata, a gelding, and then was sold by her owners.

Bint Karima ('Abeyyan) Family
1935 Grey mare from the Kafr Ibrash Farm

Obeya (root mare)
|
Karima x Dahman
|
Bint Karima x Rasheed
|
Kawthar x Mekdam
|
Ithad x Sameh
|
Hodhoda x Alaa El Din
|
Hania x *Tuhotmos   Domia x Seef   *Higran x Seef
|
*Moshira x Ibn Akhtal

Hodhoda

Hania   *Higran

*Moshira

Bint Karima Line

Obeya†/Karima (INS)/Bint Kareema/Kawthar/Ithad/Hodhoda/Domia/*Moshira

El Yasmine #0472427
'Abeyyan
Grey Mare; September 1991
Sire: Ansata Halim Shah   Dam: *Moshira
Leased from George El Cassabgui in 1994 by Ansata, Mena, Arkansas

PRODUCE
Unnamed colt by Prince Fa Moniet, March 15, 1996.

Returned to El Cassabgui

Al Kuhaylat
Kuhaylat Ajuz
(original Strain of All Arabians)

Kuhaylan Jellabi (or Saqlawi Jedran?)
*Maaroufa Family

Koheil Aguse mare, E. Volckers. Olms archive.

The Kuhaylan Jellabi Strain

The Kuhaylan Jellabi Strain or Saklawi Jedran?*

*In recent years a question has arisen as to whether the Yamama family (Negma, Mahroussa, *Maaroufa, *Fadl, *Nasr, etc.) previously recorded as Kuhaylan Jellabi tracing to El Argaa Jellabiet, actually traces to Ghazieh of the Saklawi Jedran strain instead. The debate began because Lady Anne Blunt mentions nothing in her diaries about seeing the Yamama-Jellabi line at Prince Mohamed Ali's, only the Yamama-Saklawi line. This gave rise to the question about his strain of horses and caused researchers to investigate further. DNA testing also indicates the same markers relegate the Negma/Mahroussa/Maaroufa family (known as Kuhaylan Jellabi) to the Ghazieh family (recorded as Saklawi Jedran), and causing some researchers to now consider this family as Saklawi Jedran instead of Kuhaylan Jellabi. Nevertheless, this Yamama family bred by Prince Mohamed Ali had certain traits that gave it a distinctive type, and many scholars consider it different in phenotype (and genotype) than the Ghazieh line. Regardless of which is correct, this family has been listed following the Kuhaylan strain in this book, just as it was placed in *Authentic Arabian Bloodstock* and other reference books. While there may be a question in some minds as to "strain," there is no question as to the authenticity of the line tracing to Abbas Pasha root stock.

The much heralded Jellabi family descended from the root mare Jellabiet Feisal (Feysul), a Kuhayla Jellabia, imported from the desert by Abbas Pasha I, for whom it is said he paid 7000 gold pounds.

The Kuhayla Jellabia were of the Kuhaylan Ajuz strain and were much appreciated in the desert, the line becoming most famous on the island of Bahrain among the Al Khalifa ruling family. *The Abbas Pasha Manuscript* (pp. 534-543) records this family history as follows:

> In the presence of Haizam ibn Hathleen, the Sheikh of al Ajman, those who were present at the gathering of al Hassa were asked about al Kuhayla of Ibn Jarshan.
>
> "Amir ibn Shayi' of al Hathleen, and he is a white-haired man, and Mani' ibn Mu'ajil, who is a white haired man and the paternal cousin of Haizam ibn Hathleen, and Mesaoud ibn Falah, the brother of Sirhan al 'Abd, the ra'i of the stud, replied: 'She is a Kuhayla 'Ajuz, the first of al Kuhayl and the dearest and most valuable [a'az al 'anaber] of the Arabs of Nejd. She is to be mated in the darkest night. And she originates from Ibn Jarshan of al Baqoum and has relatives at Ibn Jarshan's. And it happened there was a ghazu [a raid] on al Baqoum, and they took the flock of sheep and camels of Ibn Jarshan. And when the news reached Ibn Jarshan that his flock of sheep had been taken and al Baqoum terrorized, and the sheep were not returned, he rode his mare al Kuhayla catching up with the raiders and admonished them. And he took back his sheep from the tribe on that day.

The story further continues about the descendants of this line, stating that a grey filly named Al Hanif (the pigeon-toed) was born, and her descendants were acquired and bred by Ibn Jellab. The line became celebrated in the future as Kuhayla Jellabiya, from Ibn Jellab.

Lady Anne Blunt maintained breeding stock to this Jellabi family through the bay Yamama and the chestnut Kasida. Pictures of Kasida indicate she would not have won a most classic Arabian contest. However, her grandson was the beautiful bay stallion, Kazmeen. Although he acquired and transmitted her weakness of croup and was said to have a club foot, his positive attributes were such that the Royal Agricultural Society's representatives chose him for a breeding stallion when they purchased horses from Crabbet in 1920. Kazmeen became a significant broodmare sire for the R.A.S. and the Egyptian royal family, and his influence is felt to this day.

The Kuhaylan Jellabi strain, as it has been known in America since its introduction in 1932 by Henry Babson (*Fadl and *Maaroufa) and W. R. Brown (*Roda, *Aziza, *HH Mohamed Ali's Hamida, *HH Mohamed Ali's Hamama, *Nasr and *Zarife), was descended from Bint Yamama, the mare of H.H. Prince Mohamed Ali. There has never been a question about the Blunt-bred Kasida line of descent being Jellabi. However, this Jellabi line has not bred into the present in tail female, unless one agrees that the Bint Yamama line, from whom the Babson and Brown imports and also Ghazala I/Gazella I (Inshass) descend, is Jellabi. The current questionability about this family arises over two different mares having the same name, but different strains. It is this Bint Yamama of Prince Mohamed Ali's that has caused the recent question about whether this family is Saklawi or Jellabi. Lady Anne Blunt was a frequent visitor to Manial, Prince Mohamed Aly's estate. They were good friends, and she was an avid student of his horses. Her diaries and records indicate she only mentions seeing a mare at the Prince's named Bint Yamama, who was a half sister to Mesaoud, (<u>therefore Saklawi of the Ghazieh line</u>). She makes no mention of a Bint Yamama of the Kuhaylan Jellabi strain at his stud.

In H.H. Prince Mohamed Aly's books, *Breeding of Purebred Arab Horses* (Vol. I and II published in Cairo, 1935-36), he writes about breeding principles, and therefore, one would expect him to have mentioned the Jellabi line and to discuss the merits of certain strains. Pictures illustrating his horses traditionally known as Jellabi show no caption as to strain, nor does he make mention of it elsewhere. With reference to breeding principles, he says nothing about specific strains in relation to their value, history or characteristics, other than a broad reference as noted below [underlines mine, ed.]:

"Again, I found that some stallions and mares, who have a striking characteristic point inherited from their ancestors will transmit it to their descendants until they are crossed with another who is stronger in breed. I mention this on the authority of many books. Many people say the mare is the dominant factor in breeding, while others contend that it is the stallion. Arab logic argues that as the mare carries the foal for eleven months and gives it milk for five, she is the more important in breeding. I maintain, from personal observation, that <u>it depends upon the animal which is the stronger representative of its own strain.</u> Some stallions always sire foals of their own colour and sex; let us take a dark chestnut like Ibn Rabdan, one of the Royal Agricultural Society's stallions, for an example. He always produces dark chestnuts, no matter what the colour of the mares. This will go on until he covers a mare who produces the <u>form and type of her own strain</u>; if she is the better bred she will dominate in the formation and colouring of the foal."

He does, however, make reference to the Saklawi Jedran strain in Volume I, p. 17, referring to its courage, and on page 37 about his favorite hack who was later a stallion. Another reference is to Rabdan Saklawi Djedran from Prince Ahmed (Vol. II, p. 16). He speaks of his beloved mare, Negma as "an old white mare," but not a word about the strain of her family. In speaking about pedigrees, he mentions nothing about names, only about class. It is noteworthy that he concludes Volume II by equating "strains" with "pedigree" as follows: "I am including in this volume a translation of one of the pedigree books of my ancestor Abbas Pasha the First. I have plenty of manuscripts of such pedigrees. This translation will give all horse-lovers interested an idea of what pedigrees there are in the Orient." This was a partial translation of the complete Abbas Pasha manuscript that Gülsün Sherif and I later translated and published as *The Abbas Pasha Manuscript*. The Prince's selections were garbled, out of context, and often erroneous. Although he included the Saklawi Jedran strain's history, he left out the one he was best known for (by Raswan and American breeders) as the Kuhaylan Jellabi.

Another noteworthy reference to this strain controversy is found in the *Travelers Rest Arabian Horse Catalog*, published in 1939 by Col. J. M. Dickinson, a student of pedigrees and a notable breeder in his time. Dickinson lumps Ghazieh-Jellabiet Feysul-Selma female lines together as follows [underlines, ed.]:

"The Jellabiet Feysul line is somewhat confused owing to lack of record of early sires crossed on the line. As in the early days in Poland, some of the best known Arab breeders in Egypt failed to keep the sort of record to which we are accustomed today; or possibly their stud books have been destroyed during the various wars. However that may be, it must be remembered that the most celebrated Egyptian studs were owned by men of vast resources who spared no expense to obtain the best available breeding stock; and they had no use for other than the purest Arab blood. The proprietors of the great Crabbet Stud in England, recognized as authorities, accepted Abbas Pacha and Ali Pacha stock as pure Arab substantially upon the same basis that we accept horses from Arabia. The chart of the Jellabiet Feysul line has been prepared in the light of the best information obtainable and is believed to be correct. But whether <u>there may be some unintended error in the correct placing of names according to generations</u>, there is no doubt that all the horses of this line are pure Arabs."

The Travelers Rest Arabs from the Manial Stud tracing to the Jellabiet Feysul were: *Nasr, *HH Mohamed Ali's Hamama, *HH Mohamed Ali's Hamida, *Zarife, *Aziza, and *Roda.

In describing *Zarife, Dickinson makes a noteworthy observation: "His dam, Mahroussa, has been described as a most beautiful mare. However that may be, she has produced outstanding stock. His second dam, Negma, bred also <u>Kafifan,</u> full brother to Mahroussa. Bred in Egypt by Prince Mohamed Ali and taken to Poland, where <u>he is registered in the Polish Arabian Stud book as a Seglawi Jedran,</u> Kafifan stood among the leading half-dozen sires of winning Arab racers in that country for four years."

With reference to *Nasr (Rabdan El Azrak x Bint Yamama), he comments on his beauty and athletic ability, but not the strain: "A distinguished racehorse in Egypt, *Nasr represents a very high development of the classic type of Arabian, uniting dignity with grace. His pictures have been widely used in numerous publications both in America and abroad, where he is also known as 'Manial,' to illustrate the type for which the great Arab studs in Egypt have become famous. He is not only an example of the beautiful Arab type in itself, but is the living proof of the soundness that goes with the type. Although he raced successfully at various distances, his most notable record was made at the short distance of 6 furlongs."

Dickinson's comments on the mares are relative to the classic qualities that have continued to breed on in this family: "*Aziza (Jamil x Negma) has never been exhibited in competition in the United States. By many she is considered to represent the ultimate in the classic type of Arab for which the great studs in Egypt became famous.... *Roda [Mansour x Negma] was the Reserve Champion mare in a strong class of 12 entries in the National Arabian Show of 1933....The head of *Roda has been described by one of the most distinguished breeders of Arabs in the United States as perfectly representative of the classic type.... *HH Mohamed Ali's Hamama [Kawkab x Mahroussa] was National Champion Mare and second in the 3 gaited saddle class in the National Arabian Show of 1934, ...under saddle she has never been defeated except by stallions." With reference to *HH Mohamed Ali's Hamida [*Nasr x Mahroussa], he states: "Her dam, Mahroussa, has been described by Mr. W. R. Brown, former President of the Arabian Horse Club and owner of the internationally celebrated Maynesboro Arabian Stud, as the most beautiful Arab mare he ever saw."

When W. R. Brown decided to import horses from Egypt, he sent the respected breeding authority and gifted animal trainer, Jack Humphrey, to Cairo in 1932 to make the selections for him. Humphrey wrote careful descriptions of each horse to Mr. Brown, but <u>he never mentioned strains</u>, only individuals. Humphrey had a wonderful eye, and his breeding assessments were "right on." It is worth noting that Prince Mohamed Ali told him the "real head quality had come down <u>from Bint Yemama</u> and seemed to be dominant in her line." This coincides with a comment by Wilfrid Blunt, who said the <u>grey Yamameh [Yemama]</u> was "a very splendid mare with the finest head in the world." Yamameh bred to Aziz (a celebrated Dahman Shahwan), got Mesaoud, the most famous Saklawi Jedran of his time.

Humphrey also wrote to Brown: "The mare Negma, which the Prince is keeping as the best he has ever bred and as the key to all that spells quality and true Arab, is by Dahman (El Azrak) out of Bint Yemama, and is so simply one generation down from the key stock. His second mare is Mahrussa [Mahroussa] out of Negma by Mabrouck [Mabrouk Manial]. Mabrouck is by Farhan out of Tarfa by Dahman. This makes Mahrussa a great-granddaughter on the sire's side and a granddaughter on the dam's side of Dahman. If one works out an inbreeding co-efficient, <u>Mahrussa is inbred to Dahman .125, which is quite strong</u>." [Underline, ed.]. He also mentions that Prince Mohamed Ali told him that the stallion Dahman was a small white horse of excellent substance and extreme quality, that he transmitted good Arab type and strong bone and that he was of medium head. Humphrey believed that <u>this one stallion</u> had been the keystone of all their breeding. This assessment is valuable in reflecting on subsequent generations of this family. When the *Maaroufa line was bred to the Dahman line (*Ansata Ibn Halima, El Hilal, and derivatives thereof) the resulting get were more like Negma/Mahroussa/*Roda/*Ansata Ibn Halima in type.

Other information about the strain descent of this family comes from the 1948 English edition of *The History of the Royal Agricultural Society's Stud of Authentic Arabian Horses* by Dr. Abdel Alim Ashoub, and refers to the Jellabi strain on page 126 under the heading of Aroussa, the full sister to Negma, as follows:

Horses of the Koheilan Jellabi Strain

Aroussa: p. 126

Bint Yamama (Koheila Jellabia) by Ali Pasha Cherif's Saklawi I, out of the same owner's Yamama, Koheila Jellabia. Abbas Pasha I's Aroussa's fourth dam was Waziria K. Jellabia; she was known as Jellabiet Feisal and was sold to Abbas Pasha I by Feisal Ibn El Turki Ibn Saud for £E 7000 in gold.

The 1932 certified pedigrees for the Babson and Brown imports from the Manial Stud of Prince Mohamed Ali [as recorded by the Arabian Horse Registry of America (A.H.R.A.) show Bint Yamama as by "Saklawi El Kebir" out of Yamama, of the Jallabi strain of Ali Pasha Sherif, from "Waziria, from Abbas Pasha's stud." This is in accord with the R.A.S. Herdbook noted above. Nothing is mentioned about the sires and dams between Jellabiet Feisal and Yamama in these records, or in those of the R.A.S. Herdbook. These intermediate ancestors seem to have been supplied later by Raswan in his *Raswan Index*. Raswan was enamored of the Jellabi strain and spent quality time in Egypt at all the stud farms. Although there were opportunities to mix up the different Yemama's of that era owned by the royal family, and the Blunts, one would think an error in strain name would have been caught by someone and corrected in the records. Lady Anne Blunt was extremely observant, very mindful of strains and families, and kept detailed records of her own as well as other breeders' horses. That she mentions nothing about the Jellabi line at Manial is noteworthy, and as stated earlier, gave rise to this whole question about the strain being Saklawi instead of Jellabi.

*Maaroufa in old age. (Ibn Rabdan x Mahroussa x Negma x Bint Yamama)

Other references regarding this Jellabi subject can be found in the *Al Khamsa Arabians*, Vol. I and II; the Michael Bowling article that appeared in *Visions Magazine*, October 1998; Carol Mulder's book, *Imported Foundation Stock of North American Arabian Horses*, Vol. I-III; *The Raswan Index*; and Colin Pearson's *The Arab Horse Families of Egypt*.

The reader should be aware, however, that there can be many spellings for the name Yemameh, (e.g. Yemama, Yamama, Yamamah, etc.). Transliteration from Arabic to English results in many different spellings; HOWEVER, it is only spelled one way when written in Arabic. The same case applies for the spelling of strain names (Kuhaylan, Koheilan, Keheilan, etc.). These different English spellings of one Arabic name, plus the fact that many horses of different strains were given the same names, and that they were bred by different breeders and at varying points in time, has caused many researchers to have nightmares.

The Negma/Mahroussa/*Maaroufa Family

Much about this Negma/Mahroussa/*Maaroufa family is referenced in the book, *The Royal Arabians of Egypt and the Stud of Henry B. Babson*. The general characteristics of this line varied. *Maaroufa, whom we saw in old age shortly before she died, was a very beautiful feminine mare of harmonious proportions, medium length of head, and somewhat longer back than her full brother, *Fadl, who appeared to be very masculine, shorter-coupled and heavier-bodied than his sister.

*Maaroufa produced 12 offspring by Fay El Dine: 3 stallions, 9 mares; and two by Fa-Serr (one stallion, one mare). When bred to the Saklawi Jedran stallion, Fay El Dine (x *Bint Serra), she generally produced more refined, feminine individuals (excepting El Maar). When twice bred to Fa-Serr, the black, shorter legged, heavy-bodied full brother to Fay El Dine, who bore no resemblance to him in type or conformation, they more resembled Fa-Serr (e.g., Faaris and Serroufa). It appears that through intensification of *Fadl in the breeding program, heavier-bodied, shorter-legged more masculine individuals resulted in what later became generally known as the "Babson type." The first generation Babson mares, particularly the grey Fay El Dine daughters, were more refined and fitted more closely in type to the R.A.S./E.A.O.-bred horses we saw in Egypt in 1959.

The *Maaroufa Family

There were several branches tracing to *Maaroufa, and Ansata acquired three.

El Maar (by Fay El Dine), a first generation mare, was in the original group we obtained on lease from Mr. Babson. She was the least typey in head of the Fay El Dine-*Maaroufa crosses and did not get the refinement of body or classic head the other full sisters had, being more *Fadl in structure. Nevertheless, she was a very handsome, strong-bodied, good-legged show-winning mare that structurally would appeal to any good horseman. She was in foal to Fabah when we leased her and produced Maarqada, a sickly foal who grew into a good broodmare of pleasing proportions, but again rather plain of head that one would have expected of that mating to the stallion, Fabah. Bred to *Ansata Ibn Halima, her foals were greatly improved in type and the overall harmony of body was retained. In future generations, this proved to be a very excellent producing line.

Faye Roufa (Ansata Abbas Pasha x Bint Fay Roufa), was a granddaughter of Fay Roufa (x *Maaroufa), one of the exquisite Fay El Dine daughters. A very beautiful chestnut-roan mare with some white markings, we purchased Faye Roufa directly from the Babson Farm. One of the best Babson mares at that time, she produced several good foals for Ansata before being sold.

Shahlika (by Ansata Halim Shah x Bint Hassema), provided a line to the Maar-Ree family tracing back to Maarou (Fay El Dine x *Maaroufa). She produced two excellent fillies by Ansata Hejazi; both had very pretty heads, were well-balanced, good bodied, and of show quality.

W. Sudan's Nefisa (by Ansata Ibn Sudan x Idaa), was leased by Ansata from Baraka Arabians. She traced back to Fada (Fadaan x Aaroufa by Fay El Dine). We bred her to Ansata Omar Halim as well as to Ansata Halim Shah, producing Ansata Omari and Baraka Sulima, both of good broodmare quality. Doubling the Dahman strain through *Ansata Ibn Halima, plus a line to *Bint Bint Sabbah through Fabah, resulted in the two sisters being shorter in head, more compact, and quite different than the type produced when Fada was bred to *Ibn Moniet El Nefous resulting in Fa Moniet, who in turn was bred to The Egyptian Prince. This intensification of Saklawi blood resulted in the tall, elegant and refined stallion, Prince Fa Moniet, among others, more reminiscent of the Ghazieh tail female family and the Fay El Dine type.

Although there is much to admire in this family through its various descendants, and it is one of our favorites, it was minimal in number at Ansata. Other breeders have developed significant straight Egyptian breeding programs based on the descending lines from *Maaroufa. Many beautiful individuals as well as athletes have been perpetuated through this family. Whether it is Saklawi or Jellabi, it makes no difference; there are adequate generations documented and known for certain traits regardless of "strain." It is what it is today, and there can be no question as to its value and quality.

Faye Roufa (Ansata Abbas Pasha x Bint Fay Roufa)

Comparison of Kuhaylan Jellabi and Saklawi Jedran Strain Chart
(see discussion in the introduction to this section for details)

According to Pearson/Mol Book & Mulder

Ghazieh (d.b.)
"Waziria"
Saklawia Jedrania of Abbas Pasha
| or
or   Bint Ghazieh x **Zobeyni (per Carol Mulder)**
|
└── Yamameh (dam of Mesaoud) x **Shueyman**
     Grey mare of AP Sherif

Bint Yamama x **Saklawi I**
Grey mare, d.o.b. unknown.
bred by Khedive Abbas Pasha
Hilmi II, by Saklawi I out of Yemameh.
Saklawia Jedrania
of Ibn Sudan

Negma x **Dahman El Azrak**

Mahroussa x **Mabrouk Manial**

*Maaroufa x **Ibn Rabdan**

According to R.A.S. Studbook & Raswan

Jellabiet Feysul (Waziria) Original Arab mare
from Ibn Khalifa of Bahreyn.
Sold from Feysul Ibn Turki Ibn Saud
to Abbas Pasha I

Bint Jellabiet Feysul
    x **a horse from Abbas Pasha I**
    (per Raswan)

Bint Bint Jellabiet (El Argaa)
    x **Waziri El Auwal**
    (per Raswan)

Yamama (c. 1890) x **Aziz**
color not given
(the Blunt's bay mare per Raswan)

Bint Yamama x **Saklawi I**
no color noted
Strain recorded as Koheila Jellabia, "by Ali
Pasha Sherif's Saklawi I, out of same owner's
Yamama, Koheila Jellabia."

Negma x **Dahman El Azrak**

Mahroussa x **Mabrouk Manial**

*Maaroufa x **Ibn Rabdan**

Bint Yamama with colt, Hilal

Negma

Mahroussa

*Maaroufa

Comparison Photos of descendants of Ghazieh and the Bint Yamama branch

Ghazieh Line

Mesaoud (Aziz x Yemameh)

Bint Helwa (Aziz x Helwa)

Serra (Sahab x Jemla)

Fay El Dine (*Fadl x *Bint Serra)

Fa Serr (*Fadl x *Bint Serra)

Bint Yamama branch

Negma (Dahman El Azrak x Bint Yamama)

*Roda (Mansour x Negma)

*Fadl (Ibn Rabdan x Mahroussa)

*Maaroufa (Ibn Rabdan x Mahroussa)

El Maar (Fay El Dine x *Maaroufa)

*Aziza (Gamil Manial x Negma)

Mahroussa (Mabrouk x Negma)

*Zarife (Ibn Samhan x Mahroussa)

Fay Negma (Fay El Dine x *Maaroufa)

*Maaroufa Family Branch Chart

*Maaroufa x **Ibn Rabdan**

El Maar   Aaroufa   Maarou in background   Fay Roufa

El Maar x **Fay El Dine**   Aaroufa x **Fay El Dine**   Maarou x **Fay El Dine**   Fay Roufa x **Fay El Dine**

Maarqada x **Fabah**   Fada x **Faddan**   Maar-Ree x **Fasaab**   Bint Fay Roufa x **Fabah**

Idaa x **Fabah**   Maar-Jumana x **Disaan**   Faye Roufa x **Ansata Abbas Pasha**

W.Sudans Nafisa x **Ansata Ibn Sudan**   Maar Malika x **Maar-Rab**

Bint Haseema x **Na Ibn Moniet**

Shahlika x **Ansata Halim Shah**

Maarqada

Idaa

W. Sudans Nafisa   Bint Haseema   Faye Roufa

The El Maar Family

Traditional Jellabia Lineage:
*Jellabiet Feysul/Bint Jellabiet Feysul-El Argaa/Yamamah -Blunt /Bint Yamama/Negma/Mahroussa/*Maaroufa*

DNA Implied Lineage:
*Ghazieh/Yamameh/BintYamama/Negma/Mahroussa/*Maaroufa*

El Maar #5693

Grey Mare; March 19, 1949 – April 5, 1965
Sire: Fay-El-Dine Dam: *Maaroufa
Bred and owned by Henry B. Babson, Dixon, Illinois.
Leased by Ansata Arabian Stud, Chickasha, Oklahoma. In foal to Fabah.

PRODUCE

Maarqada #253000 April 17, 1963
Grey mare by Fabah.

1963 and 1964 serviced by *Ansata Ibn Halima.
Died after giving birth to dead filly April 5, 1965.

Maaroufa/El Maar

El Maar

Maarqada #25300

Grey Mare. April 17, 1963 – February 17, 1976
Sire: Fabah   Dam: El Maar
Bred by Mr. Henry B. Babson, Dixon, Illinois.
Transferred to Mr. and Mrs. Donald L. Forbis.

PRODUCE

Ansata El Salim #41579 March 25, 1967
Grey stallion by *Ansata Ibn Halima.
Sold to Mr. and Mrs. Norton Grow, Prosser, Washington, September 1967.

Above and below, Maarqada

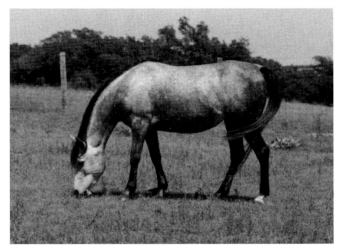
Ansata El Salim as a young horse

PRODUCE of Maarqada continued

Unnamed bay filly March 10, 1969
by *Ansata Ibn Halima. Died at birth.

Ansata El Arabi #070169 February 20, 1971
Grey stallion by *Ansata Ibn Halima. Sold to
Ned and Beth Shoell, Springville, Utah, May 1971.

Ansata Jellabia #094540 February 22, 1973
Chestnut mare by Ansata Ibn Sudan. Sold to
Cathy Downing, Chino, California.

Ansata Jasmin #105815 March 26, 1974
Chestnut mare by Ansata Ibn Sudan. Sold to
Donald and Joann Johnson, Lufkin, Texas,
January 1977.

Ansata Jellabia

Ansata El Arabi

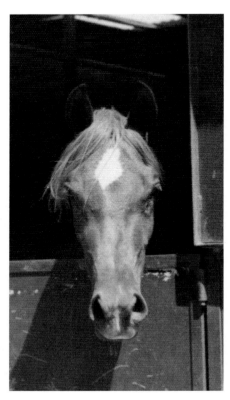

Ansata Jasmin

Faye Roufa

The Fay Roufa Family

Maaroufa/Fay Roufa/Bint Fay Roufa/Faye Roufa

Faye Roufa #100673
Roan Mare; May 3, 1973 – 1986
Sire: Ansata Abbas Pasha Dam: Bint Fay Roufa
Bred by Babson Arabian Horse Farm, Oak Brook, Illinois.
Purchased by Ansata from Babson.
Sold to Desert Arabian Bloodstock in 1985.

PRODUCE

Ansata Ali Abbas #239897 February 21, 1981
Grey stallion by *Ansata Ibn Halima. Sold to
Dorian Farms, Leverett, Massachusetts.

Ansata Faye Maara #298549 February 6, 1983
Grey mare by Ansata Ibn Shah.

Ansata El Karim #0311261 January 9, 1984
Chestnut stallion by Ansata Ibn Shah. Gelded
Transferred to John Sheppard, Mena, Arkansas.

Ansata Fay Jamila #365080 February 12, 1986
Roan mare by *Jamilll. Sold to Dr.
Hans Nagel. Exported to Germany 1990.

Above and below, Ansata Ali Abbas

Below, Ansata Fay Jamila

*Maaroufa/Fay Roufa/Bint Fay Roufa/Faye Roufa/Faye Maara

Ansata Faye Maara #298549
Grey Mare; February 6, 1983
Sire: Ansata Ibn Shah   Dam: Faye Roufa
Sold to Drs. Sparks and Berry, Jonesboro, Arkansas.

PRODUCE
DS Rawa #0387785 April 18, 1987
Grey mare by Ansata Halim Shah.
Owned by Drs. Sparks and Berry, Jonesboro, Arkansas.

The Maarou Family

*Maaroufa/Maarou/Maar-Ree/Maar Jumana/Maar Malika/Bint Hassema/Shahlika

Shahlika #0464101
Grey Mare; April 5, 1990
Sire: Ansata Halim Shah   Dam: Bint Hassema
Bred by Mr. Robert Waddell, Carlmel, Indiana.
Leased by Ansata from Robert Waddell and Ray and Karen Kasper.
Sold to Sheikh Abdullah Bin Nasir Al Thani,
Al Naif Stud, Doha, Qatar. Exported.

PRODUCE
Ansata Mahroussa #0540869 February 25, 1997
Grey mare by Ansata Hejazi. Sold to Usamah Al Kazemi
Ezzain Arabians, Kuwait. Exported to Kuwait.

1998 Bred to Ansata Hejazi.
Exported in foal. Foaled filly in Qatar.

Ansata Faye Maara

Shahlika

Ansata Mahroussa in Kuwait at Ezzain Arabians

The Aaroufa Family

*Maaroufa/Aaroufa/Fada/Idaa/W Sudan's Nafisa

■ W Sudan's Nafisa #0297880
Grey Mare; June 4, 1984
Sire: Ansata Ibn Sudan   Dam: Idaa
Leased 1992-1993 from Barbara Lewis, Baraka Arabians, Mena, Arkansas.

PRODUCE
Ansata Omari #0493701 April 10, 1993
Grey mare by Ansata Omar Halim.
Bred and owned by Ansata.

*Maaroufa/Aaroufa/Fada/Idaa/W Sudan's Nafisa/Ansata Omarai

Ansata Omari #0493701
Grey Mare; April 10, 1993
Sire: Ansata Omar Halim   Dam: W Sudan's Nafisa
Sold to H.R.H. Prince Khalid Bin Fahad Al Saud. Exported to Saudi Arabia.

PRODUCE
Ansata Omnia #0552502 April 24, 1998
Grey mare by Ansata Iemhotep
Sold to Jessie MacLean, New Brunswick, Canada.

Exported in foal to Ansata Sokar. Foaled colt in Saudi Arabia.

Above and below, W Sudans Nafisa

Ansata Omari

Ansata Omnia

Al Kuhaylat

Kuhaylan Rodan

A chestnut Arab mare. G.H. LaPorte. Musgrave Clark collection.

The Kuhaylan Rodan Strain

The Kuhaylan Rodan strain was maintained by the Shammar tribe with whom Abbas Pasha had particularly strong relations. When his emissaries set forth to learn the history of the strain, which appears in *The Abbas Pasha Manuscript* (pp. 532-533), they recorded:

> Nasser ibn-Funaykh of Al Shaalan, the Ra'i of the Stud was asked, "O Nasser, from where did Kuhayla Rodan pass to you? Do you or do you not know any history about her?
>
> And Nasser ibn Funaykh replied, "Aye, by Allah, I have knowledge. We have heard from our grandfathers that her origin is mubti - from the old times. And some time ago she passed from Beni Khaled, on the day of al Arban at Nejd, to al Sayah of Shammar, and from Sayah al Shammari she passed to Funaykh, my grandfathers, I, O Nasser. And my grandfathers said about the aforementioned mare, 'She is Kuhayla al 'Ajuz, and she is to be mated. And she is the most cherished {aziza} one of the stud, and the height {nadir} of the stud. And she is of the old stud {mubti al marabat}.

Abbas Pasha obtained some mares of this strain, but it was through the Blunt mare, Rodania, that this female line currently descends in Egypt. It has been prolific within the Crabbet breeding program, and has become influential in straight Egyptian breeding worldwide.

Rodania was a desert-bred mare imported from Arabia to Crabbet by Lady Anne and Wilfrid Blunt. She was a Kuhaylan Ajuz of the strain of Ibn Rodan and was foaled about 1869. Standing 14.2 high, she was chestnut; her near hind leg was white to above the fetlock, and she had a blaze to the mouth (with pink on the upper lip). She was deep of jowl, and her eyes showed a white oval like human eyes. Acclaimed to have extraordinary strength and style of going up to any weight, she had a somewhat uncertain temper. Wilfrid Blunt consigned Rodania to Class 2 of the Crabbet horses; however, Lady Anne apparently considered her Class 1 and wrote, "She may give sires to Class 2. She is to be covered by Seglawi's only." Many chestnuts are found through this lineage, but the degree of white isn't present in modern day Egyptian Rodania descendants in comparison to the Crabbet Rodanias who were bred "chestnut to chestnut" producing chestnuts with high white (e.g. *Rose of Sharon), because Lady Wentworth insisted on color, not liking greys until Skowronek came along.

There are two main Egyptian branches of this Kuhaylan Rodan family descending from the two mares Bint Risala (Razieh) and Bint Riyala (Risama) purchased by the Royal Agricultural Society in 1920 from Lady Wentworth and exported to Egypt along with several other horses.

The Bint Rissala descendants (see pp. 238-246, *Authentic Arabian Bloodstock*) were among the tallest and biggest horses at the E.A.O. The branch founded by the bay mare, Yashmak (by Sheikh El Arab), tended to plainer heads, but beautiful big black eyes. They were tall with relatively long backs, short croups, narrow and straight stifles, particularly noticeable in the bays such as *Rashad Ibn Nazeer (lacking in Shahloul blood), although her El Sareei daughter, Rashida, was more refined - again adding the Saklawi stallion, Shahloul, which combined well with this family. The addition of Nazeer to this family usually made a significant difference in quality, even though the dam's structure remained dominant. A descendant of this line, the magnificent chestnut mare, Sonbolah (*Serenity Sonbolah), by Sameh out of the Nazeer daughter, Bint Om El Saad, combined all the great Egyptian stallions within her pedigree. She was imported to Canada by Hansi and Bradford Heck, and shown to U.S. National Champion Mare in 1971. Those who saw her will always remember her as the epitome of classic beauty. Although she became a good broodmare, she never reproduced herself. Most of her foals were grey instead of chestnut. One wonders what the result would have been had she been bred to Ansata Ibn Sudan, her U.S. National Champion Stallion companion, who won his title the same year she did. It would have carried on the Blunt tradition of breeding the Rodan females to Saklawi stallions and matched two of the most exceptional Egyptian horses ever born.

The other Rissala branch through the grey Kateefa (by Shahloul) distinguished itself through two very refined Nazeer sons, Kaisoon (grey) and Alaa El Din (chestnut), both somewhat feminine in type. Kaisoon was not a very dominant breeding stallion, but he bred successfully in Europe. Alaa El Din, was known for his daughters in Egypt, and although his sons were not the strong breeders his daughters were, they were handsome and refined, good athletes, and were used in several programs (e.g. *Faleh, *Farazdac, *Rasheek). Although Alaa El Din was successful on the track, his elbows were somewhat tied-in (a trait found in some inbred lines to him today, with front legs set back instead of at the corners). He was somewhat narrow-chested, and he had an odd way of moving when viewed from behind. He was a tall elegant horse with a classic head; not very large, but very expressive, lively eyes, which he tended to pass on along with rather small hooves, and very good straight front legs.

Ansata did not breed any of the Rissala family in tail female, although it is incorporated through the stallion Alaa El Din who appears within some of the later Ansata pedigrees, primarily through *Jamilll, Ruminaja Ali, and Farres, and through certain mares. This line crossed well with Ansata Halim Shah blood, particularly within the Nagel program through Hanan.

The Bint Riyala branch, in general, when compared to the Risala branch was also of good size, but not as leggy; it tended to prettier and had somewhat shorter heads, shorter necks, and more compact bodies appearing more masculine and closer to the Dahman strain in type and overall harmony. The stallion Sameh also crossed extremely well with this line, producing many beautiful mares that were imported by Gleannloch and shown to multiple championships in America. The chestnut Sameh grandson, *Asadd, a U.S. National Champion at halter and in performance, was of this Riyala branch, as was the Ansata Imperial son, Imperial Imdal, a U.S. Reserve National Champion Stallion and World Champion Stallion at the Salon du Cheval, Paris; the latter also became a popular sire. Another marvelous champion of this line was the liver chestnut mare *Romanaa, a beautifully balanced most classic individual, and granddam of Imperial Imdal. Both greys and chestnuts were of equal beauty in this Riyala family.

The Bint Riyala Family

*Mawaheb (x Amani), was the first mare of this line represented at Ansata. She was on lease from Jean Kayser. Her line stemmed from Malaka and Nazeera, both very pretty headed compact mares, but tending to short necks. Amani by El Sareei was rather plain-headed with an adequate but somewhat short, heavy-based neck, wonderful body and good size. *Mawaheb took after her dam. However, when bred to *Ansata Ibn Halima, the heads were much more classic, the necks better shaped, but again not long enough.

Serrada (by Ansata El Naseri), also traces to Malaka. Serrada was a handsome and well-balanced mare, good length and shape of neck and excellent body. Ansata purchased her at a Pyramid Society Breeder's Sale to support that organization. When bred to Ansata Halim Shah, thus doubling the Ansata bloodlines, she produced the excellent colt, Ansata El Shakir, and became a successful broodmare.

Malika El Sheba (by Ansata Abu Sudan), also descended from Malaka. Her sire had been on lease to Smith Rock Arabians, and she was acquired by Ansata when they dispersed their herd. An attractive grey mare with pleasing head, of good size and overall balance, she was shown successfully and became a good broodmare.

Bint Faras Azali (by Ansata El Naseri), also traced back to Malaka. Her dam, Faras Azali, was a most classic individual, and Bint Faras Azali inherited beauty through both her sire and dam. A big white very pretty-headed mare, she was well-balanced, good-legged, and a good mover. Bint Faras Azali was acquired at age 11 by Ansata where she produced consistently good colts and fillies, several of which were exported.

The Rodania line had no major long-term impact on the Ansata program due to its late acquisition. However, many of its qualities are admirable, and it has bred on successfully for many breeders. The Rodania line has always excelled in all kinds of performance, and continues to do so.

At Ansata we leaned toward the Riyala branch finding it more consistently well-balanced and stronger-bodied with better toplines, hips and croups, than the Rissala family. While both families and sexes thereof have contributed much to Egyptian programs, the Rodan broodmares had greater impact than the stallions in Egypt.

Kuhaylan Rodan - Rodania Family Through Malaka

Ridaa

Riyala

Bint Bint Riyala

Rodania - Kuhaylan Ajuz Rodan
Rose of Sharon x **Hadban**
Ridaa x **Merzuk**
Riyala x ***Astraled**
Bint Riyala (Risama) x **Nadir**
Bint Bint Riyala x **Gamil Manial**
Malaka x **Kheir**

Malaka

Salomy

Nazeera

Nazeera x **Nazeer**
Amani x **El Sarie**
*Mawaheb x ***Ibn Hafiza**

Amani

Mamlouka

Mamlouka x **Nazeer**
Set El Wadi x **El Sarie**
*Watfa x **Nasralla**
Bint Set El Wadi x ***Ibn Hafiza**
Dal Sheba x **Dalul**
Malika El Sheba x **Ansata Abu Sudan**

*Salomy x **El Sarie**
Il Mandil x ***Moftakhar**
Moradil x ***Morafic**
Faras Azali x **Fehris**
Bint Faras Azali x **Ansata El Naseri**

**Watfa*

Dal Sheba

Faras Azali

Bint Faras Azali

KUHAYLAN RODAN STRAIN
Riyala-Malaka Family

The *Mawaheb Family

Rodania/*Rose of Sharon/Ridaa/Riyala/Bint Riyala/Bint Bint Riyala/ Malaka/Nazeera/Amani/*Mawaheb

*Mawaheb #149404
Grey Mare; February 2, 1969
Sire: *Ibn Hafiza   Dam: Amani
Bred by E.A.O., Cairo, Egypt
E.A.O. #650, Volume III
Imported on lease from Jean Kayser to Ansata, Lufkin, Texas, December 28, 1976.
Sold to Phil Petrulli and Stan Lefkowitz in June, 1984.

PRODUCE
Ansata Al Hasan #196943 February 9, 1979
Grey stallion by *Ansata Ibn Halima. Sold to John Blincoe, American Farms, Scottsdale, Arizona, May 1981.

Ansata Malima #220479 April 24, 1980
Grey mare by *Ansata Ibn Halima.

Ansata Nasr Halim #265727 March 31, 1982
Grey stallion by *Ansata Ibn Halima. Died.

Ansata El Riyal #311262 February 10, 1984
Grey stallion by Ansata Halim Shah. Gelded.

*Mawaheb

Ansata El Hassan

Ansata Malima

Ansata Nasr Halim

Ansata Mumtaza

Ansata Maliha

Serrada

Rodania/*Rose of Sharon/Ridaa/Riyala/Bint Riyala/Bint Bint Riyala/Malaka/Nazeera/Amani/*Mawaheb/Ansata Malima

Ansata Malima #0220479
(First named JKB Bint Mawaheb)
Grey Mare; April 24, 1980
Sire: *Ansata Ibn Halima   Dam: *Mawaheb
Bred by Jean Kayser, Ettelbruck, Luxembourg.
Leased to Ansata, Mena, Arkansas.
Sold to Paolo Gucci, Millfield Stables,
Yorktown, New York, March 1989.

PRODUCE

Ansata El Haddidi #328032 April 6, 1985
Grey stallion by *Jamilll. Gelded.
Transferred to Ken and Martha Young, Bethany, Oklahoma.

Ansata Maliha #0357274 March 15, 1986.
Grey mare by *Jamilll. Owned by Kayser. Sold
to Anita Wallin September 1991. Exported to Sweden.

El Malim #0396631 March 15, 1987.
Grey stallion by Ansata Halim Shah. Bred by
Jean Kayser and Mr. and Mrs. Donald L. Forbis.
Owned by Kayser. Leased to Dan and Joan Appleby,
Quitman, Texas.

Ansata Mumtaza #0424832 April 28, 1988.
Grey mare by Ansata Ibn Shah. Sold.

The *Watfa Family

Rodania/*Rose of Sharon/Ridaa/Riyala/Bint Riyala/Bint Bint Riyala/Malaka/Mamlouka/Set El Wadi/*Watfa/Bint Set El Wadi/Bint El Wadi/Serrada

Serrada #0299648
Chestnut Mare; January 5, 1984
Sire: Ansata El Naseri   Dam: Bint Elwadi
Bred by Willis or Imogene Flick, Miami, Florida.
Purchased June 5, 1985 at The Pyramid Society Breeders Sale
IV, Lexington, Kentucky by Ansata, Mena, Arkansas.
Sold to Luis Miguel Muzquiz, Mexico City, Mexico.

PRODUCE

ZF Zalima #0426550 June 28, 1988
Grey mare by Ansata Halim Shah. Bred
and owned by Mike Zicari, Suisuin, California.

Ansata El Shahkir #455946 May 29, 1990
Grey stallion by Ansata Halim Shah.
Owned by Luis Miguel Muzquiz, Mexico City, Mexico.

Above left and right Ansata El Shahkir

Rodania/*Rose of Sharon/Ridaa/Riyala/Bint Riyala/Bint Bint Riyala/
Malaka//Mamlouka/Set El Wadi/*Watfa/Bint Set El Wadi/Dal
Sheba/Malika El Sheba

Malika El Sheba #0322960
Grey Mare. January 6, 1985
Sire: Ansata Abu Sudan   Dam: Dal Sheba
Bred by Mr. Ernest M. Rawlings, Redmond, Oregon.
Transferred to Ansata, Mena, Arkansas.
Sold to Guilermo Henin, December 1990. Exported 1991 to
Argentina, South America.

PRODUCE

Ansata Tabitha #0430172 February 7, 1989
Grey mare by Ansata Halim Shah. Sold to
Guillermo Henin, December 1990. Exported
to Argentina, South America.

Ansata Al Malik #0456412 February 17, 1990.
Grey stallion by Ansata Halim Shah. Gelded.

Ansata Tabitha

The *Salomy Family

Rodania/*Rose of Sharon/Ridaa/Riyala/Bint Riyala/Bint Bint Riyala/ Malaka/*Salomy/Il Mandil/Moradil/Faras Azali/Bint Faras Azali

▪ Bint Faras Azali #0271233
Grey Mare;   March 21, 1983
Sire: Ansata El Naseri   Dam: Faras Azali
Purchased by Ansata from David and Martha Lucas, Bishopville, South Carolina 1994.

PRODUCE

Ansata Farah #529850   March 23, 1996
Grey filly by Ansata Hejazi. Transferred to David and Martha Lucas, October 1996.

Ansata Riyal #0542105   March 21, 1997
Grey stallion by Ansata Hejazi. Sold to Kuwait Arabian Horse Center
May 1998. Exported to Kuwait

Ansata Azali   #0552501   May 17, 1998
Grey mare by Ansata Iemhotep. Sold to Judith Wich, Orienta Arabians.
Exported to Germany, May 1999.

Ansata Riyadh #0568674   May 12, 1999
Grey stallion by Ansata Iemhotep. Sold to Judith Wich. Orienta Arabians. Exported to Germany.

Ansata Fayrouz #05788072 May 23, 2000
Grey filly by Ansata Hejazi. Sold 2002 to Deborah and Emil Nowak, Abraxas Arabians, Winchester, California.

Ansata Sinan Bey #589165 September 2, 2001
Grey colt by Ansata Sinan. Sold 2002 to Delyth Gamlin. Exported to England.

Above and below, Bint Faras Azali

Ansata Farah

*Ansata Riyal
EBC Champion
1998 Egyptian Event
Vesty photo.*

Ansata Azali

Ansata Riyadh

Ansata Sinan Bey

Above left and right, Ansata Fayrouz

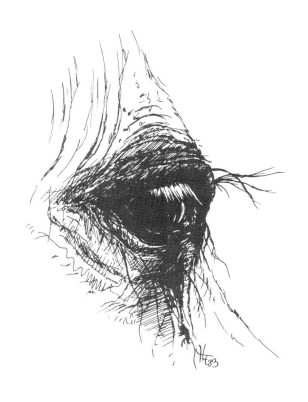

CHAPTER V
Perfecting the Dream

Dreamers! Ansata Halim Shah and young admirer, Nicole Bradbury. *Sparagowski photo*

CHAPTER V
Perfecting the Dream

The Horse Sublime

Parts near picture perfect,
the head, the neck, the legs,
a statue of ideals,
drawing applause and praise.

Many a horse may fit that mold,
yet rare is the horse to find,
with that quality transcendent,
that speaks of life sublime.

Of being past the present,
a look that says there's more,
than the weary path from birth to death,
a hint of life as pure.

Jean Hacklander

A TRIBUTE TO ANSATA HALIM SHAH

When a masterpiece has been acclaimed and something close to perfection has been achieved, it is impossible to describe it adequately. Words wouldn't come when I began to write this chapter about our beloved white stallion, Ansata Halim Shah. As if by Divine Providence there came a telephone call from Jean Hacklander who, as a teenager won Ansata's 20th anniversary essay contest in 1978. Now 24 years later, we reminisced and talked about her profession as a successful writer and film producer. Part of her essay is quoted in *Ansata Ibn Halima - The Gift*, and it seemed appropriate to ask her to write some prose or poetry for inclusion in this book. When her contribution came, it was the perfect introduction - and the perfect ending - to this chapter.

Every breeder at sometime is asked, "What is your favorite horse among all those you've bred?" In our case, this is a difficult question to answer because after forty five years of owning and breeding Egyptian horses there are many beloved and valued mares and stallions, each with their own individuality, quality, and beauty. Rarely, however, even in a long-term breeding program, does a single horse stand out within that century who could be called " a genetic giant," a horse that is phenotypically a magnificent archetype and genotypically dominant in passing on an unmistakable "look." Like a rare jewel, a great poet, an heroic warrior, or a supreme leader, he establishes a new criterion of excellence and achievement. Such a horse was Ansata Halim Shah - a breeder's dream come true. This tribute to him is just

Ansata Rosetta, dam of Ansata Halim Shah.
Sparagowski photo

Ansata Halim Shah shortly after arriving in Mena.
Forbis photo

*Ansata Ibn Halima, U.S. National Top Ten Stallion 1966, 67, 69. Sire of Ansata Halim Shah, U.S. National and international champions.
Sparagowski photo

Ansata Halim Shah as a young stallion.
Sparagowski photo

a sampling of current remembrances by some of his many admirers. And while some reminiscences may seem coincidental with others, they illustrate the consistent power and magnetism that attracted people to him. He epitomized the saying: "Behold the white horse. He is the mount of kings!"

BORN TO RULE

Ansata Halim Shah, whose name means *the kind ruler*, was born with the potential for greatness, but he was not so spectacular at birth as to be proclaimed "the dream horse." That distinctive honor he developed in due time. Like most of the Bukra family, he was slow to mature, but little by little one could see in him the balance, beauty and gentility of his sire, *Ansata Ibn Halima, the elegance, grace and spirit of his dam, Ansata Rosetta, the exceptional overall refinement of his granddam, *Ansata Bint Bukra, and the extreme pride and individuality of his dam's sire, the inbred Ansata Shah Zaman. His big black expressive eyes were a gift from all of his parents and grandparents, especially *Ansata Bint Mabrouka (Ansata Shah Zaman's dam and full sister to *Morafic). He carried like a distinctive banner a certain attitude of "royal" reserve and regal carriage enhanced by a highborne silken tail, a part of his persona. He was kind, yet somewhat haughty, a down-to-earth mischief-maker with a twinkle in his eye. He nipped at your arm - to see if he could get attention - but he never really bit down hard.

We knew he would become something special, however, when he tried to self-destruct. Most of the great ones do. As a yearling he jammed his foot through the bars of his stall door, hanging there with utter disdain until we were able to cut him free with bolt cutters and a saw. Fortunately he was intelligent, understood his precarious situation, and remained quiet, though curious and observant, throughout the entire procedure of extricating his leg. Luck was on our side and the injury to his pastern was minor. It was reminiscent of the time Ansata Ibn Sudan stuck his leg in the rim of a tire swing that hung from his favorite tree and had to be cut loose. Ibn Sudan became a U.S. National Champion Stallion, but Halim Shah was yet to make his mark.

Ansata Halim Shah
1980 - 1994

*Ansata Ibn Halima	Nazeer	Mansour	Gamil Manial	Saklawi II
				Dalal I (Al Zarka)
			Nafaa El Saghira	Meanagi Sebeli
				Nafaa El Khebira
		Bint Samiha	Kazmeyn	Sotamm II
				*Kasima (AHR 352)
			Samiha	Samhan
				Bint Hadba El Saghira
	Halima	Sheikh El Arab	Mansour	Gamil Manial
				Nafaa El Saghira
			Bint Sabah	Kazmeyn
				Sabah
		Ragia	Ibn Rabdan	Rabdan El Azrak
				Bint Gamila
			Farida	Saklawi II
				Nadra El Saghira
Ansata Rosetta	Ansata Shah Zaman	*Morafic	Nazeer	Mansour
				Bint Samiha
			Mabrouka	Sid Abouhom
				Moniet El Nefous
		*Ansata Bint Mabrouka	Nazeer	Mansour
				Bint Samiha
			Mabrouka	Sid Abouhom
				Moniet El Nefous
	*Ansata Bint Bukra	Nazeer	Mansour	Gamil Manial
				Nafaa El Saghira
			Bint Samiha	Kazmeyn
				Samiha
		Bukra	Shahloul	Ibn Rabdan
				Bint Radia
			Bint Sabah	Kazmeyn
				Sabah

Ansata Halim Shah shows off his beautiful head. Forbis photo

Being presented at the 1983 Egyptian Event, Lexington, Kentucky.

Winning U.S. Top Ten Futurity Stallion, U.S. National Horse Show, Albuquerque, New Mexico 1983.

In 1981 we left the Texas farm and moved to our new ranch in Mena, Arkansas. Nestled in the foothills of the Ouachita mountains, it was a haven of natural beauty. Several friends including Lisa Lacy and her father, Jarrell McCracken, whose Bentwood Farms became the largest Egyptian farm in the world, were on hand to meet the vans and record the arrival of *Ansata Ibn Halima and his family - among them the yearling Ansata Halim Shah. Lisa, now an international judge, vividly recalls that historical day: "My strongest memory of this regal heir to the Ansata throne is when he was a brash young colt leaping off of the van. I was videoing the arrival of the stallions, and as the younger horses disembarked .... someone said, 'The next one is Halim Shah.' I knew from the name that Judi liked this colt and had invested her future in him. He exuded quality and he knew it. He was comfortable in his own skin. He was a classic whose bearing and attitude reflected the ethereal beauty now well known to the whole world."

About this time, we made several trips abroad and went to visit Dr. Nagel's Katharinenhof Stud near Bremen, Germany. His handsome white stallion, Jamil, caught our eye. Short wedgy head, big black expressive eyes, tiny ears, long neck, well-balanced with clean bone and fine legs, he was an elegant individual, with a sire and dam of excellent quality and unique pedigree. He seemed a perfect match to incorporate with the Ansata program. Although we had met before many times, Nagel was away from the farm when we arrived; however, he was planning a trip to the States at which time we would have time to discuss our favorite subject - breeding Egyptian Arabian horses. This particular meeting would turn out to be an historical one for the breed.

"It was my first visit to Ansata in the early eighties, after Don and Judi had moved to Mena, Arkansas," Dr. Nagel recalls. "I had the opportunity to see most of the horses at that time, especially the colts and the stallions. Admiring Ansata's special look and quality, I was busy in my mind with the question, which horse might improve some features of our Arabians in Germany, and naturally, also in my own farm. My face apparently changed expression drastically when I saw the young Ansata Halim Shah coming out, and Judi apparently noticed this fact." He went on to reflect, "She asked me later about my impression of her horses, and my answer was like many of them: 'It is difficult to tell which is the one.' And Judi remarked, 'I can tell you which is the one, because I saw your reaction when Halim Shah appeared.' "

When two like-minded breeders get together and a warm friendship develops, good things are bound to happen. A cooperative agreement was struck whereby Jamil would come to America to be shown and stand at Ansata while Halim Shah would go to Europe after the Nationals, compete at the Salon du Cheval, and then stand at Nagel's farm.

SHOW HORSE

Jamil (registered in America as *Jamilll) arrived at Ansata in time to be exhibited in June on the Hall of Stallions along with Halim Shah at the 1983 Egyptian Event. Halim Shah, now age three, was eligible for the Stallion Futurity Class at the U.S. National Arabian Horse Show to be held in Albuquerque, New Mexico, that year. Richard Sanders, Ansata's resident trainer, qualified *Jamilll for the U.S. National Stallion classes easily winning several championships, and recalls: "The Egyptian Event had been very exciting that year, especially with *Jamilll's debut in this country. Even with the hoopla about *Jamilll, Halim Shah held his own and seemed to affect people with his impish regal air," Richard reminisces. "Looking back now, I recall many details as it seems like just yesterday. We had a great crew, headed by Ansata's Yvette Van Natta of Mena and Catherine Willing of Australia, and we arrived at U.S. Nationals on a quite a high note, having had a very successful summer of showing a group of straight Egyptian stallions that turned everybody's heads.

Through all of this, Halim Shah seemed to know he was on a mission. Showing never seemed to stress

Ribbons garnered by the 1983 show string trained by (l to r) Richard Sanders, groomed by Yvette Van Natta and handled by staff Pat Stodghill and James Cobb in 1983. Sparagowski photo

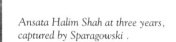

Ansata Halim Shah at three years, captured by Sparagowski.

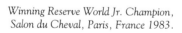

Winning Reserve World Jr. Champion, Salon du Cheval, Paris, France 1983.

him. Although he was a nipper, most of his energy came across as impish and playful. It looked good in the ring."

In those days classes were huge and highly competitive. Looking back, Richard recalls: "If I remember correctly, we had to compete against as many as 60 stallions, and as many as 3 cuts of 20 just to reach the Top Ten. The competition was exceptionally tough, but *Jamilll was chosen as a U.S. Top Ten Stallion, and Halim Shah was chosen a U.S. National Top Ten Futurity Stallion. By this time, he was getting considerable notice from breeders and Arabian enthusiasts worldwide."

Jody Cruz, young son of Dr. Felino Cruz, and now an important breeder, attended the 1983 Nationals with his dad. Both were great fans of *Ansata Ibn Halima, and two of his sons were being shown in the Futurity. Jody reminisces: "My favorite memory of Halim Shah was at the 1983 U.S. National Championship Show when he and our own El Halimaar were crowned U.S. National Top Ten Futurity colts. The two brothers made their sire proud that day." In speaking of Ansata Halim Shah, Jody reflects: "Anyone who saw him even once remembers him forever. His beauty was unmistakably 'Ansata' and his classic type unforgettable. What attracted me most was his exquisite, finely chiseled head, beautiful black eyes, and his memorable expression. Of course, one would expect such beauty from his pedigree. Surely *Ansata Ibn Halima was pleased when this son was born."

Nationals concluded on a high note, and we began immediate preparation to ship Halim Shah to Paris for the prestigious Salon du Cheval. Reflecting on that adventure, Richard remembers, "Showing in Europe was relatively new for Americans at that time. There were a few differences in the way horses were presented there. It was a challenge to get some hair to grow in Halim Shah's ears and on his muzzle in time to meet the show requirements that horses not be shaved or clipped. All the handlers were required to wear only white attire in the show ring. Additionally, much more value was put on how horses moved in hand. It was important that they move out freely with an especially fluid ground-covering trot."

Halim Shah was in peak show condition for the trip and, as Richard recalls: "We left Arkansas for Europe, flying from Houston to New York and then on to Paris after spending eight hours on the J.F.K. tarmac due to mechanical problems. However, Halim Shah did not seem to mind the delay. He sensed he was on a mission and it seemed to help keep him out of trouble. We traveled well."

THE EUROPEAN CONNECTION

Halim Shah's fame and National win had preceded him. Two other American horses had already made history at the show in 1980; the stallion, Ansata Abbas Pasha, winning Res. European Champion, and the mare, Fa Halima, also winning Res. European Champion. Both were by *Ansata Ibn Halima. Now another American invasion of his get was imminent. Reflecting on his adventure, Richard was unprepared for his first French connection and the reception committee of newspaper reporters and press that greeted him and his prized stallion as they landed in Paris. However, there was one minor detail that created anxious moments, Richard remembers. "Somehow, the arrangements to move Halim Shah from the plane across town to the Longchamp race track, where he would stay temporarily, were overlooked. My French, being very poor, forced me to draw pictures indicating the need for a ride. It was a bit traumatic, but we finally got transportation and, with a strange mare riding beside us, traveled across Paris in an open-topped two-horse trailer.

When we arrived at Port du Versailles, where the historical Salon du Cheval takes place, Mizan Taj Halim, a beautiful black-bay *Ansata Ibn Halima son from America was there with Midwest Training Stables. It was an exciting show with much class and an enthu-

At his new home in Germany, posing with Martin Nagel.
Van Lent photo

Out for a walk on Dr. Nagel's farm with Johann, his attendant.
Van Lent photo

Dr. Hans Nagel and the late Dr. Wolfgang Cranz, Director of Marbach State Stud.

Katharina Nagel leads Halim Shah from his stable.

siastic French crowd as well as spectators from around the world. When the cheering was over and the show had concluded, two *Ansata Ibn Halima sons stood out from the rest: Mizan Taj Halim was World Champion Stallion, and Ansata Halim Shah became Reserve World Junior Champion - the only horse to receive a perfect 20 in type. It was an unforgettable event."

After the show, Halim Shah was vanned to Dr. Nagel's farm in Bremen, where he was to make a legendary contribution to European breeding programs. "He remained for 18 months only, and any Egyptian horse, since that time, that does not carry his blood, ranks on a lower level," Dr. Nagel commented for this tribute. "This stallion had such a positive influence that even those breeders who were at first very skeptical about him, nearly all changed their minds, and until now, Halim Shah is still the top name in all the breeding farms concentrating on Egyptian blood." Nagel carefully selected the mares to be bred to him, not only his own, but those from the Babolna State Stud of Hungary, the Marbach State Stud of Germany, and from other private breeding farms.

In Europe, horse breeding is a serious, honorable profession, and much attention is paid to breeding horses who are conformationally sound, athletic, and have good character which allows them to be happy in performing their allotted tasks. Europeans value and live with classic art works of all kinds. Classic horses and classic horsemanship are part of their culture. The ideal classic Arabian horse, such as Halim Shah, was living confirmation to them that *beauty arises out of efficiency and significance and that beauty enhances function and produces or increases effectiveness.* At that time German breeders were still compelled to obtain official approval for their stallions when they wanted to use a horse for breeding. At the very strict German stallion licensing tests, Halim Shah was the only horse to receive a perfect 20 in type.

A letter from Stefan Walterman of the Maiworm Stud, who bred the celebrated Halim Shah son, Maysoun, explains how Halim Shah came to be even more appreciated by German breeders as time progressed:

"Dear Don, dear Judith,
Something unsurpassed happened during the 1987 German stallion licensing:
29 stallions out of 62 were licensed
4 were by Ansata Halim Shah
9 were rated "PREMIUM" (Same as 'Elite' in Tersk)
3 are by ANSATA HALIM SHAH.
The success of your stallion as a breeding horse is absolutely unique for the breeding of Arabian horses in Germany. In his article in the "Arabisch Pferde,", Dr. Wolfgang Cranz, the director of Marbach, mentioned that the four Halim Shah sons had been noticed as "superior in type, ideal neck, fine in the throatlatch, flowing movement with an extreme elasticity of the body."
We all would like to thank you for sending Halim Shah to Europe and we all congratulate you for the phenomenal success."

Dr. Nagel further comments regarding the German licensing system and Halim Shah's total record: "According to statistical figures covering 20 years of licensing, an Arabian stallion produces in all his life only one or two licensed sons. Ansata Halim Shah instead produced, in only 18 months during his short time in Germany, 9 stallions who were all approved by the governmentally guided committee."

Concludes Nagel, "His unprecedented success was definitely based on his own high breeding potential, but also on the careful selection of mares that might fit to such a horse. His striking appearance as a most classical Arabian and the quality of his ancestors listed in his pedigree, together made him a unique horse. He was also endowed with the rare ability to produce excellent sons and daughters as well."

Among the enthusiastic Egyptian horse breeders in Germany at that time were the late Dr. Tauschke and his wife, Cornelia, who recollects: "When I saw Ansata Halim Shah the first time, he was 3 years old and had just arrived at Dr. Nagel's Katharinenhof. His exotic head and balanced body impressed me very much. We were happy to have the chance to breed some mares with him. On this point we couldn't know that this young stallion would revolutionize the breeding in Europe. When the first foals were born we had the presentiment that we have something really special. Today when you see the group of El Thayeba horses, you can see Ansata Halim Shah in every horse."

Experience has taught us that most superior Egyptian breeding stallions have been grey or white. Shahloul, Mansour, Nazeer, Sheikh El Arab, *Ansata Ibn Halima, *Morafic, are legendary. All were grey. All appear in Ansata Halim Shah's pedigree. In fact, Ansata's three foundation yearlings were grey, as was *Ansata Bint Bukra, Halim Shah's granddam. And, Halim Shah proved to be a homozygous grey - siring only greys.

HOME AGAIN

When Halim Shah returned to America, he was quarantined for a month in Kentucky before traveling home. Not long after his arrival in Mena he appeared colicky, and Don rushed him to a vet clinic in Oklahoma where exploratory surgery was performed. No problems were found, and he came through the operation with his usual aplomb and serenity, surprising the vets with his rapid recovery and playful demeanor. He had never been sick a day in his life, and he was perfectly healthy again thereafter.

Having now matured into a majestic silvery-white stallion, Halim Shah made an indelible mark on people from all walks of life, many of whom were international breeders and judges. Among them was Peter Pond, who had known the Ansata horses since 1971, having imported to Australia the bay colt, Ansata El Shahwan. "I marveled at how successful a sire Halim Shah was when used in many of the Egyptian breeding programs in Europe and the Middle East," remarks Peter. " Halim Shah was truly one of the great Arabian sires of our time." Similar thoughts were echoed by two American judges, Bill Trapp and Jim Panek, who were dedicated to pinning classic type in the show ring. "Recollections of Ansata Halim Shah for me are of an exceptional individual with that elusive quality many good horses lack, a 'presence' about him, coupled with a lot of vitality," remarked Trapp. "He was the epitome of Arabian type, and, of course, had the bloodlines to 'breed on'." Panek's appraisal is similar: "If there is any one Arabian stallion that has re-awakened the world of Arabian horses to the classic Arabian, it is Ansata Halim Shah. He has perpetuated his likeness in the United States, Europe, and the Middle East. Through his offspring in each of these geographical areas, Ansata Halim Shah has become the epitome to which breeders strive. There is no doubt that as the history of the Arabian horse progresses world-wide, Ansata Halim Shah will be recognized as the standard bearer of the breed."

Writers were captivated by his extreme beauty, prideful demeanor and devilish attitude. Cynthia Culbertson, who spent a summer at Ansata in Lufkin, Texas, during her college years and is now a respected author, was one of many who fell prey to his charms: "Ansata Halim Shah, much like his legendary father, *Ansata Ibn Halima, seemed to have an inner sense of self," she muses. " In his presence, one sensed a sublime serenity and harmony. He could stand quietly, a certain stillness in his faraway gaze, and project his greatness without unnecessary fuss. He was exquisitely beautiful.... Halim Shah, like many of the greatest Arabian stallions, wasn't a tall horse, but he had extraordinary balance, and in motion could be bigger than life. He was the essence of Arabian type. Having known both his father and mother, I could see an influence from each of them in his appearance. Yet Halim Shah possessed his own unique look, beautifully distilled from

Ansata Shah Zaman, maternal grandsire of Halim Shah (left) . . .
Sparagowski photos

. . . and Ansata Halim Shah strike similar poses.

Enjoying a romp in his Mena paddock.

"The Fountain" - cover photo of Arabian Horse World, by Jerry Sparagowski, set a new standard of horse photography.

his ancestral elements. Particularly from eyes to ears, there was an expression that is typically 'Halim Shah' (and one sees it time and time again in his children and grandchildren around the world)."

Author Joe Ferriss, long the respected editor of *The Khamsat* magazine and a graphic designer, knows Ansata horses well. Looking back to when he first saw Halim Shah in the early 90's he reflects: "As I walked around him looking over his pearl-like surface, the experience was like that of viewing an Auguste Rodin sculpture. He was continuously harmonious when viewed from every angle. Knowing his pedigree, I was aware that he represented an amalgam of the key ingredients that embody the Ansata vision and type. Yet, he was not a horse that one could accurately say, 'he reminds me of....' As I looked into his deep lustrous eyes, I felt an immediate sense of quiet dignity conveying with certainty that this was a horse unto himself, a soul like no other. And at the same time, in his eyes I saw the essence of thousands of years of the indescribable character of the Arab horse. From that moment of eye contact, I was reminded what the Bedouin meant by 'a gift of Allah'."

Many stallions sire good quality in only one sex. Halim Shah sired equally good colts and fillies, and his sons and daughters began winning championships in the States: Ansata El Shahkir, Ansata Ramazan, Ansata Ranita, Ansata Majesta, Ansata Magnifica, Ansata Majesty, SM Carnaba, SM Amir Halim, SES Nafila, *Bint Amal, Ansata Selket, to name a few, as did others in Europe such as Maysoun, El Thay Ibn Halim Shah, El Thay Mansour, etc.

Breeders admired him. Halim Shah was in his element whether holding audience at home, or at the Pyramid Society's annual Egyptian Event where he was exhibited in the 80's and early 90's on the Hall of Stallions. Karen Henwood, breeder and owner of Sandybrook Farm, pays him a special tribute: "I was like a kid in a candy shop while walking through the 'Big Barn' at the Kentucky Horse Park where straight Egyptian stallions filled all the stalls. All the famous stallions that I had read about were there. What a thrill! Then, about half way around, I looked between the bars in a stall that I had yet to see the owner's or horses's name. I was stopped dead in my tracks. Before me was a vision I thought only possible in ancient paintings. I had never seen anything like this in the flesh. I looked at the nameplate on the door. This was Ansata Halim Shah. He was a true masterpiece, a true living piece of art. I thought he looked like a piece of the finest china; or perhaps a porcelain statue that had been made from the artist's imagination. Ansata Halim Shah was the most exotic horse I had ever seen. He was exotic type personified. He had an air of royalty about him. I believe he knew he was a King. His skin was like a fine, white silk. There was no other horse in that barn that gave me the feeling that Halim Shah did."

"At first we saw him in his huge double stall, his coat opalescent as if he were cast from fine bone china; huge dark eyes glanced briefly then he returned to the order of his day," Joan Skeels and Sue Burnham of Hope Farm in Arkansas memorably recall. "Regal, a king in his demeanor, were our thoughts of him. Once on a lead shank, he became animated; his tail a plume over his back claiming full attention from those observing him, yet reflecting kindness and docility. Standing there, each point of his construction fluidly created a model of perfection."

Photographers worldwide considered him a perfect model. Because of his ideal bone structure, he was very photogenic, and every photo session produced excellent pictures. Jerry Sparagowski, who began his horse photography career at Ansata in the early 70's, recorded Halim Shah throughout his life in America, and his "fountain" photograph - Halim Shah looking over his back at his high-plumed tail - set a new stage for Arabian horse photography. Harking back to those formative days, he remembers: "In the early seventies, Ansata Arabian Stud had three important stallions: *Ansata Ibn Halima, Ansata Ibn Sudan and Ansata

"Grey Monarch"
pencil sketch by Barbara Lewis
when Halim Shah was age 7.

Inspiration of artists, writers, photographers, poets. Sparagowski photo

German photographer Erwin Escher takes a shot...
"we never before saw Beauty like him!"

Shah Zaman. They were very different in style, and one could easily pick out which foals were by which stallion. As time moved on and Ansata's breeding program continued to mature, a distinctive 'look' started to appear. In my eye, that 'look' was Ansata Halim Shah. He had *Ansata Ibn Halima's beauty, Ansata Ibn Sudan's strength of elegance, and Ansata Shah Zaman's flair for life. Without question he was a magnificent stallion. To my mind, he had become the Ansata horse."

Internationally famous equine photographer Erwin Escher of Germany became an enthusiastic fan of Ansata Halim Shah when he and his wife, Annette, came to America in 1993. "We first visited Ansata Arabians, and this has had much influence on our future breeding program," Erwin remembers. "The first horse we saw there was Ansata Halim Shah! Don showed him to us outside his barn, and we were totally impressed by his incomparable Arabian beauty and charisma...white, dry in type, elegant - simply a real ARABIAN HORSE, we never before saw a Beauty like him! At this moment we knew - the only horses we wanted to buy this time could be some of his offspring."

Another young photographer, who also began his career at Ansata, was Gigi Grasso of Italy. He has since published numerous books portraying his artistic photographs of Arabian horses. "Ansata Halim Shah will always be a special horse to me because he was the very first Arabian that I had the chance to photograph." Gigi fondly remembers of his first visit to Ansata in the early 90's: "Even after I have had the privilege of photographing many of the world's most beautiful and famous Arabians, I still cannot forget Ansata Halim Shah. After I had seen several of his offspring, I said to myself, 'Just once in my life I would like to have a horse of this blood!' His charisma, his eyes, his fine skin, his splendid quality -all of this made Ansata Halim Shah simply spectacular."

From time immemorial, artists have been inspired by beautiful Arabian horses, and Halim Shah touched the soul of many in this era. Barbara Lewis of Baraka Farm and Studio in Mena portrayed him in a charming pencil sketch entitled "Grey Monarch" when he was seven years old. "On occasion, it would seem that God sends us an Arabian horse of such beauty and elegance that it becomes an ideal by which all others are to be measured," she reminisces. "They remain a goal for breeders to accomplish, and for lovers of perfection to own. Though always apart as an individual, they pass a special, ongoing quality to their children, and the generations to follow. Ansata Halim Shah was such an Arabian horse. He had that indefinable, ethereal quality. His demeanor made it known that he did not question his importance. If I were to compare him, it would have to be with the finest Meissen china....delicate, and refined, yet with a strength that is enduring....he was perfection in elegance. Oh, that God would grant us all such a lovely animal as we strive to share the obvious delight that He must feel when giving life to such perfection. It is the hope of this achievement that drives us forward against all odds."

When celebrated equine sculptress, Karen Kasper, was commissioned to portray him in bronze, she came to Ansata and spent several days studying his structure, measuring him, and observing his mischievous habits. She chose to depict him in a rearing position, a pose she didn't realize was also common to his sire, *Ansata Ibn Halima. "Although a stallion of exquisite refinement and beautiful type, I remember most his unique and expressive energy that projected a powerful charisma, attitude of superiority, and an inherent sense of mischief," she says in retrospect. " I was among those who sensed that he was profoundly important in some way beyond our present understanding, and I felt that Halim Shah himself sensed this as well. I sculpted him rearing proudly yet with a playful eye on his admirers, and this is how I will always remember him."

Sculptress Karen Kasper captured his playful essence, remarking: "I was among those who sensed that he was profoundly important in some way beyond our present understanding."

This is the attitude used by artists to depict the horses on which gods and heroes ride.
Xenophon

Father (*Ansata Ibn Halima), above, and son (Ansata Halim Shah) right. Sparagowski photos

RENAISSANCE STALLION

During the late 80's and early 90's Ansata was not only receiving guests from across America, but a renaissance in Arabian horse breeding was about to begin in the Arabian Gulf and the Arabian countries as the result of a visit to Ansata by Sheikh Abdulaziz Bin Khaled Bin Hamad Al Thani, a young man from the ruling Al Thani family of the State of Qatar. Arriving at Ansata with an entourage of friends, and minus their luggage, they uncomplainingly made themselves comfortable with clothes hastily purchased from Wal-Mart until their bags arrived two days later. Rain or shine, they walked through the pastures and looked carefully at every horse. The young Sheikh had the keen eye of his beloved falcons, and he missed nothing. He knew exactly what he wanted - whether or not it was for sale. Mindful of his extreme enthusiasm, remarkable pedigree knowledge, and his potential to lead the Arab world to new heights in breeding Arabian horses, we agreed to sell him two exceptional young mares: the beautiful chestnut, Ansata Splendora (by *Jamilll) and the white Ansata Majesta (by Ansata Halim Shah), the latter who became the first National Junior Mare Champion of Qatar and was subsequently National Senior Champion mare, among countless other wins. Trotting into the ring projecting all the majesty her name represented, those who saw her win that first championship will never forget her. She lit the torch that would carry Halim Shah's flame into the Arab world.

Sheikh Abdulaziz today recalls: "When I first saw Ansata Halim Shah in photographs I had high expectations. When a few months later visiting the Ansata Arabian Stud, I realized this stallion was still more special! Halim Shah has made such an impression on me that from that day I gladly have dedicated my own breeding program towards him, hoping that one day I will be honored with a colt foal possessing his incomparable Arabian horse type, quality and charm."

It wasn't long after that importation that the Heir Apparent and now current Emir of the State of Qatar, H.H. Sheikh Hamad Bin Khalifa Al Thani, wanted to purchase Halim Shah and a select group of mares for his Al Shaqab Stud. Much thought was given to the pros and cons of letting Halim Shah leave the States again, but it seemed a wonderful opportunity for him to become the "light on the hill" that could help spark an Arabian horse renaissance in the Arab world. Therefore, we made the decision to let him go. Breeder Christie Metz of Silver Maple Farm remembers her visit to Ansata just before that fateful decision: "I saw Halim Shah, along with Omar Halim, Manasseh and Prince Fa Moniet, the first time I visited your farm . I remember thinking how special it was to see these beautiful silver talismans of the breed and thinking that I wanted to breed Sahbine to Halim Shah. The next year you sold him and the rest is history. A lesson for me in follow your instincts immediately, do not wait." These sentiments were shared by many others who wished they had bred to him while they had the chance. Nevertheless, "the moving Finger writes, and having writ, moves on," and a special farewell party was held in the show barn, complete with carrot cake, to wish Halim Shah bon voyage. Soon thereafter Don and Dr. Craig Bullock, D.V.M., were accompanying him on his flight to the mystical faraway land of Qatar, a tiny peninsula jutting off Saudi Arabia into the turquoise waters of the Arabian Gulf.

The trip was undertaken without mishap. The long journey ended, Halim Shah deplaned into the bright white glare of the desert sands and was taken to his new home at Al Shaqab Stud. Sheikh Hamad Bin Ali Al Thani, the farm manager , recalls: "From the beginning of the foundation of Al Shaqab Stud, we selected the best stallions which became the basis of our present elite horses. I am remindful Ansata Halim Shah is one stallion that I chose and I have never forgotten him; he was the main pillar at Al Shaqab Stud."

Halim Shah at the Egyptian Event 1988 with Judi and Don Forbis. Polly Knoll photo

Sheikh Abdulaziz Bin Khaled Bin Hamad Al Thani of Qatar visits with his favorite Ansata stallion in Mena, Arkansas.

Ansata Majesta (Ansata Halim Shah x Ansata Malika) wins Qatar National Jr. Champion Mare; Sheikh Abdulaziz and Ansata's trainer, David Bradbury accepting the trophy.

Bon Voyage party and carrot cake in the show barn.

The harem awaits Ansata Halim Shah at Al Shaqab Stud, Doha, Qatar. Rik Van Lent Jr. photo

Posing in the Qatar desert for photographer Rik Van Lent, Jr.

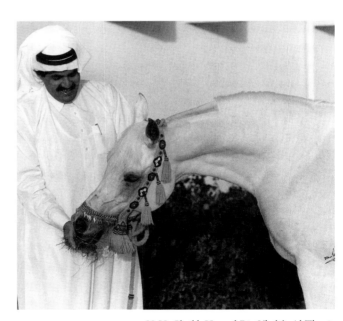

H.H. Sheikh Hamad Bin Khalifa Al Thani, Emir of Qatar, greets the new Al Shaqab herd sire. Rik Van Lent Jr. photo

A select harem of straight Egyptian mares, as well as Arabians of other bloodlines awaited him, and Halim Shah promptly went about his mission of getting them in foal. The following year promised to be an exciting one - to see the new foals and his first production in the Arab world. Then tragedy struck, an accident, an irreparable accident! And the beautiful white light went out, to shine only in reflection through his get and their descendants.

"Ansata Halim Shah was the best Egyptian horse; he comprised all the great things which all breeders love," Sheikh Hamad nostalgically reminisces. "Although it is not difficult to find a beautiful stallion, it is very difficult to get one with the specifications of Ansata Halim Shah. He incarnates beauty, magnificence, elegance and endurance and was much beloved. I have never seen any other like him nor one that had something of his fascination; he was a unique stallion that cannot come back.

I tried to find suitable words that can describe my lovely stallion, but I couldn't. He is gone, though he lives in my heart and mind. All breeders over the world from U.S. and Europe to the Middle East are missing him. I can say Al Shaqab Stud is very lucky because the blood of Ansata Halim Shah is running into new generations of stallions born here, and Al Aadeed Al Shaqab is one of them. He is not Ansata Halim Shah, but he takes most characteristics of his sire. Thus, I can say he is a unique stallion. I love him, just as I loved his sire."

As this is written, much has changed since the young men from Qatar first visited Ansata fourteen years ago. Halim Shah has become a legend, some of his get are living legends around the world, and Qatar has become a major force in the international Arabian horse community. Halim Shah influenced many lives, leaving each person with a treasured remembrance of him. The last artist with a camera to portray him was H.H. the Emir of Qatar's official photographer, Rik Van Lent, Jr., whose words and photo portrait opposite provide an everlasting image: "Halim Shah never died. He will always live on and his remarkable influence will always be there."

Horse of a Different Stature

By Jean Hacklander

Of all animals domesticated is not the horse the most magical? None can match the romance of the horse, on whose back came the founding of nations, by whose sweat came the building of civilizations, from whose beauty sprang the inspiration of fairytales.

Beguiled by this magic....I developed a keen appreciation – no, a reverence – for the rare horse whose beauty surpasses the sum of its parts. Like a masterpiece on the wall of the Louvre, image becomes essence; flesh and bone give way to inspiration and light; and viewers cannot look away. Is it the intelligent eyes? The graceful movement? The glistening hide? Elusive, yet it is there. A transcendent beauty, a glimpse into eternity. Such a horse may not be the most picture perfect assemblage of parts, but to be in the presence of such a horse is to be in some small way changed. To believe in the wonder and magic of life. To believe in the presence and triumph of good. To believe in the power of beauty to transform.

Ansata Halim Shah
1980-1994

Rik Van Lent Jr. photo

THE HALIM SHAH LOOK

A random collection of sons, daughters and grandchildren illustrating his consistency as a sire.

Ansata Hejazi (Ansata Halim Shah x Ansata Sudarra), Dahman Shahwan. Multi-Champion Stallion. Owned by Al Ajmal Stud, Kuwait. Sparagowski photo

Above, Al Aadeed Al Shaqab (Ansata Halim Shah x Sundar Alisayyah), Hadban Enzahi. Multi-Champion Stallion. Owned by Al Shaqab Stud, Qatar. Gigi Grasso photo

Salaa El Dine (Ansata Halim Shah x Hanan), Abeyyan Om Jurays. Sire of champions. Owned by Katharinenhof, Germany. Escher photo

Maysoun (Ansata Halim Shah x Maysouna), Saklawi Jedran. Champion Stallion. Owned by Maiworm Stud, Germany. Escher photo

Dalia Halim (Ansata Halim Shah x AK Bint Dalia II). Hadban Enzahi. Champion Stallion. Owned by Al Muntaha Stud, Germany. Escher photo

Assad (Ansata Halim Shah x Arussa). Abeyyan Om Jurays. A sire used by the Babolna State Stud, Hungary. Gigi Grasso photo.

The doubling of Ansata Halim Shah has been very successful as is demonstrated by two of his grandget:

Above, Ansata Malik Shah (Ansata Hejazi by Ansata Halim Shah x Ansata Malaka by Ansata Halim Shah). Double Halim Shah grandson. Dahman Shahwan. Champion Stallion. Owned by Ansata, USA. Sparagowski photo.

At right, Safir (Salaa El Dine by Ansata Halim Shah x Aisha by Ansata Halim Shah). Double Halim Shah grandson. Abeyyan Om Jurays. Owned by Al Rayyan, Qatar. Van Lent Jr. photo.

Ansata Sekhmet (left) and Ansata Selket (right), Dahma Shahwania; full sisters by Ansata Halim Shah x Ansata Samarra. Sekhmet is the dam of valued stallions and mares. She was exported to Nejd Stud, Saudi Arabia and died there. Ansata Selket is a multi-champion, now owned by Al Rayyan Farm, Qatar. Sparagowski photo.

Ansata Majesta (Ansata Halim Shah x Ansata Malika). Dahma Shahwania. Champion Mare. Still a Queen in the show ring at age 12. Owned by Al Rayyan Farm, Qatar, Judith Wich photo.

Ansata Malaha (Ansata Halim Shah x Ansata Malika). Dahma Shahwania. Champion Mare. Owned by Al Shaqab Stud, Qatar. Judith Wich photo.

Ansata Nefertiti (Ansata Halim Shah x Ansata Sudarra). Dahma Shahwania. Dam of Champion Ansata Nefer Isis. Owned by Ajmal Stud, Kuwait. Gigi Grasso photo.

Ansata Majesty (Ansata Halim Shah x Ansata Malika). Dahma Shahwania. Champion Mare. Owned by Ansata, USA. Sparagowski photo

Ansata Taya (Ansata Halim Shah x JKB Nehaya).
Saklawia Jedrania. Escher photo

Malaka (Ansata Halim Shah x Kis Mahiba).
Saklawia Jedrania. Escher photo

Ken Bint Bint Mahiba (Ansata Halim Shah x Bint Mahiba). Saklawia Jedrania. Escher photo

Amar (Ansata Halim Shah x Ameera). Abeyya Om Jurays.
Deceased. Owned by Al Rayyan, Qatar. Van Lent Jr. photo

Aischa (Ansata Halim Shah x Ghazala). Abeyya Om Jurays. Champion Mare.
Owned by Al Rayyan, Qatar. Van Lent Jr. photo

Ansata Manasseh (Ansata Halim Shah x Ansata Aliha). Dahman Shahwan. Champion Stallion. Owned by El Miladi Farm, U.S.A. Richard Bryant photo

El Thay Ibn Halim Shah (Ansata Halim Shah x Mahameh) Dahman Shahwan. Champion Stallion (deceased). Owned by El Thayeba, Germany. Escher photo

Lohim (Ansata Halim Shah x Lohelia) Hadban Enzahi. Champion Stallion. Owned by Classic Arabian Stud, France. Escher photo

Nigmh (Ansata Halim Shah x *Nasbah). Hadban Enzahi. Champion Stallion. Owned by W. Pontrello. Javan photo.

Al Aadeed Al Shaqab winning the highly coveted Supreme Stallion Championship at the Nation's Cup, Aachen, 2002. Ansata Halim Shah get and his descendants were major factors in the Qatari team taking home the Nation's Cup this year. Left to right: Judith Forbis, Sheikh Hamad Bin Ali Al Thani, Al Aadeed Al Shaqab, his trainer Ahmed, and Don Forbis. Bert Van Lent photo.

Nineteen years after Ansata Halim Shah won Res. World Jr. Champion Stallion at Paris in 1983, Al Aadeed Al Shaqab avenges his sire by winning the Supreme Champion Stallion title for Qatar at the Salon du Cheval World Championships, Paris, 2002, with 479 points, alleged to be the highest score ever recorded. Ansata Halim Shah bloodlines were also represented in the Reserve Champion Stallion, Hlayyil Ramadan, owned by H.R.H. Princess Alia Bint Hussein of Jordan, and Johara Al Naif, Junior Champion Mare, owned by Sheikh Abdullah Bin Nasir Al Thani of Qatar. Six of the eight 2002 World Champions carried Ansata bloodlines, a unique accomplishment in the annals of this historical show.

At right and below left: Al Aadeed Al Shaqab and handler Michael Byatt admiring his new trophy saddle that was presented to the Supreme Champion Stallion by H.R.H. King Mohamed IV of Morocco. Al Aadeed stands up for the judges, who rewarded his exceptional quality with 20's in type during his presentation by U.S trainer, Michael Byatt. Below right: Junior World Champion Filly, the double Halim Shah-bred Johara Al Naif, by the Halim Shah son, Ansata Shalim, and out of Al Johara, by Prince Fa Moniet x Ansata Majesta, Al Rayyan Farm's multi-champion Ansata Halim Shah daughter. At right are her trainer, Glenn Jacobs, and Amid Abdelhamid, manager of the Royal Stables of Morocco, presenting the trophy. Escher photos.

Fatima and groom, Victor Adam. Courtesy Bibliothequé Nationale.

CHAPTER VI

Sharing the Dream

"Horse photography has been elevated to a fine art."   Rik Van Lent, Jr. photo, Al Rayyan mares, Qatar.

CHAPTER VI
Sharing the Dream

*Composer, sculptor, painter, poet, prophet, sage
these are the makers of the afterworld, the architects of heaven.
The world is beautiful because they have lived; without them,
laboring humanity would perish.*

James Allen, *As A Man Thinketh*

INTRODUCTION

No man is an island, nor is any breeder sufficient unto himself. Sooner or later each must find another soul who can provide a missing piece to a breeding program puzzle. Without sharing knowledge and achievements, nothing prospers.

In the early days of our lifelong pursuit of the classic Egyptian Arabian horse, there were few photographs and little information available to make family bloodline comparisons and decisions. Each trip to Egypt was a gathering process. The Egyptian directors at the E.A.O., such as Dr. Marsafi, Dr. Khalil Soliman, and Dr. Ibrahim Zaghloul, generously shared their knowledge. They provided access to books and prints from the Prince Mohamed Ali library and to old scrapbooks and photos from the R.A.S./E.A.O. collection. They also exhibited great patience by allowing countless horses to be photographed - the grooms parading then posing them one by one, day after day, to be recorded on film for posterity. Such studies enabled us to see patterns in pedigrees emerge, as well as physical shapes and forms that persisted in certain families and how they were, or were not, overcome in the breeding program over time. While the camera can lie, it can also show subtle nuances that may be missed by the naked eye. Composing a picture takes absolute mental focus, and with such concentration, one sees features about a horse that might easily be overlooked by casual observation. Acquiring historical photographs and taking photographs of current horses are of inestimable reference value, as are herdbooks and notes kept by stud managers of important studs.

In the 1800's breeders relied on, and were inspired by, artists' conceptions of Arabian horses, such as those by Vernet, DeDreux, Adam, Gericault, Gros, and others. In the 1900's and in the new millennium, Arabian horse photography has been elevated to a fine art through the exceptional talents of certain professional Arabian horse photographers. Both artistic as well as ordinary photographs are valuable records, because all photographs allow us to look back as much as six to eight decades and see how dedicated breeders have made lemonade from many lemons over time.

As the craze for Egyptian horses gained momentum in the 60's, the thirst for knowledge grew. Data gathered in Egypt was shared through magazine articles and other books, but *Authentic Arabian Bloodstock* has become the constant companion and most-used reference guide by breeders of Egyptian horses worldwide. However, that book and this one could not have been written had others not shared in the Arabian horses' development by contributing information, photographs and additional support. Magazines, for example, provide useful current information on many horses and trends in the breeding world. However, they do not fulfill the purpose that books can for archival sources of information, sources that are ready references for the breeder. Books last. They are loving friends - companions along the way, day or night, at home or abroad. Reference books are family!

In using Chapter VI, readers should also refer to analyses of strains and families in Chapters II - IV, and to *Authentic Arabian Bloodstock* for additional definitive information. For visual comparisons and easy reference, this chapter is also arranged according to strains and families through the TAIL FEMALE LINE, as they have been recorded throughout the centuries. Each strain begins with a poem or short history excerpted from *The Abbas Pasha Manuscript,* the primary record about the strains from which modern Egyptian and other Arabian horses descend. A few matriarchs who influence families within the strain have been chosen for reference and their personal story value. Many other special mares have also founded or are founding dynasties of their own, as will be noted from their prominence throughout the succeeding pages. Regrettably, neither space nor time allows for a detailed coverage of each, even though many are as worthy.

The title of this chapter, "Sharing The Dream," aptly says what it means and means what it says. A very special thank you is extended to all those benefactors and patrons who made this particular chapter a sharing experience for everyone. Some may have incorporated or started with Ansata bloodlines in making their dreams come true; others may have been inspired or influenced by different breeders, past or present, having the same love for Egyptian Arabian horses. By whatever means, we are led to become breeders, in the universal scheme of things we are all sharing in the Divine Plan as interim custodians of our Creator's beloved creations.

"Sharing in the Divine Plan." Van Lent Jr. photo, Al Rayyan mares, Qatar.

The Dahman Shahwan Strain

Shahwan's Poem About Al Dahma

Oh men who are skilled among hunters
    know ye that I excel you.
My hunting grounds are the lofty heights
    attainable by none
      but my mare and I.

How swift is she
    of the beautiful countenance.
No horse can surpass her in flight
    save only her reflection.

Laudable is her size,
    she whose origin traces to al Abhar.
Look how marvelous her legs.
Behold her pleasing shape
    which attracts the eyes of all who see her.

Heavy burdens she bears lightly
    over distances near or far.
Broad is her forehead,
    her intelligence keen.

Prepared is she, day or night for the battle.
She has borne me and my armor,
    my brother and my son as one,
      while her heart beat heavily within her.
Such is her endurance and strong will.

Ye are aware that most hunters wound their prey,
    but know ye that mounted upon my swift mare,
      I stay my quarry drawing not its blood.

For my Dahma carries me to the realms
    accessible to none other.

Sharing the dream - Dahman

Matriarchs of the Sabah/Bukra Family: Ansata Delilah

		Nazeer	Mansour
	*Morafic		Bint Samiha
		Mabrouka	Sid Abouhom
Ansata Shah Zaman			Moniet El Nefous
		Nazeer	Mansour
	*Ansata Bint Mabrouka		Bint Samiha
		Mabrouka	Sid Abouhom
Ansata Delilah			Moniet El Nefous
		El Moez	Ibn Fayda
	Sameh		Bint Zareefa
		Samira (INS)	El Deree
*Ansata Bint Misr			El Samraa
		Nazeer	Mansour
	*Ansata Bint Bukra		Bint Samiha
		Bukra	Shahloul
			Bint Sabah

The grand matron, Ansata Delilah in old age at one of the Ansata seminars.

To those who share a love for horses of the Dahman strain, Shahwan's poem that introduces this section strikes a resonant chord as it did to the emissaries of Abbas Pasha I. Traversing all of Arabia to interview Bedouins in whose keeping the strain had been entrusted, they traced its history back to "our Lord Suleiman, may peace be upon him," thus assuring their august master of its veracity and antiquity. It became the Pasha's favorite strain, and he gave it first preference among all the strains listed in his priceless *Abbas Pasha Manuscript*. As popular now as it was in his time, the Dahman Shahwan strain became the foundation for other renowned stud farms over the past two centuries. While this strain was first in the heart of Abbas Pasha, it is also first in the hearts of breeders today, as evidenced by its listing as number one in *The Pyramid Society's Manual of Straight Egyptian Arabian Horses* with 5994 registrations tracing to the celebrated El Dahma of Ali Pasha Sherif as of the close of 2000.

To select just a few special matriarchs from this incredible and extensive strain is like choosing among bright stars in the firmament. Some sparkle more prominently at appointed times. Ansata Delilah and her descendants are among such luminaries in recent years, particularly among the Bukra family, thus her "star quality" is one of those worthy of special mention.

A regal attitude, a proud demeanor and an independent soul marked Delilah from the day she was born. Standing in the pasture apart from her mother, she would look off into the distance at something sensed but not seen. Her full sister, Ansata Damietta, who also established a significant family, was a more beautifully conformed show champion, but Delilah transformed herself into a charismatic charmer when she entered a show ring - and she too emerged a winner. It was an "I'm special attitude" that propelled her into stardom of her own. However, superior productivity as a broodmare far outweighed her looks. She wore nobility like a crown. She had lively big black "apple" eyes and a captivating expression. Her head was somewhat long and slightly heavy of foreface, her nostrils capable of great expansion. Her well-set lengthy but shapely ears were ever listening to zephyrs chasing over the hills. Her neck of good length, though somewhat straight, was attached to good withers and deep well-laid back shoulders. Her body and topline were commendable, and her croup was adorned by a high set tail which she carried with aplomb. She stood on four good legs, and she carried herself with pride, whether standing or in motion. During her lifetime, she produced 12 foals for Ansata; 7 sons, 5 daughters. Her first foal was the incomparable stallion, Ansata Imperial (by Ansata Ibn Sudan), a halter champion whose contribution as a sire of international magnitude is legendary. His full brother,

Ansata Delilah in old age.

the chestnut Ansata Amon Ra, was also a successful sire for his owners, the Spitzers. Her sons by *Ansata Ibn Halima (Ansata Emir Halim and Ansata Haji Halim) were handsome and good sires. Another son, Ansata Haji Jamil (by *Jamilll) was presented to Morocco's H.R.H. King Hassan II as a special gift from his loyal countrymen. Bred to *Ansata Ibn Halima, she produced Ansata Samantha and Ansata Bint Halima. Both were excellent broodmatrons, but Ansata Samantha's family has grown into one of the largest, most popular and successful worldwide.

Ansata Al Murtajiz (Ansata Hejazi x Ansata Samsara). J. Al Kazemi photo

Ansata Samantha as a filly

The "girl named Sam," as Ansata Samantha was affectionately called, captivated everyone with her huge black eyes, beautiful head, and silken coat of deep chestnut hue. A bit short on her correct legs, and somewhat long in body, she had an incredible topline and high-set tail that waved like a banner whenever she moved. Her disposition was as sweet as her countenance was beautiful, and she always appeared to be in a good humor. In her lifetime, she produced 10 foals: 7 fillies and 3 colts. Bred to *Jamilll came the bay Ansata Samarra, a broodmare supreme, who was eventually exported to Saudi Arabia after producing Ansata Selket and Ansata Sekhmet - two superior mares who also found their way to Al Rayyan in Qatar and the Nejd Stud in Saudi Arabia, respectively. Samarra's full sister, Ansata Samaria, was of "cover girl" quality as a foal and after producing the handsome colt, Ansata El Sharaf (by Ansata Halim Shah), she and eventually Samantha's next daughter, the graceful crowd-pleasing beauty, Ansata Sharifa (by Ansata Ibn Shah) were sent to Qatar to become foundation mares in the Al Rayyan program. Ansata El Salaam, Sharifa's good-looking chestnut brother, was exported to Europe where he won championships and made his name as a credible sire.

Bred to Prince Fa Moniet, Samantha produced the show winning grey mare, Ansata Samsara, who was sold to the Arabian Horse Center in Kuwait. Samsara tragically died before shipment; however, her excellent son and fine sire, Ansata Al Murtajiz, and her only daughter, Ansata Suleyma (both by Ansata Hejazi), now reside at Ezzain Arabians in Kuwait to carry on their mother's valued traits. Other commendable daughters out of Samantha by Prince Fa Moniet followed: Ansata Samai, who was acquired by Dr. Chess Hudson, Ansata Samiha, who is retained by Ansata, and Ansata Shalimar, now at the Ajmal Stud, Kuwait.

Ansata Samsara as a filly (Prince Fa Moniet x Ansata Samantha)

Ansata El Salaam (Ansata Ibn Shah x Ansata Samantha). Escher photo

Ansata Sokar (*Imperial Madheen x Ansata Samantha)

Sharing the dream - Dahman

The full sisters Ansata Samantha (left) and Ansata Bint Halima (right), by *Ansata Ibn Halima and out of Ansata Delilah. Sparagowski photo

Samantha's last foal, Ansata Sokar (by *Imperial Madheen) arrived wearing a bright chestnut coat and a fortuitous white star emblazoned on his forehead. A handsome and powerful stallion much resembling his dam, he has been a progenitor of all colors (in contrast to the many homozygous grey Ansata sires) and his progeny have been exported to Europe, the Gulf, and Saudi Arabia. Having delivered him, Samantha was scheduled for service by frozen semen to Ansata Halim Shah. However, her time was waning and it was not to be. During a hot summer evening in Kentucky, her light dimmed and then ceased to shine. Thankfully her influence multiplies with each succeeding generation, bringing to others the beauty she brought to Ansata.

Not to be overshadowed by her sister's early popularity, as often happens in life between competing sisters, Ansata Bint Halima brought forth some stellar progeny of her own: a spectacular filly by Ansata Ibn Shah, who was trampled in the pasture and had to be destroyed, then came two attractive Prince Fa Moniet daughters, the grey sisters, Ansata Neoma and Ansata Nawarra (the latter now at Al Rayyan in Qatar). Neoma and her daughters, Ansata Neamat, Ansata Nadra and Ansata Nariyah (by Ansata Hejazi) remain in America where they are founding families of their own. Bint Halima was eventually purchased by the Shaqab Stud in Qatar, but after a short time at her new home she passed on, leaving her attractive grey daughter, Halima Al Shaqab (by Ansata Manasseh) to carry on her legacy at the stud.

Another star in Delilah's crown was her daughter, Ansata Sudarra (by Ansata Abu Sudan). She was elegant and taller than her dam, and she had more size and scope than her half-sisters. Sudarra also inherited that special independent attitude and those big black soulful eyes that indicated she knew exactly who she was and where she stood in the universal scheme of things. She gave birth to two exceptionally beautiful mares, Ansata Nefara and Ansata Nefertiti (by Ansata Halim Shah) as well to as their world-famous brother, Ansata Hejazi, who became a herd sire for Ansata before his export to Kuwait where he reigns supreme at the Ajmal Stud. Nefara bred to Prince Fa Moniet became the dam of champions Ansata Sinan, Ansata Iemhotep, and Ansata Nafisa. Sinan was exported to Switzerland and won consistently in Europe, establishing a dynasty there. Ansata Iemhotep, a senior herd sire for Ansata, won U.S. National honors and numerous class A halter and performance championships, while sister Nafisa won in Egypt after being imported to the land of her ancestors by Omar Sakr. Ansata Nefertiti claimed to her broodmare credits the stunning European and Qatari champion Ansata Nefer Isis, the lovely Ansata Queen Nefr and the elegant Ansata Nefr - full sisters of special merit all sired by Prince Fa Moniet. And Sudarra produced two more admirable grey

Ansata Nawarra (Prince Fa Moniet x Ansata Bint Halima). Van Lent Jr. photo

The renowned Ansata Sudarra, (Ansata Abu Sudan x Ansata Delilah). Van Lent Jr. photo

Ansata Shalim (Ansata Halim Shah x Ansata Nefertari), Ansata Sudarra's grandson. Filsinger photo

Jamila Al Rayyan (Al Aadeed Al Shaqab x Ansata Selket), a Delilah great great granddaughter. Van Lent Jr. photo.

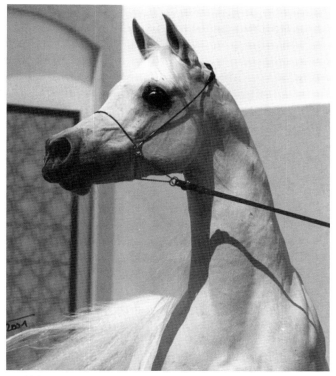

World-famous sire, Ansata Hejazi (Ansata Halim Shah x Ansata Sudarra). Filsinger photo

daughters, Ansata Nefertari (by Prince Fa Moniet), whose son Ansata Shalim is making a name as a sire in Qatar, and Ansata Nefri (by Ansata Manasseh), who is establishing herself as a good broodmare at Ansata. Having given Ansata these exceptional progeny, Sudarra joined her herdmates at Al Shaqab in Qatar where in the desert lands of her ancestors she spent her final days.

Another classic Delilah filly was Ansata Deborah (by Ansata Halim Shah), born in 1984. She brought a high-selling yearling price at the 1985 Classic Arabian Horse Sale in Scottsdale. Pretty Deborah was later purchased by Paolo Gucci. After the dispersal of his herd, she was sold to Al Shaqab Stud and exported to Qatar. Ansata Delilah's last foal, born in 1990, was the tall and distinctively stylish Ansata Sarai, (by Prince Fa Moniet), who resembles her most of all, yet is more beautiful of face. After producing her closest replica, Delilah decided she had earned time out from broodmare duties and would spend her final days at pasture. When her appointed time on earth had come she was laid to rest in Arkansas beside *Ansata Ibn Halima and other friends whose lives were testimony to the aphorism:

> *Beauty is Truth, Truth is Beauty*
> *that is all you know on earth*
> *and all you need to know.*
>
> Keats

Sharing the dream - Dahman

Combining Three Bukra Families
Rhodora - Delilah - Rosetta

Bint Rebecca
1993 Grey Mare
Sire: Salaa El Dine
(Ansata Halim Shah x Hanan)
Dam: Ansata Rebecca
(Ansata Abu Sudan x Ansata Rhodora
x *Ansata Bint Bukra)
Bred by Dr. Hans Nagel
Supreme Female Champion
at Wels International
B-Show, Austria.

Samara SQR
1999 Black Mare
Sire: Imperial Madori
(Imperial Madheen x Imperial Orianah)
Dam: Ansata Shahnaz (Prince Fa Moniet x
Ansata Sekhmet tracing to Ansata Delilah.)
Bred by Omar Sakr, Egypt

Gigi Grasso photos

NK Hafid Jamil
1996 Grey Stallion • Sire: Ibn Nejdy (Nejdy x Ghazala)
Dam: Helala (Salaa El Dine x Ansata Gloriana tracing to Ansata Rosetta)
Bred by Dr. Hans Nagel • Co-Owned by Al-Nakeeb and Dr. Nagel

AL-NAKEEB ARABIANS
Mr. Hassanain Al-Nakeeb • Gillingham Hall • Gillingham, Norfolk NR34 0ED, U.K.

Descending from Sabah through Bukra

Radames II
1992 Grey Stallion
Sire: Salaa El Dine
(Ansata Halim Shah x Hanan)
Dam: Ansata Rebecca
(Ansata Abu Sudan x Ansata Rhodora
x *Ansata Bint Bukra)
Bred by Dr. Hans J. Nagel, Germany

Escher photo

Rayhanah Bint Rebecca
2001 Grey Mare
Sire: Mangalee Ibn Shadwan
(Classic Shadwan x Bint Bint Mahiba)
Dam: Bint Rebecca
(Salaa El Dine x Ansata Rebecca)
Bred by Dirnhofer Family

BIRKHOF STUD
Dirnhofer Family • Birkhof 3 • Burglengenfeld, Germany 93133

Sharing the dream - Dahman

Descending from Bukra through Rhodora

Judy Perry photo

"Ibriz — the purest gold."

Ansata El Ibriz
1988 Grey Stallion
Sire: Ansata Halim Shah
(*Ansata Ibn Halima x
Ansata Rosetta)
Dam: Ansata Rebecca
(Ansata Abu Sudan x
Ansata Rhodora)

A Most Classic Arabian winner.

IBRIZ ARABIANS
Wilfrid & Nancy Bourque • 2271 Route 890 • Cornhill, New Brunswick, Canada E4Z 1B5

Sharing the dream - Dahman

Descending from Bukra through Bint Sudan

Gigi Grasso photo

Moniet El Dine
1994 Grey Stallion
Sire: Salaa El Dine
(Ansata Halim Shah x Hanan)
Dam: Ansata Moniet Sudan
(Prince Fa Moniet x
Ansata Bint Sudan)
Bred by Anne-Katrin Drewes

NEDSCHD ARAB
Dr. Ferdinand Denzinger & Gabriele Schuster • Hans-Watzlik-Straße 21 • D-92539 Schönsee • Germany

Sharing the dream - Dahman

Descending from Bukra through Rosetta

Ansata Ken Ranya
1992 Grey Mare
Sire: Salaa El Dine (Ansata Halim Shah x Hanan)
Dam: Ansata Prima Rose (*Jamilll x Ansata Rosetta)
Bred by Ansata Arabian Stud and Sylvie Eberhardt.

*"From fairest creatures we desire increase,
that thereby beauty's rose might never die."*
                                     Shakespeare

Jamal El Dine
2001 Grey Stallion
Bred by Dr. Hans Nagel
Sire: Hafid Jamil (Ibn Nejdi x Helala)
Dam: Ansata Ken Ranya

Ansata Ken Ranya and Jamal El Dine are co-owned by Dr. Hans Nagel and Mahmood Al Zubaid

Gigi Grasso photos

EL ADIYAT ARABIANS
Mahmood Al Zubaid • P.O. Box 270 • Surra, 45703 Kuwait

Sharing the dream - Dahman

Ansata Ken Rashik
1993 Grey Stallion
Full brother to Ansata Ken Ranya
Bred by Ansata Arabian Stud
and Sylvie Eberhardt

Cathy Childs photo

NILE BAY ARABIANS
Gary & Tracy Davis • 1241 Crabapple Road • Ozark, Missouri 65721

Sharing the dream - Dahman

Descending from Bukra through Rosetta

NK Yasmin
1999 Grey Mare
Bred by
Dr. Hans Nagel

Jaber Al Kazemi photo

	Salaa El Dine	Ansata Halim Shah
		Hanan
Adnan		Ghazal
	Ghazala	Hanan
NK YASMIN		Ansata Halim Shah
	Salaa El Dine	Hanan
Helala		*Jamilll
	Ansata Gloriana	Ansata Ghazia
		x Ansata Ghazala
		x Ansata Rosetta

Shall I compare thee
to a Summer's day?
Thou art more lovely
and more temperate.
   Shakespeare

EZZAIN ARABIANS
Usamah Zaid Al Kazemi • P.O. Box 30 • Safat 13001, Kuwait

Sharing the dream - Dahman

Descending from Bukra through Delilah

Ansata Al Murtajiz
1997 Grey Stallion
Sire: Ansata Hejazi (Ansata Halim Shah x Ansata Sudarra)
Dam: Ansata Samsara (Prince Fa Moniet x Ansata Samantha x Ansata Delilah)

Jaber Al Kazemi photos

"He is a presence to be known in darkness and in light."
Shelley

EZZAIN ARABIANS
Usamah Zaid Al Kazemi • P.O. Box 30 • Safat 13001, Kuwait

Sharing the dream - Dahman

Sparagowski photo

Descending from Bukra through Delilah

Ansata Serqit
1997 Grey Mare
Class A Res. Champion Mare &
Egyptian Event Res. Champion Futurity &
Multiple Top Tens

But what is higher beyond thought than thee?
More strange, more beautiful,
more smooth, more regal,
What is it? And to what shall I compare it?
It has a glory, and nought else can share it.
Keats

	Prince Fa Moniet	The Egyptian Prince
Ansata Iemhotep		Fa Moniet
	Ansata Nefara	Ansata Halim Shah
ANSATA SERQIT		Ansata Sudarra
	Ansata Halim Shah	*Ansata Ibn Halima
Ansata Selket		Ansata Rosetta
	Ansata Samarra	*Jamilll
		Ansata Samantha

CLAUDIA QUENTIN
Buenos Aires, Argentina

Descending from Bukra through Delilah

Van Lent Jr. photo

Ansata Samantha (right) and her granddaughter, **Ansata Selket**

Sparagowski photo

Ansata Osiron 1998 Grey Stallion
Full brother to Ansata Serqit
Now owned by Sheikha Sarah Al Sabah, Kuwait

This page is in celebration of
Ansata Samantha 1979-1999
(*Ansata Ibn Halima x Ansata Delilah)
and her exotic granddaughter,
Ansata Selket
(Ansata Halim Shah x Ansata Samarra)
now owned by Sheikh Abdulaziz Al Thani,
Al Rayyan Farm, Qatar,
and of their descendants around the world.

CLAUDIA QUENTIN
Buenos Aires, Argentina

Sharing the dream - Dahman

Descending from Bukra through Delilah

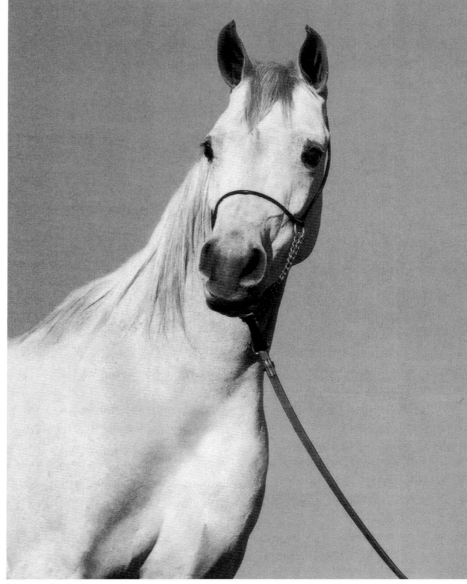
Sparagowski photo

Ansata Neamet
1998 Grey Mare

A strange stillness dwells in the eye of the horse, a composure that appears to regard the world…from the depths of a dream.
                Hans Heinrich Isenbart

		Ansata Halim Shah	*Ansata Ibn Halima
			Ansata Rosetta
	Ansata Hejazi		
		Ansata Sudarra	Ansata Abu Sudan
ANSATA NEAMET			Ansata Delilah
		Prince Fa Moniet	The Egyptian Prince
	Ansata Neoma		Fa Moniet
		Ansata Bint Halima	*Ansata Ibn Halima
			Ansata Delilah

CLAUDIA QUENTIN
Buenos Aires, Argentina

Sharing the dream - Dahman

Descending from Bukra through Delilah

Sparagowski photo

Ansata El Shahraf
1992 Grey Stallion
Jr. Champion Stallion
Class A, Canada;
Res. Champion Stallion, Class A;
Liberty Champion, Class A;
Top 5 Liberty - Egyptian Event.

		*Ansata Ibn Halima	Nazeer
	Ansata Halim Shah		Halima
		Ansata Rosetta	Ansata Shah Zaman
ANSATA EL SHAHRAF			*Ansata Bint Bukra
		*Jamilll	Madkour I
	Ansata Samaria		Hanan
		Ansata Samantha	*Ansata Ibn Halima
			Ansata Delilah

BARAKA FARM
Barbara & Tyrone Lewis • 5223 Hwy 71 South • Cove, Arkansas 71937

Sharing the dream - Dahman

Descending from Bukra through Damietta

Susan McAdoo photo

Zandai Ibn Omar
1994 Grey Stallion
Sire: Ansata Omar Halim
(*Ansata Ibn Halima x
Ansata Rosetta)
Dam: Ansata Dia Halima
(*Ansata Ibn Halima x
Ansata Damietta)
Bred by Dr. W. Hudson
Class A Halter Champion
and Egyptian Event
Top Ten Stallion in Halter

ZANDAI ARABIANS
Dr. William Hudson • 6010 Riley Road • Cumming, Georgia 30041

Sharing the dream - Dahman

Descending from Bukra through Regina and Damietta

Richard T. Bryant photos

"Who hath not proved how feebly words essay
To fix one spark of beauty's heavenly ray?
Who doth not feel, until his failing sight
Faints into dimness with its own delight,
His changing cheek, his sinking heart, confess
The might, the majesty of loveliness?

Lord Byron

Ansata Manasseh
1989 Grey Stallion
Sire: Ansata Halim Shah
(*Ansata Ibn Halima x Ansata Rosetta)
Dam: Ansata Aliha
(*Jamilll x Ansata Regina x *Ansata Bint Bukra)

Ansata Starletta
1989 Grey Mare
Sire: Ansata Halim Shah
(*Ansata Ibn Halima x Ansata Rosetta)
Dam: Ansata Marietta
(Ansata Ibn Sudan x Ansata Damietta)

EL MILADI ARABIAN STUD
Cynthia Culbertson • P.O. Box 928 • Vista Del Malpais Ranch • Carrizozo, New Mexico 88301

Sharing the dream - Dahman

Descending from Bukra through Delilah

Frasera Futura
1996 Grey Mare
Bred by La Frasera
Reserve Champion
Fillies - C Show

		Maysoun	Ansata Halim Shah
	Shahil		Maysouna
		Shahila	Ibn Galal
FRASERA FUTURA			Salima
		Prince Fa Moniet	The Egyptian Prince
	Ansata Nefer Isis		Fa Moniet
		Ansata Nefertiti	Ansata Halim Shah
			Ansata Sudarra
			x Ansata Delilah

LA FRASERA
Francesco & Serenella Santoro • Via Borneo • Roma, Italy 00144

Sharing the dream - Dahman

Frasera Ramses Shah
1999 Grey Stallion
Full brother to Frasera Futura
Bred by La Frasera

Gigi Grasso photos

LA FRASERA
Francesco & Serenella Santoro • Via Borneo • Roma, Italy 00144

Sharing the dream - Dahman

Descending from Bukra through Rosetta and Delilah

Ansata Helwa
1995 Grey Mare
Sire: Salaa El Dine
(Ansata Halim Shah x Hanan)
Dam: Ansata Gloriana (*Jamilll x Ansata Ghazia to Rosetta)
Bred by Ansata Arabian Stud.
Reserve Champion Mare,
Egypt Breeders Open Show.

C. Toischel photo

Ansata Nafisa
1995 Grey Mare
Sire: Prince Fa Moniet
(The Egyptian Prince x Fa Moniet)
Dam: Ansata Nefara
(Ansata Halim Shah x Ansata Sudarra)
Bred by Ansata Arabian Stud
Full sister to Ansata Sinan and Ansata Iemhotep.
Reserve Champion Mare, Egypt Breeders Open Show.
Dam of halter champions.

SAKR ARABIANS
Sherifa & Omar Sakr • 6 Hassan Sabry Street • Zamalek, Cairo, Egypt

Sharing the dream - Dahman

Descending from *Bint Bint Sabbah through Fay-Sabbah

Alidarra
1989 Grey Mare
Bred by Glorieta Ranch
Reserve Champion Mare, Egyptian Nationals
Produce of Dam, Egypt Breeders Open Show
Dam of champion mares and stallions in Egypt.

Escher photos

	Alidaar	Shaikh Al Badi	*Morafic
			*Bint Maisa El Saghira
ALIDARRA		Bint Magidaa	*Khofo
			*Magidaa
	Glorieta Serima	*Ansata Ibn Halima	Nazeer
			Halima
		Glorieta Shahlima	Ansata Shah Zaman
			Ansata Sabiha
			to Fay-Sabbah

Ghazal Sakr
1997 Grey Stallion
Sire: Shaheen
(El Haddiyah x Bint Bint Hamama)
Dam: Alidarra
Bred by Sakr Arabians
Junior Champion Colt,
Most Classic Head,
Egypt Breeders Open Show.

SAKR ARABIANS
Sherifa & Omar Sakr • 6 Hassan Sabry Street • Zamalek, Cairo, Egypt

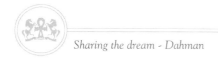

Sharing the dream - Dahman

Matriarchs of the Sabah / *Bint Bint Sabbah Family: Ansata Sabiha

			Mansour	Gamil Manial
		Nazeer		Nafaa El Saghira
			Bint Samiha	Kazmeyn
	*Ansata Ibn Halima			Samiha
			Sheikh El Arab	Mansour
		Halima		Bint Sabah
			Ragia	Ibn Rabdan
Ansata Sabiha				Farida
			*Fadl	Ibn Fayda
		Fabah		Bint Zareefa
			*Bint Bint Sabbah	Baiyad
	Sabrah			Bint Sabah
			Fa-Serr	*Fadl
		Serrasab		*Bint Serra
			Fay-Sabbah	Fay El Dine
				*Bint Bint Sabbah

Ansata Sabiha as a filly

The birth of Ansata Sabiha at the Ansata stud in Chickasha, Oklahoma, was a celebratory occasion. She was the first offspring of the black mare Sabrah, who had arrived at Ansata, sight unseen, from the Babson Farm in exchange for the stallion, Ansata Abbas Pasha. Unsuccessful at getting her granddam, Fay-Sabbah, in foal when she was on lease to Ansata, she was a hopeful choice to incorporate that line into the herd. Sabrah was serviced as a three- year-old to *Ansata Ibn Halima. A well-balanced sweet filly, with a pretty face, though far from exotic, resulted. She was named after Sabiha, the young girl who came daily to the Diyarbakir race track in Turkey to watch us training our horses. Her father, opposed to having a woman jockeying race horses in this patriarchal society, would often beat her for skipping out of her mud brick house against his orders, but she waited patiently until he was gone and then came anyway. Ansata Sabiha had the stuff great broodmares of her family are made of. Her half-brother, Ansata El Reyhan (by Ansata Ibn Sudan), became a U.S. National performance champion, and her full sister, Fa Halima, became the Res. European Champion at the Salon du Cheval and the 1981 U.S. National Champion Mare; thus both made a name in the show ring for their family. Regrettably, Fa Halima passed away after having produced only six foals; 5 colts (Shaikh Halim, Al Khalid, AK Nazaar, AK El Maleek and Sar Fa Rafic) and a filly who died young. Sweet Sabiha, a Dam of Significance, left a legacy of champions through both sexes.

Sabiha's twelve progeny were divided evenly: 6 sons, 6 daughters. She was first serviced to *Morafic and produced Ansata El Emir, a handsome grey colt of good quality who became a champion and head sire for Arabesque Farm. Next came her magnificent and charismatic chestnut daughter, Ansata Sabrina (by Ansata Ibn Sudan) a U.S. Top Five Futurity winner and performance champion. Sabrina in turn produced a beautiful chestnut daughter Ansata Splendora (by *Jamilll) who was purchased by Sheikh Abdulaziz for his Al Rayyan Farm at the time he bought Ansata Majesta. Splendora bred on well, but unfortunately died young. Sabiha also produced the U.S. National Futurity Champion Stallion, Ansata Shah Zam (by Ansata Shah Zaman), who became an influential sire of the 70's and 80's for Downing Arabians and did much to popularize Egyptian horses in California and on the West Coast. Mike and Kiki Case of Glorieta Ranch in Texas had been watching Sabiha and her production, and knowing her daughter, Sabrina, could be retained by Ansata

Ansata Sabiha in her prime. Polly Knoll photo.

Sharing the dream - Dahman

*Far left, Glorieta Serima (*Ansata Ibn Halima x Glorieta Shahlima x Ansata Sabiha), left Glorietasayonaara (Ansata Abu Nazeer x Glorieta Shahlima) Johnny Johnston photo.*

*Three generations of the Sabiha family: left, Ansata Splendora (*Jamilll x Ansata Sabrina x Ansata Sabiha), center, RN Safia (Ali Kamal x Ansata Splendora) and right, Naif Al Rayyan (Al Aadeed Al Shaqab x RN Safia). L/R: Sparagowski, Van Lent and Filsinger photos.*

as her replacement, they asked to buy Sabiha for their foundation broodmare. Their choice brought them good fortune particularly through her daughters Glorieta Shahlana, Glorieta Sabieta, Glorieta Sahlima, Glorieta Saafrana and Glorieta Sabdana.

When Glorieta dispersed their stud, Sabiha's granddaughter, Glorieta Serima (by *Ansata Ibn Halima) became the object of a bidding war between Paolo Gucci and another buyer, Bob Hutton, the latter paying $155,000, just $5000 more than Gucci had authorized Ansata to bid in his absence. Considering that problems developed and the Gucci herd was dispersed under adverse circumstances, it was a stroke of luck he did not get her. Serima was in foal to Alidaar and produced Alidarra, who has become one of the most important broodmares in Egypt today and a foundation mare of the Sakr Stud.

A valuable line remaining in America is that of Sabiha's descendants through Glorietasayonaara (by Ansata Abu Nazeer x Glorieta Shahlima) a line which Silver Maple has used as a foundation so successfully. Meanwhile, at age 20, Glorieta Saafrana (by Ansata Abu Nazeer) made her way to the Ajmal Stud in Kuwait and has been blessed with foals at her new home. Could little Sabiha in Turkey witness what her namesake has left to posterity, she would be very proud indeed.

Travel.
Thou wilt find a friend
in the place of him thou leavest;
and fatigue thyself;
for by labour are the sweets of life obtained.
            Tales of the Arabian Nights

Sharing the dream - Dahman

Descending from *Bint Bint Sabbah through Fay-Sabbah

Polly Knoll photo

Glorietasayonaara
1983 Grey Mare
Sire: Ansata Abu Nazeer
(*Ansata Ibn Halima x *Ansata Bint Zaafarana)
Dam: Glorieta Shalima (Ansata Shah Zaman x Ansata Sabiha x Sabrah x Serrasab x Fay-Sabbah)
Bred by Glorieta Ranch, Texas
1991 Egyptian Event Aged Mares Champion;
Region XVIII Top Five;
Park Champion Class A.
Dam of Multi-Champions.

Gigi Grasso photo

Sahbine • 1990 Grey Mare   *Polly Knoll photo*
Sire: Ruminaja Ali (Shaikh Al Badi x Bint Magidaa)
Dam: Glorietasayonaara
Bred by Dr. Charles & Judy Jones, Atallah Arabians
Third, Egyptian Event Two Year Old Futurity;
Top Five Yearling Region XI

Shaboura • 1998 Grey Mare
Sire: PVA Karim (Imperial Imdal x BKA Rakiisah)
Dam: Sahbine
Bred by Silver Maple Farm
Egyptian Event Top Ten

SILVER MAPLE FARM, INC.
Christie & Henry Metz • 5125 Happy Canyon Road • Santa Ynez, California 93460

Sharing the dream - Dahman

Descending from *Bint Bint Sabbah through Fay-Sabbah

Ali Saroukh
1994 Grey Stallion
Sire: Ruminaja Ali
(Shaikh Al Badi x Bint Magidaa)
Dam: Glorietasayonaara
Bred by Silver Maple Farm, Inc.
Res. Champion Egyptian Event Yearling ;
Region X Res. Champion Yearling,
Res. Champion Egyptian Event Two Year Old Colts.
Sire of Champions.

When we contemplate Nature it suggests to us that Beauty is the vestment and expression of the Creator; that He made the pursuit of the Beautiful the Supreme Law of the Universe.
                                            van Loon

Gigi Grasso photos

A collection of Silver Maple's beautiful foundation mares.
(Left to right)
Shaboura
Glorietasayonaara
Sahbine

SILVER MAPLE FARM, INC.
Christie & Henry Metz • 5125 Happy Canyon Road • Santa Ynez, California 93460

Sharing the dream - Dahman

Descending from *Bint Bint Sabbah through Fay-Sabbah

Samura
(Pictured with her filly **Bint Bint Sayo**
by Ali Saroukh)
1995 Bay Mare
Sire: Anaza El Farid
(Ruminaja Ali x Bint Deena)
Dam: Glorietasayonaara
Bred by Silver Maple Farm, Inc.
Egyptian Event Top Five;
Class A Halter Championships

Darryl photos

Suleiman SMF
2001 Bay Stallion
Sire: Ali Saroukh
Dam: Samura
Bred by Silver Maple Farm, Inc.
Egyptian Event Top Five Yearling Colts

Sarjah SMF
2000 Bay Stallion
Full brother to Suleiman SMF
Bred by Silver Maple Farm, Inc.
Region XVIII Top Five,
Egyptian Event Res. Champion
Yearling Colts, & Top Ten Two Year Old Colts

SILVER MAPLE FARM, INC.
Christie & Henry Metz • 5125 Happy Canyon Road • Santa Ynez, California 93460

Sharing the dream - Dahman

Descending from *Bint Bint Sabbah through Fay-Sabbah

Ansata Halisha
1998 Grey Mare
Bred by Ansata Arabian Stud

	Prince Fa Moniet	The Egyptian Prince
Ansata Iemhotep		Fa Moniet
	Ansata Nefara	Ansata Halim Shah
ANSATA HALISHA		Ansata Sudarra
	Ansata Halim Shah	*Ansata Ibn Halima
Sundar Alisha		Ansata Rosetta
	BKA Alisabbah	Ruminaja Ali
		Glorieta Sabdana
		x Ansata Sabiha x Sabrah

*Nature is more beautiful than art;
and in a living creature freedom of
movement makes nature more beautiful.*
Buffon

HALYPA - AL DUHAYM
Annalisa & Pasquale Monticelli Berloco • Via Luca Spaziante No. 20 • Altamura - Bari, Italy 70022

Sharing the dream - Dahman

Descending from *Bint Bint Sabbah through Fay-Sabbah and descending from Farida through Inas

Almaas Al Sabah
(Pictured with her colt
Azda Hani Sokari by Ansata Sokar)
1991 Grey Mare
Sire: Prince Fa Moniet
(The Egyptian Prince x Fa Moniet)
Dam: Ansata Asmarra
(Ansata Shah Zaman x Ansata Sabrina
x Ansata Sabiha x Sabrah)
Bred by Evergreen Arabians, Tennessee

Bahiya Al Nour
1998 Grey Mare
Sire: Imperial Al Kamar (El Hilal x Imperial Sonbesjul)
Dam: Almaas Al Sabah
Bred by Kimberly McGill, Florida

***Fortun** 1993 Grey Stallion
Sire: Ibn Bint Inas
(Ansata Halim Shah x Bint Inas)
Dam: Ruala Farha
(Golmoud el Ahmar x Ruala Farida x
Isis x Bint Inas)
Bred by Dr. Erwin Piduch, Germany

This page is in memory of Dr. Erwin Piduch

AZDA ARABIANS
Kimberly McGill • P.O. Box 522 • Hobe Sound, Florida 33475

Sharing the dream - Dahman

Matriarchs of the *Bint Bint Sabbah family:
Falima – progenitress of the Nile family

			Mansour	Gamil Manial
		Nazeer		Nafaa El Saghira
			Bint Samiha	Kazmeyn
	*Ansata Ibn Halima			Samiha
			Sheikh El Arab	Mansour
		Halima		Bint Sabah
			Ragia	Ibn Rabdan
Falima				Farida
			Ibn Rabdan	Rabdan El Azrak
		*Fadl		Bint Gamila
			Mahroussa	Mabrouk
	Fa-Habba			Negma
			Baiyad	Mabrouk
		*Bint Bint Sabbah		Bint Gamila
			Bint Sabah	Kazmeyn
				Sabah

When Fa-Habba came to Ansata on lease from the Babson Farm, she was the last choice of the four mares selected. She turned out to be the best choice. Bred to *Ansata Ibn Halima, she produced the much revered bay mare, Falima - an exemplary Dahmah Shahwania in type; beautiful head, large lively eyes, well-set ears, average neck and good mitbah, strong body, topline and high tail set, clean legs and superb movement - a characteristic passed on continually to the Nile family line that she established.

Falima was a true matriarch. She was the herd boss, but a kindly one - always rushing to the rescue with a worried look on her face whenever another member of the herd was in trouble, or a foal needed attention. Out of her seven progeny, 3 stallions and 4 mares, daughters were her most valued assets. Her grey filly, Ansata Nile Jewel (by *Morafic) was a charmer - very classic, petite, with a short beautiful head, well-shaped neck and good body and topline. Although one of her eyes was kicked out as a foal, she was of such quality that when she entered the show ring wearing a black sequinned eye-patch, she won. Jewel went on to become a valued foundation mare for the Hacklander's Hadaya Stud and from her came the exceptional grey mare Ansata Nile Gift (by Ansata Ibn Sudan). Two of her daughters are the beautiful GiftoftheNile (by Ruminaja Bahjat) and Ansata White Nile (by Prince Fa Moniet), the latter a champion, and both credits to the Nile family as broodmares.

Ansata Nile Queen (by Ansata Ibn Sudan), was an ethereal bay filly with a shimmering white star that illumined her beautiful face. Even more attractive than her mother, and with a longer neck, she was the benchmark by which the Nile family

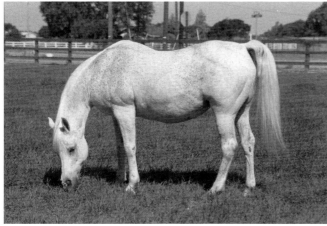

Ansata Nile Gift (Ansata Ibn Sudan x Ansata Nile Jewel x Falima)

Ansata Nile Dream (Ansata Ibn Sudan x Ansata Nile Queen)

was measured. Shown successfully as a filly, she went on to become a valued broodmare for Ansata, her descendants becoming first-class foundation stock in Qatar and Saudi Arabia. Full sister, also an exquisite mare, was the bay Ansata Nile Mist, who became a foundation mare for Imperial Egyptian Stud and left a prominent legacy through her many daughters and granddaughters such as Imperial Mistry, Imperial Impress, Imperial Mistilll, Ansata Misty Nile, etc.

Falima's grey daughter, Ansata Nile Star (by Ansata Shah Zaman), was an elegant and special mare too, but unfortunately she died young from a kick.

The Nile family has come into prominence in Europe through Ansata Nile Pasha (by *Jamilll) a "most popular" winner and crowd pleaser because of his extreme classic type. Imperial Kammill, a grandson of Ansata Nile Mist, as well as RN Hejazia, a daughter of Ansata Blue Nile, have recently arrived in England, the latter at Northbrook Farm Arabians, adding more Nile bloodlines to those already in that country through Nile Pasha's earlier sojourn at the Maxwell Farm.

The Arab world also acquired a number of the Nile family: the beautiful Nile Allure, and her progeny brought this line to Egypt; Ansata White Nile went to Kuwait, producing by Ansata Hejazi two special full sisters. She had already produced the outstanding colt, Ansata Nile Echo (by Ansata Hejazi), who was exported from Ansata to Al Naif Stud in Qatar. There he became a champion as well as in Europe where he exemplifies his family's noble traits and is known as "Mr. Trot." Another full sister, Ansata Nile Pearl, had previously gone to Saudi Arabia. Ansata Nile Gypsy (by Ansata Hejazi x Ansata Nile Starr), is a valued foundation mare at Al Naif Stud in Qatar, and the dam of Jendeh Al Naif (by Ansata Shalim), also an international winner. Nile Gypsy's full sister, Ansata Nile Jade, is also producing in Saudi Arabia having been exported to the Nejd Stud. The Nile family, despite many exports, still boasts significant numbers in America, where its legacy for classic type and superb movement is much appreciated.

Regrettably many breeders stopped using the word Nile within the name of Falima's direct female line descendants, but the quality flows on, much like the life-giving river after whom this precious family was named.

Ansata Nile Mist (Ansata Ibn Sudan x Falima). Sparagowski photo.

I have observed that the stagnation of water corrupteth it;
if it floweth, it becometh sweet; but otherwise, it doth not.
The grains of gold upon their native bed are regarded as mere dust;
and the aloeswood, where it groweth, is a kind of firewood:
If exported, it becometh an object of high demand; but if not,
it attaineth no kind of distinction.

            Tales of the Arabian Nights

*Imperial Mistilll (*Jamilll x Ansata Nile Mist)*

An extended Nile family: from right to left, Ansata Nile Mist (Ansata Ibn Sudan x Falima), Imperial Mistry (Mosry x Ansata Nile Mist), Imperial Im Tiarah (Ansata Imperial x Ansata Mistry), Imperial Tali (Ruminaja Ali x Imperial Im Tiarah), and Imperial Kamaala (Imperial Al Kamar x Imperial Tali). Scott Trees photo.

Ansata Nile Echo (Ansata Hejazi x Ansata White Nile). Gigi Grasso photo.

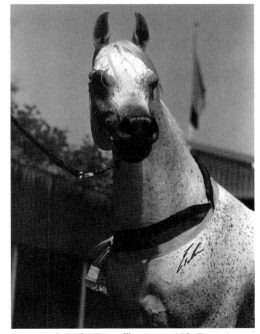

*Ansata Nile Pasha (*Jamilll x Ansata Nile Dream). Escher photo.*

Sharing the dream - Dahman

Descending from *Bint Bint Sabbah through Fa-Habba

Ibn El Nil
1999 Grey Stallion
Sire: Montasir
(El Hadiyyah x Maysa)
Dam: Nile Allure
Bred by Rahim Arabians

Nile Allure
1991-2002 Grey Mare
Sire: Ansata Halim Shah
(*Ansata Ibn Halima x
Ansata Rosetta)
Dam: Ansata Nile Magic
(*Jamilll x Ansata Nile Rose x
Ansata Nile Charm)
Bred by Ansata Arabian Stud
EAHBA: 1998 Event Res.
Champion Mare
& Most Classic Head Mares
and Fillies; 1999 Class Winner
and 2nd Produce of Dam Group.
Dam of Champion **Bint El Nil**
and Class Winner **Aruset El-Nil.**

RAHIM ARABIANS
Dr. Aly Abdel Rahim • 439 Pyramids Street • Giza, Egypt 11511

Sharing the dream - Dahman

Descending from *Bint Bint Sabbah through Habba

Polly Knoll photos

Antara Shalima
2000 Grey Mare
Sire: Ansata Hejazi
(Ansata Halim Shah x Ansata Sudarra)
Dam: Sha Latif (Shaikh Halim x Latifa Raqqasa x
Raalima x Asal Sirabba)
Bred by Jerry Gates and Nancy Gates, Arkansas

2001 Egyptian Event: Res. Jr. Champion
Res. Champion Futurity Yearling
Top Four World Class Yearling

2002 Egyptian Event: Most Classic Head Winner
2 Year Old Futurity Champion
Amateur "Shoot Out" Champion Filly
Junior Champion Filly
Reserve Supreme Champion Mare
"Breeder's Award" Highest Placing Straight Egyptian
Mare of the Show

Then to the perfect horse He spake:
"Fortune to thee I bring,
Fortune, as long as rolls the earth
Shall to thy forelock cling."
    B. Taylor from the Arabic

This page is dedicated with love and respect
to the late Gerald F. Mayden

ANTARA EGYPTIAN ARABIANS
Jerry & Nancy Gates • 550 Gavin Lane • Rison, Arkansas 71665

Sharing the dream - Dahman

Descending from *Bint Bint Sabbah through Fa-Abba

LA Selket (and her daughters)
1988 Grey Mare
Sire: Ansata Abu Sudan
(Ansata Ibn Sudan x
Ansata Bintmisuna)
Dam: Marah Abadiyah
(Waseem Kumaal x Sereneh x
Serr Abba x Fa-Abba)
Bred by P.R. Luecken, U.S.A.

I am the eye with which the Universe
Beholds itself and knows itself divine…
   Shelley

Photos by Gigi Grasso

Serene Ciai Dii
2001 Grey Mare
Sire: Salaa El Dine (Ansata Halim Shah x Hanan)
Dam: La Selket
Bred by Serene Egyptian Stud
Owned by Paolo Damilano & Gigi Grasso

Serene Desire • 2002 Grey Mare
Sire: Teymur B (Assad x 214 Ibn Galal 1)
Dam: La Selket
Bred by Serene Egyptian Stud

SERENE EGYPTIAN STUD
Anna & Stefano Galber • Via Coletto 2, Loc. SERENE • Affi (VR), Italy 37010

Sharing the dream - Dahman

**Descending from
*Bint Bint Sabbah
through Fa-Abba**

*HS Tadeusz
1997 Grey Stallion
Imported son of Simeon Sadik
Bred by Shirley Watts of Halsdon
Egyptian Event Top Ten
World Class of 1997

Stuart Vesty photo

		Asfour	Malik
	Simeon Sadik		Hanan
		Simeon Safanad	Sankt Georg
*HS TADEUSZ			27 Ibn Galal-V
		Imperial Imperor	Ansata Imperial
	Ma Tatia		Imperial Mistry
		AK Tatima	Moniet El Sharaf
			SRA Nasaddi
			x Fa El Hannah x Bint Hannah

MAISANO ARABIANS
Karen Maisano • 9188 Eagle Run Drive • Brighton, Michigan 48116

Sharing the dream - Dahman

Descending from Bukra Delilah

Ansata Sharifa
1987 Grey Mare
Sire: Ansata Ibn Shah
(Ansata Shah Zaman x Ansata Jezebel)
Dam: Ansata Samantha
(*Ansata Ibn Halima x Ansata Delilah
 x *Ansata Bint Misr)
Bred by Ansata Arabian Stud
Produce of Dam Winner Qatar 1999 National &
International Shows

Gigi Grasso photos

Al Wajba Al Rayyan
1996 Grey Mare
Sire: Safir
(Salaa El Dine x Aisha)
Dam: Ansata Sharifa
Bred by Al Rayyan Farm
Res. Jr. Champion Filly Jordan International Show 1998

Sinan Al Rayyan
2002 Grey Colt
Sire: Ansata Sinan (Prince Fa Moniet x
Ansata Nefara)
Dam: Al Wajba Al Rayyan

AL RAYYAN FARM
Sheikh Abdul Aziz Bin Khaled Bin Hamad Al-Thani • P.O. Box 375 • Doha, Qatar

Sharing the dream - Dahman

Descending from Bukra through Delilah

G Shafaria
1992 Grey Mare
Sire: Prince Fa Moniet
(The Egyptian Prince x Fa Moniet)
Dam: Ansata Sharifa
Bred by Paolo Gucci, U.S.A.
Egyptian Event Res. Champion Mare

Gigi Grasso photos

Ansata Selman
1999 Grey Stallion
Sire: Ansata Hejazi
(Ansata Halim Shah
x Ansata Sudarra)
Dam: G Shafaria
Bred by
Ansata Arabian Stud
Qatar National Jr. Champion Colt 2000
Qatar International Jr. Champion Colt 2002

Jalila Al Rayyan
2000 Grey Mare
Full sister to Ansata Selman

AL RAYYAN FARM
Sheikh Abdul Aziz Bin Khaled Bin Hamad Al Thani • P.O. Box 375 • Doha, Qatar

Sharing the dream - Dahman

Descending from Bukra through Delilah

Gigi Grasso photos

Ansata Selket
1987 Grey Mare
Sire: Ansata Halim Shah
(*Ansata Ibn Halima x Ansata Rosetta)
Dam: Ansata Samarra
(*Jamilll x Ansata Samantha x Ansata Delilah)
Bred by Ansata Arabian Stud
Most Classic Arabian Wins, Class A
Championships U.S.A.
Qatar National Champion Mare

Ansata Nefer Isis
1991 Grey Mare
Sire: Prince Fa Moniet
(The Egyptian Prince x
Fa Moniet)
Dam: Ansata Nefertiti
(Ansata Halim Shah x
Ansata Sudarra
x Ansata Delilah)
European Res. Champion Mare 1996, Vichy/France;
Qatar International Res. Champion Mare 2001 and 2002

Dana Al Rayyan
2000 Grey Mare
Sire: Alidaar
(Shaikh Al Badi x Bint Magidaa)
Dam: Ansata Samaria
(*Jamilll x Ansata Samantha)
Bred by Al Rayyan Farm
Champion Filly Foal Qatar National Show 2000

AL RAYYAN FARM
Sheikh Abdul Aziz Bin Khaled Bin Hamad Al Thani • P.O. Box 375 • Doha, Qatar

Sharing the dream - Dahman

Descending from Farida through Halima

Ashhal Al Rayyan
1996 Grey Stallion
Sire: Safir (Salaa El Dine by
Ansata Halim Shah x Aisha
by Ansata Halim Shah)
Dam: Ansata Majesta
(Ansata Halim Shah x Ansata Malika)
Bred by Al Rayyan Farm
Qatar National Jr. Champion Colt 1998
Qatar International Shows:
Jr. Champion Colt 1999 and Best Qatari Bred;
Res. Jr. Champion Colt 2000 and
Res. Sr. Champion Stallion 2002

Gigi Grasso photos

Naama Al Rayyan
1998 Grey Mare
Full sister to Ashhal Al Rayyan
Bred by Al Rayyan Farm
Class Winner Qatar National Show 2000

Al Aangha Al Rayyan
2000 Grey Mare
Sire: Alidaar (Shaikh Al Badi x Bint Magidaa)
Dam: Ansata Majesta
Bred by Al Rayyan Farm
Qatar National Jr. Champion Filly 2002

AL RAYYAN FARM
Sheikh Abdul Aziz Bin Khaled Bin Hamad Al Thani • P.O. Box 375 • Doha, Qatar

 *Sharing the dream - Dahman*

Descending from Farida through Nefisa

RN Farida
1995 Grey Mare
Bred by Al Rayyan Farm out
of the mare Noha while she was
leased in Germany by special arrangement
from the Marbach State Stud.
Jordan International Jr. Champion; Middle
East Jr. Champion 1998
Qatar National
Champion Mare 1999
Qatar International Shows:
Res. Champion Mare 2000; Champion
Mare 2001
and Best Qatari Bred Horse

Gigi Grasso photo

		Ansata Halim Shah	*Ansata Ibn Halima
	Salaa El Dine		Ansata Rosetta
		Hanan	Alaa El Din
RN FARIDA			Mona
		Hadban Enzahi	Nazeer
	Noha		Kamla
		Nadja	Nazeer
			Nefisa

AL RAYYAN FARM
Sheikh Abdul Aziz Bin Khaled Bin Hamad Al Thani • P.O. Box 375 • Doha, Qatar

Matriarchs of the Sabah/Kamar family: *Pharrah

			Nazeer	Mansour
		*Morafic		Bint Samiha
	Farag		Mabrouka	Sid Abouhom
				Moniet El Nefous
		Bint Kateefa	Sid Abouhom	El Deree
				Layla
*Pharrah			Kateefa	Shahloul
				Bint Rissala
		*Tuhotmos	El Sareei	Shahloul
				Zareefa
	Tamria		Moniet El Nefous	Shahloul
				Wanisa
		Kamar	Nazeer	Mansour
				Bint Samiha
			Komeira	Nabras
				Layla

Those who saw the Nazeer daughter, Kamar, at home in Egypt will always remember her exquisitely chiseled head, extreme refinement, feminine demeanor and dryness typical of true desert horses. One of the Sabah family's special treasures, Kamar's values are shared through her progeny, and particularly through her daughters Tamria 9 (by *Tuhotmos), imported to Babolna State Stud, Hungary, and *Kahramana (by Antar). The beautiful-headed *Kahramana, who was more substantially bodied than her dam, achieved renown for Gleannloch through her U.S. National Futurity Champion son, *Ibn Morafic (the sire of good broodmares), his full brothers, champion Shah Nishan, herd sire for Dorian Farm, and the handsome sire, Fehris, as well as through her daughters Kamaara, Kahleela, Shirzada and Bint Kahramana. While this family is prominent in breeding programs worldwide including Tamria II (by Ansata Halim Shah x 211 Zohair) now in Germany and dam of 10 progeny, the special story of Tamria's daughter, *Pharrah, is one tale among thousands that Scherezade might have told.

Tamria II (Ansata Halim Shah x 211 Zohair). Toischel photo.

The first time we saw Tamria she was a vision in white lying against the golden straw in her stall at Babolna. She didn't rise; her feet hurt too much from laminitis. Beauty crowned her sculptured face. Her big black eyes gave only a hint of the pain she bore with dignity, but she nevertheless enjoyed having visitors admire her. Among her daughters, the much praised *Pharrah (by Farag) ascended as a shining star. Born in January 1972 at Babolna, *Pharrah was imported five years later to America and in 1978 she became one of Barbara Griffith's cherished broodmares at Imperial Stud in Maryland. Of royal demeanor, she carried herself with grace, dignity and pride, her beauty, refinement and sweet character endearing her to everyone. One was drawn immediately to her extreme head with very long shapely foreface and teacup muzzle, a trait probably inherited from the double Shahloul through *Tuhotmos up close her in her pedigree. Some of her granddam's physical characteristics were also evident, but *Pharrah was much better conformed.

*Pharrah (Farag x Tamria). Sparagowski photo.

In America *Pharrah claimed broodmare notoriety through her champion sons, Royal Jalliel (by Ansata Imperial), an extremely classic and well-balanced stallion, and his taller, bigger brother, Royal Mikhiel (by Nabiel), who eventually was exported to the Al Jaffar Stud in Jordan, and through her daughters, many of whom became champions and champion producers: AK Komeira (by Ibn Galal), Imperial Phanadah (by *Ibn Moniet El Nefous - and the dam of World Champion Imperial Phanilah) Imperial Pharalima (by *Ansata Ibn Halima), Royal Bint Pharrah (by Ansata Imperial) and Royal Pharriel (by Nabiel).

After *Pharrah had produced admirable daughters for Imperial, Barbara Griffith, a staunch supporter of the Pyramid Society, reluctantly agreed to consign her to its 1983 sale held at the Egyptian Event in Lexington, Kentucky. Auction sales and prices for Egyptian horses were reaching record heights and the Society wanted only the best of the best to offer its discriminating buyers. Anyone in attendance that day will remember when stately *Pharrah stepped into the sales ring the bidding went to $100,000 and upwards - fast. The buyer who had come from California just to bid on *Pharrah was tenacious - and successful. When the gavel fell, Vincent Fortuna, had purchased her for $255,000 - a precious gift to his beloved wife, Diane. As a result, Barbara had not lost a mare, she had gained two lifetime friends.

*Imperial Phanadah (*Ibn Moniet El Nefous x *Pharrah). Scott Trees photo.*

Arriving at the Fortuna's scenic valley farm nestled in the hills near Santa Barbara, *Pharrah became "family." In this idyllic setting, her beautiful foals with the prefix "royal" were born....until one fateful day in 1990 flames appeared out of nowhere, sweeping across the hills turning the peaceful haven into a valley of death. The Fortuna's barely managed to rescue a few of the horses, including *Pharrah, but fifteen, including her youngest daughter by El Halimaar, were lost. Vincent poignantly remembers how friends offered their vacant ranch 40 miles away to house the surviving horses, and *Pharrah was taken there. She had foundered prior to the fire and was recovering under her regular farrier's capable hands. Attention by a different blacksmith resulted in a tragic reversal, and her feet became worse....and worse. Vincent placed her at a special clinic and daily drove 30 miles to hold her head in his lap, encouraging her to survive this challenge as well. She listened, but she heard a different call - and leaving behind her legacy, galloped free from earthbound pain, to join her family once again.

Oh - the horses!
Vibrant flames they seemed,
Who gave mute testimony to the purity of their ancient heritage.
Though the several families showed subtle differences,
The common bond was clear in the dryness, the spareness of line,
the total absence of anything indistinct or 'smudgy'
- and in their huge eyes,
the look of Time remembered!

Linnell Smith

*Royal Jalliel (Ansata Imperial x *Pharrah). Jeff Little photo.*

Sharing the dream - Dahman

Descending from Sabah through Kamar

Filsinger photo

Imperial Phanilah
1984 Grey Mare
Bred by Imperial Egyptian Stud,
U.S.A.
Phanilah not only won
championships in America,
but she became 1994
World Champion Mare,
Salon du Cheval, Paris;
Qatar Champion Mare;
and Middle East Champion Mare

	Ansata Ibn Sudan	*Ansata Ibn Halima
Ansata Imperial		*Ansata Bint Mabrouka
	Ansata Delilah	Ansata Shah Zaman
IMPERIAL PHANILAH		*Ansata Bint Misr
	*Ibn Moniet El Nefous	*Morafic
Imperial Phanadah		Moniet El Nefous
	*Pharrah	Farag
		Tamria

*...the best produce is that which proceeds from
a sire and a dam both of pure race.
In this case, it is gold allying itself with gold.*

Amir Abd El Kader

AL SHAQAB STUD
P.O.Box 90055 • Doha, Qatar

Sharing the dream - Dahman

Descending from Sabah through Kamar

Amira El Shaqab
1998 Grey Mare
Sire: Al Aadeed Al Shaqab
(Ansata Halim Shah x Sundar Alisayyah)
Dam: Imperial Phanilah
Bred by Al Shaqab Stud
Qatar National Shows: 1998 Class Winner
& Champion Yearling Filly
Qatar International Show: 1998 Champion Female,
1999 Champion Yearling Filly,
Res. Champion Jr. Female; 2000 two Year Old Class
Winner and Champion Filly;
2001 Class Winner and Res. Jr. Champion Mare

Gigi Grasso photos

AL SHAQAB STUD
P.O. Box 90055 • Doha, Qatar

Sharing the dream - Dahman

Descending from Sabah through Kamar

Irina Filsinger photo

Aliaa Al Shaqab
1995 Grey Mare
Sire: Ansata Halim Shah
(*Ansata Ibn Halima x Ansata Rosetta)
Dam: Imperial Im Pharida
(Ansata Imperial x Imperial Phanadah)
Bred by Al Shaqab Stud
Reserve Champion Filly, Amman 1996
Qatar National Shows:
1997 Class Winner 2 Year old Filly
1999 Class Winner Mares 4-6 years old

ALIAA AL SHAQAB	Ansata Halim Shah	*Ansata Ibn Halima	Nazeer
			Halima
		Ansata Rosetta	Ansata Shah Zaman
			*Ansata Bint Bukra
	Imperial Im Pharida	Ansata Imperial	Ansata Ibn Sudan
			Ansata Delilah
		Imperial Phanadah	*Ibn Moniet El Nefous
			*Pharrah

AL SHAQAB STUD
P.O. Box 90055 • Doha, Qatar

Sharing the dream - Dahman

Descending from Bukra through Delilah

Halima Al Shaqab
1994 Grey Mare
Sire: Ansata Manasseh
(Ansata Halim Shah x Ansata Aliha)
Dam: Ansata Bint Halima
(*Ansata Ibn Halima x Ansata Delilah)
Bred by Ansata Arabian Stud, U.S.A.

Gigi Grasso photos

AL SHAQAB STUD
P.O. Box 90055 • Doha, Qatar

Sharing the dream - Dahman

Descending from Bukra through Rosetta

Gigi Grasso photo

Sultana Al Shaqab
1997 Grey Mare
Bred by Al Shaqab Stud
In 1999 Sultana was shown
to Second Place 2 Year Old Fillies
at the Qatar International Show
& Third Place 2 Year Old Fillies
at the Qatar National Show

		Imperial Al Kamar	El Hilal
	Imperial Pharouk		Imperial Sonbesjul
		Imperial Phanadah	*Ibn Moniet El Nefous
SULTANA AL SHAQAB			*Pharrah
		*Jamilll	Madkour
	Ansata Prima Rose		Hanan
		Ansata Rosetta	Ansata Shah Zaman
			*Ansata Bint Bukra

AL SHAQAB STUD
P.O. Box 90055 • Doha, Qatar

Descending from Farida through Halima

Filsinger photo

Ansata Malaha
1993 Grey Mare
Sire: Ansata Halim Shah
(*Ansata Ibn Halima x Ansata Rosetta)
Dam: Ansata Malika
(*Jamilll x *JKB Masouda x Maymoonah)
Bred by Ansata Arabian Stud, U.S.A
1st Place Yearling and Champion Filly Middle East Championships 1994;
Res. Champion Amman, Jordan; First Qatar 2 Year Old Filly 1996;
First 4-6 Year Mares 1998; Senior Female Champion
& Qatar National Res. Champion Mare 1999

Gigi Grasso photo

Gigi Grasso photo

AL SHAQAB STUD
P.O. Box 90055 • Doha, Qatar

Sharing the dream - Dahman

Descending from Farida through Helwa

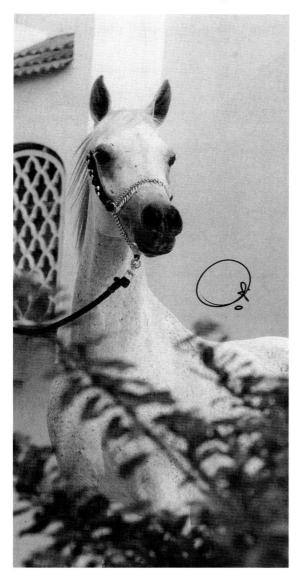

Rabaab Al Shaqab
1995 Grey Mare
Bred by Al Shaqab Stud

Gigi Grasso photos

```
                                          *Ansata Ibn Halima    Nazeer
                        Ansata Halim Shah                       Halima
                                          Ansata Rosetta        Ansata Shah Zaman
      RABAAB AL SHAQAB                                          *Ansata Bint Bukra
                                          Adl                   Ikhnatoon
                        Rahamat                                 Enayah
                                          Rola                  Raki
                                                                Bint Hakeema
```

AL SHAQAB STUD
P.O. Box 90055 • Doha, Qatar

Sharing the dream - Dahman

Descending from Farida through Halima and Bukra through Delilah

Filsinger photo

Ansata Meryta
1988 Grey Mare
Sire: Ansata Halim Shah
(*Ansata Ibn Halima x Ansata Rosetta)
Dam: *JKB Masouda
(Madkour x Maymoonah x Malikah)
Bred by Jean Kayser, Luxembourg,
while Masouda was on lease to Ansata

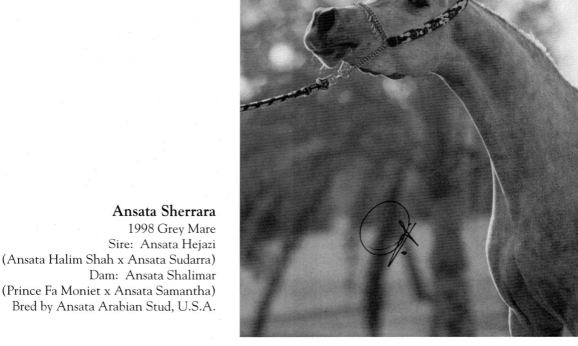

Ansata Sherrara
1998 Grey Mare
Sire: Ansata Hejazi
(Ansata Halim Shah x Ansata Sudarra)
Dam: Ansata Shalimar
(Prince Fa Moniet x Ansata Samantha)
Bred by Ansata Arabian Stud, U.S.A.

Gigi Grasso photo

ARABIAN HORSE CENTER
P.O. Box 22436 • Safat 13085, Kuwait

Sharing the dream - Dahman

Descending from Sabah through Bukra and *Bint Bint Sabbah through Fa-Habba

Latiefa
1993 Grey Mare
Sire: Hamasa Khazzan (Kasr El Nil x Nafteta x Moneera)
Dam: Hamasa Tarifa (Farag x Menha
x Nagwa x Bint Bukra/EAO)
Bred by Arabian Horse Center

Gigi Grasso photo

Filsinger photo

Ansata White Nile
1990 Grey Mare
Sire: Prince Fa Moniet
(The Egyptian Prince x Fa Moniet)
Dam: Ansata Nile Gift (Ansata Ibn Sudan x
Ansata Nile Jewel x Falima x FaHabba.
Bred by Leonor Romney/Ansata while
Ansata Nile Gift was on lease to Ansata
White Nile won Multi-Top Tens at The Egyptian Event,
U.S.A. and was a Class A Sr. Champion Mare

Wafaa Elkuwait
1998 Grey Mare
Sire: Ansata Hejazi
(Ansata Halim Shah x Ansata Sudarra)
Dam: Ansata White Nile
Bred by Ansata Arabian Stud, U.S.A.

Gigi Grasso photo

ARABIAN HORSE CENTER
P.O. Box 22436 • Safat 13085, Kuwait

Sharing the dream - Dahman

Descending from Bukra through Delilah

Ansata Nefertiti
1987 Grey Mare
Full sister to Ansata Hejazi
Bred by Ansata Arabian Stud, U.S.A.
Dam of **Ansata Nefer Isis**, winner of European and Qatar Championships

Filsinger photo

Ansata Samari
1997 Grey Mare
Sire: Ansata Hejazi
(Ansata Halim Shah x Ansata Sudarra)
Dam: Ansata Samarra (*Jamilll x Ansata
Samantha x Ansata Delilah)
Bred by Ansata Arabian Stud, U.S.A.
Egyptian Event EBC Res.Champion
Yearling Filly and World Class Top Ten

A thing of beauty is a joy forever;
It's loveliness increases; it will never
Pass into nothingness.
      Keats

Gigi Grasso photo

AJMAL ARABIAN STUD
Mohammed Jassim Al Marzouk • Area 11, Farm No. 1 & 2 • Wafra, Kuwait 3131

Sharing the dream - Dahman

Descending from Bukra through Delilah

Gigi Grasso photo

Ansata Hejazi 1991 Grey Stallion
Sire: Ansata Halim Shah (*Ansata Ibn Halima x Ansata Rosetta x *Ansata Bint Bukra)
Dam: Ansata Sudarra (Ansata Abu Sudan x Ansata Delilah x *Ansata Bint Misr x *Ansata Bint Bukra)
Bred by Ansata Arabian Stud, U.S.A.
Class A Reserve Supreme Champion U.S.A.; Sr. Champion & Res. Supreme Champion
& Liberty Class Top 5 at Egyptian Events, U.S.A. Sire of Multi-Champions including **Ansata Samari, Antara Shalima, Ansata Malik Shah, Ansata Selman, Ansata Nile Echo, Ansata Haisam**, and other winners

AJMAL ARABIAN STUD
Mohammed Jassim Al Marzouk • Area 11, Farm No. 1 & 2 • Wafra, Kuwait 3131

Sharing the dream - Dahman

Descending from Farida through Futna and *Bint Bint Sabbah through Fay-Sabbah

Gigi Grasso photo

Ali Jamila
1987 Black Mare
Sire: Ruminaja Ali
(Shaikh Al Badi x Bint Magidaa)
Dam: Ansata Justina
(Jamil x Ansata Judea x Ansata Jamila)
Bred by Ansata Arabian Stud
Dam of **The Atticus, Shabab Jamal**

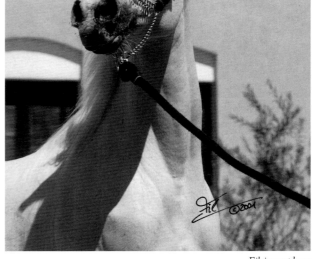

Glorieta Saafrana
1982 Grey Mare
Sire: Ansata Abu Nazeer
(*Ansata Ibn Halima x *Ansata Bint Zaafarana)
Dam: Ansata Sabiha
(Sabrah x Serrasab x Fay-Sabbah)
Bred by Glorieta Ranch

Filsinger photo

AJMAL ARABIAN STUD
Mohammed Jassim Al Marzouk • Area 11, Farm No. 1 & 2 • Wafra, Kuwait 3131

Sharing the dream - Dahman

Descending from Farida through Halima

Ansata Magnifica
1991-2002 Grey Mare
Sire: Ansata Halim Shah
(*Ansata Ibn Halima x Ansata Rosetta)
Dam: Ansata Malika (*Jamilll x JKB Masouda x Maymoonah x Malikah)
Bred by Ansata Arabian Stud
Class A Champion Mare U.S.A; Wins at Qatar National and International Championships.
Dam of **R.N. Mezna**, Qatar National Res. Jr. & Res. Sr. Champion & Most Classic Head Winner

Gigi Grasso photos

Ajmal Maghrebeia
2000 Grey Mare
Sire: Salaa El Dine (Ansata Halim Shah x Ansata Rosetta)
Dam: JKB Majida (Ansata Halim Shah x *JKB Masouda)
Bred by Mohammed J. Al Marzouk

Quality is a treasure that cannot be touched,
An inspiration that cannot be lost,
A source of pride that never loses its lustre.
     Anonymous

AJMAL ARABIAN STUD
Mohammed Jassim Al Marzouk • Area 11, Farm No. 1 & 2 • Wafra, Kuwait 3131

Sharing the dream - Dahman

Descending from Bukra through Rosetta

Ansata Exotica
1986 Grey Mare
Sire: *Jamilll (Madkour 1 x Hanan)
Dam: Ansata Ghazala
(Ansata Ibn Sudan x Ansata Rosetta)
Bred by Ansata Arabian Stud, U.S.A.

Gigi Grasso photos

Aarak Al Safinat
2002 Grey Mare
Sire: Alidaar (Shaikh Al Badi x Bint Magidaa)
Dam: Ansata Exotica
Bred by Khaled Ben Shokr

AL SAFINAT FARM
Khaled Ben Shokr • P.O. Box 158 • Sura, Kuwait 45702

Descending from Bukra through Delilah

RN Rayana
1994 Grey Mare
Sire: Prince Fa Moniet
(The Egyptian Prince x Fa Moniet)
Dam: Ansata Sharifa
(Ansata Ibn Shah x Ansata Samantha)
Bred by Ansata Arabian Stud, U.S.A.

Filsinger photo

Rana Al Safinat
2001 Grey Mare
Sire: Ansata Hejazi
(Ansata Halim Shah x Ansata Sudarra)
Dam: RN Rayana
Bred by Khaled Ben Shokr

Gigi Grasso photo

AL SAFINAT FARM
Khaled Ben Shokr • P.O. Box 158 • Sura, Kuwait 45702

Sharing the dream - Dahman

Descending from Farida through Nefisa

Aliikat
1989 Grey Mare
Sire: Ruminaja Ali
(Shaikh Al Badi x Bint Magidaa)
Dam: Katourah
(Mohssen x Nourah x *Bint Nefisaa x Nefisa)
Bred by Charles & Judy Jones, Atallah Arabians

Filsinger photo

Adjbah Al Safinat
2000 Grey Mare
Sire: Ansata Iemhotep
(Prince Fa Moniet x Ansata Nefara)
Dam: Aliikat
Bred by Khaled Ben Shokr

Nefisa Al Safinat
2001 Grey Mare
Sire: Ansata Hejazi
(Ansata Halim Shah x Ansata Sudarra)
Dam: Aliikat
Bred by Khaled Ben Shokr

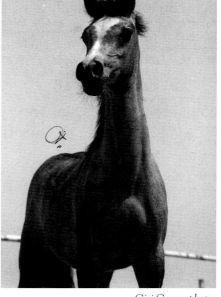

Gigi Grasso photos

AL SAFINAT FARM
Khaled Ben Shokr • P.O. Box 158 • Sura, Kuwait 45702

Sharing the dream - Dahman

Descending from Bukra through Delilah,
Farida through Helwa,
and Sabah through Kamar

Ansata Selma
1998 Grey Mare
Sire: Ansata Hejazi (Ansata Halim Shah x
Ansata Sudarra)
Dam: G Shafaria (Prince Fa Moniet x
Ansata Sharifa x Ansata Samantha)
Bred by Ansata Arabian Stud, U.S.A.
1999 Qatar National Top Five

Gigi Grasso photo

Saada Al Rayyan
2000 Grey Mare
Sire: Alidaar (Shaikh Al Badi x Bint Magidaa)
Dam: RN Azbah
(Ansata Halim Shah x Izeema x Icomb)
Bred by Al Rayyan Farm, Qatar

Filsinger Photo

Jamala Al Zamet
Sire: Salaa El Dine
(Ansata Halim Shah x Hanan)
Dam: Abbas Pasha I-12
(Abbas Pasha I x 211 Zohair x 28 Farag)
Bred by Susanne Schreibvogel, Germany

N. Sachs photo

AL JAZIRA ARABIAN STUD
Mr. Talal Abdullah Al Mehri • P.O. Box 16991 • Qadisiyah, Kuwait 35860

Sharing the dream - Dahman

Escher photos

Descending from Farida through Helwa and from *Bint Bint Sabbah through Habba

Kafrawi
1992 Bay Stallion
Sire: Gad Allah
(Adeeb x Omnia)
Dam: Bint Misr
(Anas x Basant x Hayfaa x Abla x Helwa)
Bred by the E.A.O., Cairo, Egypt

Pretty Woman
1990 Grey Mare
Sire: The Minstril
(Ruminaja Ali x *Bahila)
Dam: Our Kibriya
(Shaikh Al Badi x Raalima
x Asal Sirabba x Habba)
Bred by Carousel I Joint Venture

ABD-EL GAWAD GAWDAT
23 Kafr Abdou Street • Roushdy, Alexandria, Egypt

Descending from Bint El Bahreyn
through Maisa and through Tifla

Sharing the dream - Dahman

Karima GAD
1997 Grey Mare
Sire: Mourad
(Gassir x Mabrouka)
Dam: Bint Imdara
(Pasha Nabilahh x Imdara x Bukhayt x
Rihahna x *Bint Maisa El Saghira)
Bred by Gawdat Stud

Escher photos

Black Halim
1987 Black Stallion
Sire: Halim El Nefous
(*Ansata Ibn Halima x Monadena)
Dam: *Bint Soheir II
(Gubran x *Soheir II x Tifla)
Bred by Hamilton Bank

ABD-EL GAWAD GAWDAT
23 Kafr Abdou Street • Roushdy, Alexandria, Egypt

Sharing the dream - Dahman

Descending from Farida through Futna

Al Rayyan
1993 Grey Stallion
Bred by Imperial Egyptian Stud, U.S.A.
Middle East Jr. Champion Colt, Jordan 1997;
Supreme Champion Saudi Arabian Western Province Championships and Most Classic Head, 2000.

AL RAYYAN	*Imperial Madheen	Messaoud	Madkour
			Maymoonah
		Madinah	Ibn Galal
			Mona II
	Imperial Naffata	Moniet El Nafis	*Ibn Moniet El Nefous
			*Hoyeda
		Imperial Fanniya	*Faleh
			*Deenaa

AL-AADEYAT ARABIAN STUD
Khalid Saad Al-Haddad • Madina Road • P.O. Box 54110 • Jeddah, Saudi Arabia 21514

Sharing the dream - Dahman

Descending from Farida through Futna, and from Farida through Nefisa

With black eye, wide nostrils,
Clean limbs, and a faithful heart!
Nothing at present is equal to my horse of pure blood.
  Amir Abd el Kader

Al Rayyan

Salma
1993 Grey Mare
Sire: Adl (Ikhnatoon x Enayah)
Dam: AK Nouasha (Ansata Abbas Pasha x AK Nouara x Noha x Nadja x Nefisa)
Bred by Saqr Farm, Egypt
Salma is pictured with her 2002 filly,
AA Qana Al Aadeyat by
Shaikh Al Shahwan
(Shaikh Al Badi x Dahmah Shahwaniah)
Salma's dam, AK Nouasha, has been many times a Champion Mare in Egypt

AL-AADEYAT ARABIAN STUD
Khalid Saad Al-Haddad • Madina Road • P.O. Box 54110 • Jeddah, Saudi Arabia 21514

Sharing the dream - Dahman

Van Lent photo

Descending from Bint El Bahreyn through Tifla and through Maisa

Madinah Bint Saariyah
Sire: *Imperial Madheen
(Messaoud x Madinah)
Dam: PLF Saariyah
(Shaikh Al Badi x AK Samiha x
Soheir 11 x Tifla)
Grand Champion Mare (unanimous)
Egyptian Event U.S.A. 1997

Al Maymoon Al Aadeyat
2000 Grey Stallion
Sire: Salaa El Dine (Ansata Halim Shah x Hanan)
Dam: Madinah Bint Saariyah

Stuart Vesty photo

Simeon Sitri
1993 Bay Mare
Sire: Simeon Sadik (Asfour x Simeon Safanad)
Dam: Maardassa (El Halimaar x Hadassa x
Nafairtiti x *Bint Maisa El Saghira)
Bred by Simeon Stud
Shown to Top Ten win in Australia 1997.

AL-AADEYAT ARABIAN STUD
Khalid Saad Al-Haddad • Madina Road • P.O. Box 54110 • Jeddah, Saudi Arabia 21514

Matriarchs of the Farida Line
Moheba and her descendants: Malikah and Moheba II

			Gamil Manial
		Mansour	Nafaa El Saghira
	Nazeer		Kazmeyn
		Bint Samiha	Samiha
Ghazal			Ibn Rabdan
		Shahloul	Bint Radia
	Bukra		Kazmeyn
		Bint Sabah	Sabah
Malikah			Ibn Rabdan
Moheba II		Shahloul	Bint Radia
	El Sareei		Kazmeyn
		Zareefa	Durra
Malacha			El Deree
		Sid Abouhom	Layla
	Moheba		Sheikh El Arab
		Halima	Ragia

The splendid Malikah (Ghazal x Malacha) full sister to Moheba II. Filsinger photo.

To look at the early Moheba family's photos in the Marbach Studbook, one would never think these mares could play such an important role in history. Who would buy such mares today, when "exotic beauty" is a criteria that governs the choices of certain purchasers, regardless of genetic potential! Here again, blood will tell, along with selective breeding, and it has.

A bay Sid Abouhom daughter out of Halima, Moheba was in foal to El Sareei when she was imported to Germany from Egypt by Count Knyphausen during the time Von Szandtner was manager of El Zahraa. She produced the bay mare, Malacha, and founded an incredible family that became the only source of Halima blood in tail female line.

For those who saw Moheba's granddaughter, the beloved matriarch, Malikah, she remains an indelible memory. Typical of the Dahman strain into which she was born, Malikah was one of the few straight Egyptian daughters of the regrettably

Moheba II (Ghazal x Malacha). Note: this picture has been incorrectly identified as Mona III in other publications.

Mona III (Hadban Enzahi x Moheba II)

little-used Bukra son, Ghazal (by Nazeer). Both her tail female and sire's tail female were of the Dahman strain (Sabah through Ghazal, Farida through Moheba). Additionally, a line to the Bint El Bahreyn Dahman strain came through her dam's sire, El Sareei.

Born in 1962, Malikah became fleabitten grey, the color of many superior broodmatrons of the breed. Dr. Filsinger proudly introduced us to her against the background of a white winter wonderland, a portrait of old world type and grace with very feminine classic head, expressive eyes, and quite typical of that Dahman family in structure. Next came her grandson, Maymoon (by Kaisoon), playfully trotting in the new fallen

Sharing the dream - Dahman

Malik (Hadban Enzahi x Malikah).

Mahomed (Hadban Enzahi x Malikah)
All photos this page, Filsinger photos.

Maymoon (Kaisoon x Maymoonah x Malikah)

Mahyubi (Nabil Ibn Nazeefa x Maisa x Malikah)

Marani El Malikah (Maymoon x Malikah) Malikah's last foal

Sharing the dream - Dahman

Ansata Malika (*Jamilll x *JKB Masouda x Maymoonah x Malikah)

Ansata Majesta (Ansata Halim Shah x Ansata Malika) and trainer, Bart Van Buggenhout.

Ansata Marha, (granddaughter of Ansata Malika) a multi-champion mare in Qatar. Filsinger photo.

snow. A very refined and elegant young stallion, he bore a striking resemblance to his sire and was an exceptionally classic individual. His dam, Maymoonah (x Malikah) was considered to be one of the most beautiful Hadban Enzahi daughters in Germany, and she also produced the prestigious sire, *Messaoud. Regrettably, Maymoon died young, but fortunately Messaoud, one of the all time high scoring stallions at the German Licensing trials, carried on. Other influential Malikah progeny were her most successfully shown son, Manal (by Anchor Hill Halim), who was supreme Champion and "Best of Show" at the Asil Cup International 1988, Champion at the Stallion Licensing Show, and Best in National Stallions Performance Test; the well-known sires, Mahomed (by Hadban Enzahi), his full brother, Malik, and their celebrated full sister, Maisa, a dam of significance who remained at Filsinger Stud.

Without a doubt the most recent descendants of this Malikah line are the celebrated daughters of her namesake, Ansata Malika, (*Jamilll x *JKB Masouda), who produced four champions by Ansata Halim Shah, the first being the regal Ansata Majesta, who won every National and International Champion mare title in Qatar after also becoming a champion in America; Ansata Malaha, a multi-champion in Qatar and class winner for the Qatari team at the Nation's Cup, 2002; Ansata Magnifica, also a champion in America and Qatar and dam of the beautiful champion mare, RN Mezna, Ansata Majesty, a champion in America at Ansata, and her full sister, Ansata Malaka, dam of the young champion stallion, Ansata Malik Shah (by Ansata Hejazi). Ansata Majesta has become one of the most celebrated mares of this family in recent years. Her multi-champion son, Ashhal Al Rayyan, recently stood at stud in Europe during 2001, thereafter returning home to Qatar to take up stud duties at Al Rayyan. Another breathtaking beauty of this line, and a granddaughter of Ansata Majesta, is the champion filly, Johara Al Naif (by Ansata Shalim), bred and owned by Al Naif Stud, and truly "a jewel" who is living up to her name.

Ashhal Al Rayyan (Safir x Ansata Majesta). Judith Wich photo

Sharing the dream - Dahman

Madkour I (Hadban Enzahi x Moheba II). Forbis photo.

El Thay Ibn Halim Shah (Ansata Halim Shah x Mahameh). Carola Toischel photo

Bred by Count Knyphausen, Malikah spent her lifetime owned by the Filsinger family in Germany, the pride of their life, and deservedly so. Discriminating breeders considered her a Grand Lady and a Mare of Distinction in the Arabian breed, indeed many called her "the Queen." She produced 13 foals who have left their mark in breeding programs around the world. To name every individual worthy of mention in this short synopsis is quite impossible. One could write a book about her life, and her many distinguished descendants. Perhaps one day someone will.

Another venerated mare of this "M" family was the well-known Moheba II, full sister to Malikah, and also fleabitten. She too was bred at the Knyphausen Stud in Lutetsburg. Born in 1960, in her 27 years she produced ten foals, among them the important stallion Madkour I (by Hadban Enzahi), sire of *Jamilll. Moheba II also produced the lovely Mona III (by Hadban Enzahi).

El Thayeba stud began close linebreeding with this family, producing El Thay Bint Kamla and Kamla II, as well as several prominent stallions such as El Thay Khemal Pasha, and El Thay Mashour. Mona III was also the dam of the Gross's pretty Mahameh (by Ibrahim), dam of the captivating white stallion, El Thay Ibn Halim Shah, a champion stallion who quite mirrored his legendary sire, but met a tragic death in a trailer accident before his true impact as a breeding stallion could be measured. The M line through Mona III, Kamla II, and their descendants, brings another favorable dimension to this distinguished family. Like Malikah, Moheba II's progeny are numerous and her descendants are scattered around the world. Both of the well-bred and much revered sisters have brought honor and nobility to the Marbach foundation mare, Moheba, and to her dam, Halima, the treasured original matriarch of this ever-valued line.

> *If you work with a creative and kind eye,*
> *you will bring forth beauty.*
> Meister Eckhart

Mahameh (Ibrahim x Mona III x Moheba II). Nicole Sachs photo

Sharing the dream - Dahman

Descending from Farida through Halima

El Thay Mashour
1994 Grey Stallion
Sire: Madkour I (Hadban Enzahi x Moheba II)
Dam: El Thay Bint Kamla
(El Thay Mansour x Kamla II)
Bred by El Thayeba, Cornelia Tauschke
Jr. Champion Stallion Show; Supreme Champion,
Champion Stallion Show, Aachen 1997;
Supreme Champion Egyptian Event,
Germany 2001

Nicole Sachs photo

El Thay Bint Kamla
1989 Grey Mare
Sire: El Thay Mansour
(Ansata Halim Shah x Ansata Rosetta)
Dam: Kamla II (Ansata Halim Shah x
Mona III x Moheba II x Malacha)
Bred by El Thayeba, Cornelia Tauschke
Premium Mare German Arab Horse Society

Nichole Sachs photo

El Thay Kamla
1999 Grey Mare
Sire: El Thay Mashour
(Madkour x El Thay Bint Kamla)
Dam: Kamla II (Ansata Halim Shah x Mona III)
Bred by El Thayeba, Cornelia Tauschke
Jr. Champion and Overall Show Champion, Tillburg, Holland 2002
Jr. Champion, National Championships, Germany 2002
Jr. Champion Pyramid Society Europe,
Egyptian Event Europe 2002

Melanie Groger photo

EL THAYEBA
Cornelia Tauschke • Kneter Sand 1 • 26197 Grossenkneten, Germany

Sharing the dream - Dahman

Descending from Farida through Halima

Mudira
1997 Grey Mare
Bred by Vollblutaraber
Gestüt Seidlitz, Germany

Carola Toischel photo

MUDIRA	Ansata Amir Zaman	Ansata Shah Zaman	*Morafic
			*Ansata Bint Mabrouka
		Ansata Bint Sudan	Ansata Ibn Sudan
			*Ansata Bint Bukra
	Matthana	Montasar	Madkour
			Maymoonah
		Mohebba Bint Maymoonah	Madkour
			Maymoonah

ISABELL BOERSING & REINHOLD SZABO
Obergruenthal 6 • 83064 Raubling, Germany

Sharing the dream - Dahman

Descending from Farida through Halima

Judith Wich photo

Ansata Mital
1999 Grey Stallion
Bred by Ansata Arabian Stud

*Love horses, tend them well,
for they are worthy of your tenderness.
Treat them like your own children,
nourish them like friends of the family,
clothe them with care!*

              Sidi Aomar

		Prince Fa Moniet	The Egyptian Prince
	Ansata Iemhotep		Fa Moniet
ANSATA MITAL		Ansata Nefara	Ansata Halim Shah
			Ansata Sudarra
		Ansata Halim Shah	*Ansata Ibn Halima
	Ansata Majesty		Ansata Rosetta
		Ansata Malika	*Jamilll
			*JKB Masouda

ORIENTA ARABIANS
Judith Wich • Eichenbuehl 26 • Wilhelmsthal, Germany 96352

Sharing the dream - Dahman

Descending from Farida through Futna

Scott Trees photo

Jubilllee
1989 Bay Mare
Sire: The Minstril (Ruminaja Ali x *Bahila)
Dam: Ansata Justina (*Jamilll x Ansata Judea)
Bred by Ansata Arabian Stud
(Justina purchased by Arabians Ltd.)

Gigi Grasso photo

Insignia DeSha
2000 Bay Colt
Sire: Thee Desperado
(The Minstril x AK Amiri Asmarr)
Dam: Jubilllee
Full brother to Thee Cappuccino
and DeSha Java
Bred by Hank or Sandra DeShazer
2001 & 2002: Egyptian Event Futurity Champion;
2002: Egyptian Event Jr. Champion Stallion and Reserve Supreme Champion Stallion;
Scottsdale Top Ten

These pages are in tribute to Ansata Justina, a most valued broodmare, and to her owners, Phyllis & the late Hank Browning, who loved her dearly.

DESHAZER ARABIANS
Hank & Sandra DeShazer • 17025 Shaw Road • Cypress, Texas 77429

Sharing the dream - Dahman

Descending from Farida through Futna

Thee Cappuccino
1994 Bay Mare
Full sister to Insignia De Sha & DeSha Java
Bred by Hank or Sandra DeShazer
Egyptian Event Futurity Champion 1996

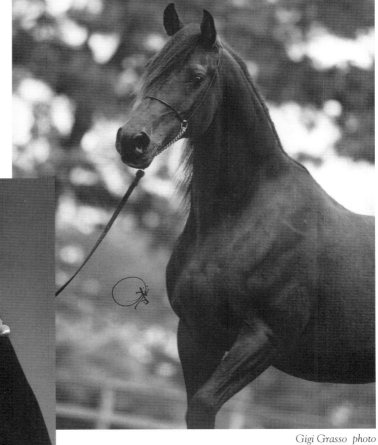

Gigi Grasso photo

DeSha Java
1998 Black Mare
Full sister to Thee Cappuccino & Insignia De Sha
Bred by Hank or Sandra DeShazer
An Egyptian Event Futurity Res. Champion

Scott Trees photo

DESHAZER ARABIANS
Hank & Sandra DeShazer • 17025 Shaw Road • Cypress, Texas 77429

Sharing the dream - Dahman

Descending from Farida through Futna

Van Lent photo

Heirogance
1997 Grey Stallion
David Houseknecht, U.S.A.
Winner of Egyptian Event
Champion in Europe
and Most Classic Arabian Head
in U.S.A.

	El Hilal	Shaikh Al Badi
VP Regal Heir		Bint Magidaa
	Bint Sammara	Mohssen
HEIROGANCE		*Sammara
	AK Bafadi	Shaikh Al Badi
BSA Bataarha		AK Fariha
	Taarifa	*Moatasim
		Bint Deenaa

HANAYA ARABIANS
Nayla Hayek • Expohof •Schleinikon, Switzerland CH 8165

Sharing the dream - Dahman

Descending from Farida through Futna

Van Lent photo

Thee Brigadier
1992 Bay Stallion
Bred by Phyllis Browning, U.S.A.
Champion Wels, Austria & Champion Egyptian Event Europe

In the race-course of valour
May Allah bless the noble courser!
His chest is of steel, and his flanks of iron;
And when he gallops, he puts the thunder to shame
                Amir Abd el Kader

		Ruminaja Ali	Shaikh Al Badi
			Bint Magidaa
	The Minstril		Ibn Galal 1
		*Bahila	Bakria
THEE BRIGADIER			
		*Jamilll	Madkour I
			Hanan
	Ansata Justina		Ansata Abu Sudan
		Ansata Judea	Ansata Jamila

HANAYA ARABIANS
Nayla Hayek • Expohof •Schleinikon, Switzerland CH 8165

Sharing the dream - Dahman

Escher photos

Descending from Farida through Futna

Endow Blu
1989 Grey Mare
Sire: Ansata Halim Shah
(*Ansata Ibn Halima x Ansata Rosetta)
Dam: Ansata Diana (Ansata Ibn Sudan x Mossa RSI x Soja RSI x Bint Dahma x Dahma II x Futna)
Bred by T. Patterson & M. Albertini, U.S.A.
Res. Championesse Gieboldehausen 1997

AL MUNTAHA
Christian Kesseböhmer • Waldstrasse 56 • Bad Essen 49152 Germany

Sharing the dream - Dahman

Descending from Farida through Futna

Filsinger photo

Farres
1995 Grey Stallion
Bred by
David and Kathy Teitrick, U.S.A.
2001 Egyptian Event
Champion of Class,
Stallions 1996 & before;
Region VI Res. Champion Stallion;
2002 German Champion Licensing
Show - Awarded Gold Medal
Breeding Stallion
Sire of Champions
including **Ansata Mouna**,
& **Ansata Qasim**, 2002 Egyptian
Event Top Ten in Futurity and World
Yearling Class

	Ruminaja Ali	Shaikh Al Badi
Anaza El Farid		Bint Magidaa
	Bint Deenaa	*Ansata Ibn Halima
FARRES		*Deenaa
	MFA Mareekh Amir	*El Mareekh
Shameerah		MFA Monien
	AK Shahniya	AK Shah Moniet
		AK Fanniya x Narimaan x *Deenaa

KARIJA ARABIANS
Michael & Valerie Resch • P.O. Box 525 • Emerson, Manitoba, Canada R0A 0L0

Sharing the dream - Dahman

Descending from Farida through Futna

Stine photo

Imperial Baarakah
1990 Chestnut Mare
Bred by Imperial Egyptian Stud
2000 Res. Champion Filly, Region 15;
2000 Champion Futurity Yearling Filly, Egptian Event;
2002 Champion Extended Specialty 3 Year Old Mares.

She exemplifieth that which is called beautiful.
Salim Abdulla Haj

		PVA Kariim	Imperial Imdal
	Imperial Baarez		BKA Rakiisah
		BB Ora Kalilah	*Orashan
IMPERIAL BAARAKAH			PH Safina
		Imperial Al Kamar	El Hilal
	Imperial Karmah		Imperial Sonbesjul
		Imperial Janaabah	Imperial Im Jasim
			Imperial Fanniya x *Deenaa

AA ARABIANS
Brooke & Candi Weeks • 14528 Lime Kiln Road • Grass Valley, California 95949

Sharing the dream - Dahman

Descending from Bint El Bahreyn through Maisa

J. Little photo

Hilala Mystique
1990 Grey Mare
Bred by Allen &
Donna Clemenger
Class A Halter Champion
1992 Egyptian Event West
Champion Futurity Filly

*"Ageless,
Timeless Beauty
and Eloquence"*

		Ansata Imperial	Ansata Ibn Sudan
	Imperial Imperor		Ansata Delilah
		Imperial Mistry	Mosry
HILALA MYSTIQUE			Ansata Nile Mist
		El Hilal	*Ansata Ibn Halima
	Moon Mystique		*Bint Nefisaa
		Nafairtiti	*Morafic
			*Bint Maisa El Saghira

AA ARABIANS
Brooke & Candi Weeks • 14528 Lime Kiln Road • Grass Valley, California 95949

Sharing the dream - Dahman

Descending from Farida through Futna and from Bint El Bahreyn through Zareefa

Gigi Grasso photo

CN Jericho
1993 Chestnut Stallion
Sire: Thee Desperado
(The Minstril x AK Amiri Asmarr)
Dam: Jubilllee (The Minstril x Ansata Justina)
Multiple Egyptian Event Top Tens, Regional Ribbons Western Pleasure & Dressage; Leading Juvenile Stallion 2001; Top Junior Sire of Halter Champions 2001;
Sire of Jr. Res. Champion Filly, Egyptian Event 2002
Owned by Second Wind & Dundee Training Stable

….the desert bred has something characteristic in its look, in the peculiar metallic shine of its hair, in the firmness of its muscles, in the steel of its tendons and hooves, in the special animation of its temperament..
                                                    J. v. Vugel

J. Little photo

AJ Ahlams Delight
1984 Black Mare
Sire: Moniet El Sharaf
Dam: Il Bint Ahlam
(Bay Halima x *Cleopatra x
Ahlam II x Bint Zareefa)

…The beauty of…head, eyes, ears, jaw, mouth and nostrils should be seen to be appreciated…The eye is peculiarly soft and intelligent with a sparkle characteristic of the breed…The neck is a model of strength and forms a perfect arch…The build of the Arab is perfect…
                                                    Homer Davenport

SECOND WIND
M. Kent Mayfield & John R. Ford, Jr. • 5653 State Road 130 • Dodgeville, Wisconsin 53533

Sharing the dream - Dahman

Descending from Farida through Halima and from Bukra through Ansata Bint Sudan

Montasar 1981 Grey Stallion
Sire: Madkour (Morafic x Maisa)
Dam: Maymoonah (Hadban Enzahi x Malikah x
Malacha x Moheba x Halima)
Bred by Seidlitz
Gold Medalist German Stallion Licensing 1984
Sire of **Maareesa**, Class Winner and
Top Ten Egyptian Event Europe 1989

Marqueesa
1991 Grey Mare
Sire: Ansata Amir Zaman (Ansata Shah Zaman x
Ansata Bint Sudan) Dam: Maareesa (Montasar x
Maamounah x Maymoonah)
Bred by Seidlitz • Top Ten Asil World Cup 2000

Ansata Amir Zaman 1984 Grey Stallion
Sire: Ansata Shah Zaman
(*Morafic x *Ansata Bint Mabrouka)
Dam: Ansata Bint Sudan
(Ansata Ibn Sudan x *Ansata Bint Bukra)
Bred by Ansata Arabian Stud, U.S.A.
Overall Champion of the German Stallion
Licensing 1990. Sire of **Mahdeenah**,
Champion Mare, Paderborn 2000

This page is dedicated to the memory of Günter W. Seidlitz

VOLBLUTARABER-GESTÜT FAMILIE GÜNTER W. SEIDLITZ
Gut Muggenbach • Sesslach, Germany D-96145

Sharing the dream - Dahman

Descending from Sabah through Bukra and Sabah through Kamar

Escher photos

Ansata Aya Maria 1985 Chestnut Mare
Sire: Ansata Shah Zaman
(*Morafic x *Ansata Bint Mabrouka)
Dam: Ansata Aya Halima (*Ansata Ibn Halima x
*Ansata Bint Misr x *Ansata Bint Bukra)
Bred by Ansata Arabian Stud, U.S.A.
Dam of **Ali Zaman**, 1997 Jr. Res. Champion, Gieboldehausen,
Liberty Champion Kaub, Class Winner Pyramid Cup.
1998 Jr. Champion Gieboldehausen

Pasha Farid 1996 Black Stallion
Sire: Anaza El Farid (Ruminaja Ali x Bint Deenaa)
Dam: Shariha (Moniet El Sharaf x AK Tashiha x
Hegrah x Hagir x Kamar)
Breeder: Marian Christiansen
Silver Medal Winner Stallion Licensing Show,
Aachen, Germany 1999

GR Aya Farida 1998 Bay Mare
Sire: Pasha Farid
Dam: Ansata Aya Maria
Bred by Rothenberg Stud, Annette & Erwin Escher

ROTHENBERG STUD
Annette & Erwin Escher • 86653 Monheim, Germany

Sharing the dream - Dahman

Descending from Sabah through Kamar

Gigi Grasso photo

Abbas Pasha I-12
1988 Grey Mare
Bred by Babolna State Stud, Hungary
Several show wins in Halter classes. In performance completed 2000 km. ride. Dam of **Muniah**, a Sr. Champion Mare & **Abbas Sinan,** Jr. Res. Champion, Tilbourg 1997 & Jr. Res. Champion "Arabian Horse Cup", Saudi Arabia 1999.

From the past, to the present...for the future.

	Ansata Abbas Pasha	*Ansata Ibn Halima
Abbas Pasha I		*Ansata Bint Mabrouka
	2 Ghalion-11	Ghalion
ABBAS PASHA I-12		12 Bint Inas
	Zohair	Alaa El Din
211 Zohair-2		Zebeda
	28 Farag-3	Farag
		9 Tamria (*Tuhotmos x Kamar)

AL-ZAMET
Susanne & Ingo Schreibvogel • Zum Jugendheim 4 • D21220 Seevetal, Germany

Sharing the dream - Dahman

Descending from Farida through Inas

Bint Amer
1992 Grey Mare

...purebred mares whose far-removed ancestors were numbered among our tribes from ancient times, They are as the gazelles grazing in the valley; to see them is to forget the authors of one's days!
From an Ancient Arabic Song

Gigi Grasso photo

		Ansata Halim Shah	*Ansata Ibn Halima
	Salaa El Dine		Ansata Rosetta
		Hanan	Alaa El Din
BINT AMER			Mona
		Amer	Mohafez
	236 Amer		Arussa
		2 Ghalion-11	Ghalion
			12 Bint Inas x Inas x *Ghazalahh

ALFABIA STUD
Gigi Grasso & Paolo Damilano & • via Pittamiglio 1 • Cherasco, Italy 12062

Sharing the dream - Dahman

Descending from Sabah through Kamar

Taghira B
1995 Grey Mare
Sire: El Thay Mameluk
(Ibn Nazeema x El Thay Mansoura)
Dam: 211 Zohair -2
(Zohair x 28 Farag-3 x Tamria x Kamar)

Alfabia Mameluka
2002 Grey Mare
Sire: ZT Faaiq
(Anaza El Farid x ZT Jamdusah)
Dam: Taghira B

Alfabia Sohar
2001 Grey Colt
Sire: Salaa El Dine
(Ansata Halim Shah x Hanan)
Dam: Taghira B

Gigi Grasso photos

ALFABIA STUD
Gigi Grasso & Paolo Damilano & • via Pittamiglio 1 • Cherasco, Italy 12062

Sharing the dream - Dahman

Descending from Sabah through Kamar

Sally Richerson photo

RA Jehan
1988 - 2000 Chestnut Mare
Sire: AK El Zahra Moniet
(*Ibn Moniet El Nefous x Maarena)
Dam: Bint Bint Tuhotmos
(*Tuhotmos x Bint Tuhotmos x
*Gazbeya x Hagir x Kamar)
Bred by Dr. A.B. Melton & Sheryl Melton
All of Jehan's foals that have been shown
have been Top Five Halter winners.

*For all that fair is, is by nature good;
that is a sign to know the gentle blood.*
                                    Edmund Spenser

Sally Richerson photo

Abiebi
1996 Chestnut Mare
Sire: El Hadiyyah (*Ansata Ibn Halima x Ansata Jellabia)
Dam: RA Jehan
Bred by Sharon Litizzette
Abiebi was exported to Australia in 1997
where she was shown to 1st Place in her age group.

Taher Sihr
1998 Chestnut Stallion
Full brother to Abiebi
Bred by Sharon Litizzette
Multiple Class A Top 5's,
3rd Place EBC Egyptian Event

This page is dedicated to RA Jehan, our beloved mare who left us much too soon and took a piece of our hearts with her.

WINDAMERE ARABIANS
Ed and Sharon Litizzette • 9121 Aspen Drive • Weed, California 96094

Sharing the dream - Dahman

Descending from Sabah through Kamar

Love horses and take care of them,
Spare no trouble;
By them comes honor,
by them comes beauty.
   Ben-el-Abbas

MB Sataarka
1996 Grey Mare
Bred by Montebello
Egyptian Bloodstock

Gotcha photo

		*Ibn Safinaz	Seef
	Imperial Saturn		Safinaz
		Imperial Mistilll	*Jamilll
MB SATAARKA			Ansata Nile Mist
		*Zaghloul	Gassir
	Kaaramal		Gharbawia
		Kahleela	*Faleh
			*Kahramana x Kamar

ROYAL SHAHARA ARABIANS
Michael & Leslie Nord • 13517 S. Greyhawk Lane • Spokane, Washington 99224

Sharing the dream - Dahman

Descending from Farida through Helwa

Farressa CA
2001 Black-Bay Mare
Sire: Farres
(Anaza El Farid x Shameera)
Dam: Queen Sheeba
(Thee Desperado x Minstrils Gabriela)
Bred by Steve & Mauri Chase

Scott Trees photos

We began in Arabians in memory of my deceased son, Weston.

ROYAL SHAHARA ARABIANS
Michael & Leslie Nord • 13517 S. Greyhawk Lane • Spokane, Washington 99224

Sharing the dream - Dahman

Descending from Farida through Helwa

Queen Sheeba
1996 Black Mare
Sire: Thee Desperado
(The Minstril x AK Amiri Asmarr)
Dam: Minstrils Gabriela
Bred by Steve & Mauri Chase

J. Little photo

Izara Blue CA
1998 Grey Mare
Sire: Ansata Hejazi (Ansata Halim Shah x Ansata Sudarra)
Dam: Minstrils Gabriela
Bred by Steve & Mauri Chase
and Owned by Danny & Miry Miro

J. Little photo

Bint Gabriela CA
2000 Grey Mare
Full sister to Izara Blue CA
Bred by Steve & Mauri Chase

Sparagowski photo

CHASE ARABIANS
Steve & Mauri Chase • 4641 New Hope Road • Aubrey, Texas 76227

Sharing the dream - Dahman

Descending from Farida through Helwa

Princess Shamira
1983 Grey Mare
Sire: The Egyptian Prince
(*Morafic x *Bint Mona)
Dam: AK Shahgat
(*AK Shahm x *Nagat x Abla x Helwa)
Bred by Janice Abrams Ferreyra

J. Little photo

Minstril's Gabriela
1990 Grey Mare
Sire: The Minstril (Ruminaja Ali x *Bahila)
Dam: Princess Shamira
Bred by Interfund Corporation
Class A Halter Champion; 1992 Egyptian Event,
Third Place World Class Fillies.

*We are shaped and fashioned
by what we love.*
                Author unknown

J. Little photo

CHASE ARABIANS
Steve & Mauri Chase • 4641 New Hope Road • Aubrey, Texas 76227

Sharing the dream - Dahman

Descending from Bint El Bahreyn through Maisa

J. Little photo

Shaamisa Mystique
1987 Grey Mare
Sire: Nabiel (*Sakr x *Magidaa)
Dam: Moon Mystique
(El Hilal x Nafairtiti x *Bint Maisa El Saghira)
Bred by John R. & Joyce L. Hurd
Dam of multiple Egyptian Event World Class &
Futurity Champions and **Entebbe CA**.

Blessed are the Broodmares…
For they shall bear foals.
  M. Phyllis Lose

Entebbe CA
2000 Grey Stallion
Sire: Thee Desperado
(The Minstril x AK Amiri Asmarr)
Dam: Shaamisa Mystique
2002 Egyptian Event Res. Jr. Champion Colt
and Res. Champion Futurity Colts

Sparagowski photo

CHASE ARABIANS
Steve & Mauri Chase • 4641 New Hope Road • Aubrey, Texas 76227

Sharing the dream - Dahman

Matriarchs of the Bint El Bahreyn Family: *Bint Maisa El Saghira

The Bint El Bahreyn line is relatively small in number compared to the El Dahma family of Ali Pasha Sherif. Bred by Aissa Ibn Khalifah, Sheikh of Bahreyn, she was given to Khedive Abbas Pasha II in 1903 and later purchased by Lady Anne Blunt in 1907. Although unquestionably of Dahman origin, she had her own family and, as mentioned earlier, this Bint El Bahreyn line has generally been taller and bigger-bodied than the El Dahma line. Tabulated in the Pyramid Society's *Manual of Straight Egyptian Arabian Horses*, at the close of 2000, this family totalled 1107 registered in contrast to 5884 of the El Dahma of Ali Pasha Sherif. In addition to the descendants of Maisa (by Shahloul), one also finds Bint El Bahreyn family lines to Elwya through her daughter, *Ansata Bint Elwya (by Antar), and another family through Tifla, which is increasing in number. The Maisa line has been internationally prominent through two particular individuals: her tall grey son, Madkour (by *Morafic) in Europe, and her daughter *Bint Maisa El Saghira in America.

If ever there were a beloved Egyptian mare who helped popularize Egyptian Arabian horses in America during its rise in the 60's and 70's, it was *Bint Maisa El Saghira. Imported by Gleannloch Farms as a filly, she grew into a tall, elegant and noble mare with an attractive head, large eyes, and a strong body. A mare who could win in both halter and performance, she began her show career in concert with *Ansata Ibn Halima, and together they brought classic performances into the show ring. Both won championships at halter, English Pleasure and Park, much to the delight of spectators who loved to cheer them on.

Zareefa (Kazmeyn x Durra)

Bint Maisa El Saghira	Nazeer	Mansour	Gamil Manial
			Saklawi II
			Dalal Al Zarka
		Nafaa El Saghira	Meanagi Sebeli
			Nafaa El Kebira
	Bint Samiha	Kazmeyn	Sotamm
			*Kasima
		Samiha	Samhan
			Bint Hadba El Saghira
	Maisa	Shahloul	Ibn Rabdan
			Rabdan El Azrak
			Bint Gamila
		Bint Radia	Mabrouk Manial
			Radia
		Zareefa	Kazmeyn
			Shahloul
			Zareefa
		Durra	Saadun
			Dalal Al Hamra

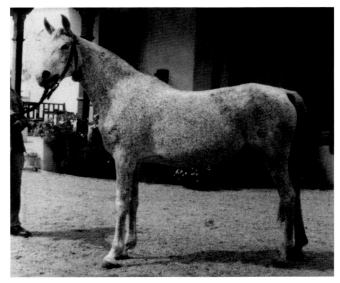

Maisa (Shahloul x Zareefa)

**Bint Maisa El Saghira (Nazeer x Maisa). R. McNair photo*

Sharing the dream - Dahman

Shaikh Al Badi (*Morafic x *Bint Maisa El Saghira). J. Johnston photo

Nafairtiti (*Morafic x *Bint Maisa El Saghira) J. Little photo.

Radia (*Morafic x *Bint Maisa El Saghira). R. McNair photo

Amaal (*Morafic x *Bint Maisa El Saghira)

Rihahna (*Morafic x *Bint Maisa El Saghira).

Sharing the dream - Dahman

Shamruk (*Ibn Hafiza x *Bint Maisa El Saghira). Bickle photo.

*Bint Maisa endeared herself to audiences across America while she was on the campaign trail, not only because of her kind temperament, but because she performed and entertained so well in different disciplines. When she was retired to become a broodmare, her first foal was sired by her traveling companion, *Ansata Ibn Halima. Together they produced an especially noble bay filly named Dahmah Shahwaniah, whose big black beautiful eyes captivated all who saw her. She was shorter headed and prettier than her dam, with good well-shaped neck, strong short-coupled body, good topline and prideful carriage. In addition to Dahmah Shahwaniah, *Bint Maisa El Saghira produced eight more foals: three daughters were by *Morafic; the handsome greys Rihahna and Nafairtiti, and the good-looking Radia, a bay with white markings. The very beautiful chestnut, Dahma Il Ashekwar was sired by Fahidd, and the grey Mashallah was by *Moftakhar. However, *Bint Maisa el Saghira became especially renowned through her sons who became popular breeding stallions. Two were by *Morafic: the white Amaal, much appreciated for his exotic head and huge black eyes, and his handsome grey full brother, Shaikh El Badi, a U.S. National Futurity Champion Stallion and sire of many champions including U.S. Res. National Champion, Ruminaja Ali, and multi-champions Alidaar and Ruminaja Bahjat, among others. Her last son was the bay Shamruk (by *Ibn Hafiza), who made his mark in Canada.

The treasured values of *Bint Maisa El Saghira can be found in her many descendants around the globe. Gone now, one likes to imagine that she and her traveling companion are reaping their rewards in realms beyond - peacefully grazing on pastures green, having sown fertile seeds on earth.

Dahmah Shahwaniah (*Ansata Ibn Halima x *Bint Maisa El Saghira). J. Johnston photo

Ibn Dahman (*Ibn Hafiza x Dahmah Shahwaniah). J. Johnston photo

To every thing there is a season, and a time to every purpose under the heaven:
a time to plant, and a time to pluck up that which was planted...
   Ecclesiastes

Sharing the dream - Dahman

Descending from Bint El Bahreyn through Maisa
and from *Bint Bint Sabbah through Fa-Habba

Van Lent Jr. photo

Abraxas Halimaar
1990 Grey Stallion
Sire: El Halimaar
(*Ansata Ibn Halima x RDM Maar Hala)
Dam: SF Moon Maiden (Nabiel x Kachina Moon x
Nafairtiti x *Bint Maisa El Saghira)
Bred by Felino Cruz, M.D.
U.S. & Canadian National Top Ten Stallion
U.S. & Canadian National Champion Stallion,
A.O.T.H. East Coast Champion Stallion,
Region VII Champion Stallion,
Region XII Champion Dressage. Egyptian Event
Supreme Champion Stallion. I.A.H.A. High Point
Halter Horse of the Year. Sire of Regional, Egyptian
Event and International Winners.

Hadaya El Tareef
1986 Grey Stallion
Sire: Imperial Imdal
(Ansata Imperial x Dalia)
Dam: Hadaya Nile Merytaten
(Ansata Halim Bey x Hadaya Nile Nefertiti x
Ansata Nile Jewel x Falima)
Bred by John & Marianne Hacklander
U.S. & Canadian National Top Ten Stallion,
Egyptian Event Supreme Champion;
East Coast Champion Stallion,
Regional Championships;
International Sire of Champions.
Twice Egyptian Event Get of Sire Winner.
Sire of Champion **Tareefs SA Diyyah**

Stuart Vesty photo

CHAPEL FARMS ARABIANS
Robert & Christine Fauls • 3129 Smokey Road • Newnan, Georgia 30263

Sharing the dream - Dahman

Descending from Farida through Nefisa and from Bint El Bahreyn through Maisa

Abraxas Moonique
2000 Grey Mare
Sire: Abraxas Moonstruk
(ET Crown Prince x SF Moon Maiden)
Dam: Cashai
Bred by Abraxas Arabians
Jr. Champion Filly

Abraxas Nejd Moon
2001 Bay Stallion
Full brother to Abraxas Moonique
Bred by Abraxas Arabians

J. R. Little photos

Abraxas Moonstruk 1992 Bay Stallion
Sire: ET Crown Prince
(The Egyptian Prince x RDM Maar Hala)
Dam: SF Moon Maiden (Nabiel x Kachina Moon x Nafairtiti x *Bint Maisa El Saghira)
Bred by Abraxas Arabians • Egyptian Event Top Four

ABRAXAS ARABIANS
Emil & Debra Nowak • 32415 Yates Road • Winchester, California 92596

Sharing the dream - Dahman

Descending from Farida through Nefisa

Cashai
1991 Grey Mare
Sire: *Simeon Shai
(Raadin Royal Star x Simeon Safanad)
Dam: Nourah (*Ansata Ibn Halima x
*Bint Nefisaa x Nefisa)
Bred by Dr. Charles & Judy Jones,
Atallah Arabians

J. R. Little photo

Polly Knoll photo

Patinaa
1995 Grey Mare
Sire: VP Regal Heir
(El Hilal x Bint Sammara)
Dam: Sidra (El Hilal x Nourah x *Bint Nefisaa)
Bred by Dr. Charles & Judy Jones,
Atallah Arabians

J. R. Little photo

Abraxas Prince 1999 Grey Stallion
Sire: ET Crown Prince (The Egyptian Prince x
RDM Maar Hala) Dam: Cashai • Bred by Abraxas Arabians

ABRAXAS ARABIANS
Emil & Debra Nowak • 32415 Yates Road • Winchester, California 92596

Descending from Bint El Bahreyn through Tifla

Hawaa
1985 Grey Mare
Sire: Amaal
(*Morafic x *Bint Maisa El Saghira)
Dam: Haseenah
(*Zaghloul x Hebet Allah x *Soheir II x Tifla)
Bred by Gleannloch Farms
Class A Halter Champion
Dam of **Zourrak**, Reserve National Champion, Chile;
The Mistral, Supreme Champion Gelding, Egyptian Event;
Shah al Hawaa, Supreme Champion Gelding, Egyptian Event;
Ma-Ajmala, International Senior Champion, Top Five All Nations Cup Aachen, Class Winner European Championship Verona, Italy

Escher photo

Jenni May photo

DHS Halima Bint Hawaa
2001 Grey Mare
Sire: Thee Brigadier
(The Minstril x Ansata Justina)
Dam: Hawaa
Bred by Mr. & Mrs. F. J. Schwestermann, Dowdstown House Stud

DOWDSTOWN HOUSE STUD
Mr. & Mrs. F. J. Schwestermann-Lawless • Dowdstown • Maynooth, Co. Kildare, Ireland

Sharing the dream - Dahman

Descending from Bint El Bahreyn through Tifla

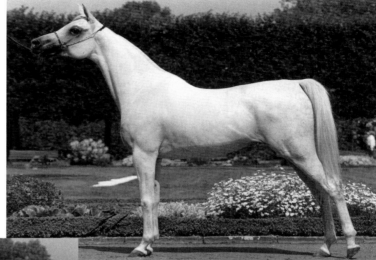

Ma-Ajmala
1993 Grey Mare
Sire: The Minstril
(Ruminaja Ali x *Bahila)
Dam: Hawaa
Bred by Mrs. S. Remond, Texas, USA
International Senior Champion,
Top Five All Nations Cup Aachen,
Class Winner European
Championship Verona, Italy

Escher photo

DHS Miraya El Nizr
1998 Grey Mare
Sire: Anaza El Nizr (Ruminaja Ali x Bint Deenaa)
Dam: Ma-Ajmala
Bred by Mr. & Mrs. F. J. Schwestermann,
Dowdstown House Stud
Top Five All Nations Cup Aachen

Escher photo

DHS Mabrouk
2002 Grey Colt
Sire: Al Lahab (Laheeb x Thee Vision HG)
Dam: Ma-Ajmala
Bred by Mr. & Mrs. F. J. Schwestermann,
Dowdstown House Stud
Son of European Champion Stallion, Verona, Italy

Jenni May photo

DOWDSTOWN HOUSE STUD
Mr. & Mrs. F. J. Schwestermann-Lawless • Dowdstown • Maynooth, Co. Kildare, Ireland

Sharing the dream - Dahman

Ron Shimer photo

Descending from Bint El Bahreyn through Tifla and from *Bint Bint Sabbah through Fay-Sabbah

Fayalia Nile Moon
Sire: Anaza El Farid (Ruminaja Ali x Bint Deenaa)
Dam: Senisa Fayali
(Ruminaja Ali x *Naadya x Tifla x Elwya)
Bred by Karen Henwood

*Your mother was taken from you too soon,
But in you, she left her Beauty of Spirit, Mind and Body.*
  Karen Henwood (in memory of Senisa Fayali)

Scott Trees photo

Three Lovely Dahma Mares tracing to Fay Sabbah: (Left to right)
Aleseri (Ibn El Naseri x Alecsis) 1992 Mare, Bred by Karen Henwood
Glenglade Lecsi (full sister to Aleseri) 1990 Mare, Bred by Mr. & Mrs. Willis Flick
Alecsis S (Ansata Iemhotep x Glenglade Lecsi x Alecsis x Glenglade Dahira x Bint Shebaa x Shebaa x Fay Sabbah), 1999 Mare bred by Karen Henwood

This page celebrates the memory of Ansata El Naseri, Glenglade Dahira, Alecsis and Senisa Fayali

SANDYBROOK EGYPTIAN ARABIAN FARM
Karen Henwood • 13885 N.W. Highway 27 • Ocala, Florida 34482

Sharing the dream - Dahman

For More Than 40 Years –
Arabian Horse World
and Ansata Arabians –
The Standard for Beauty.

ARABIAN HORSE WORLD
1316 Tamson Drive, Suite 101, Cambria, CA 93428 U.S.

Al Rayyan mares at the seaside, Al Wakrah, Qatar. Gigi Grasso photo.

The Saqlawi Strain

Part 2. Al Sudaniyat

Dabbi ibn Shutaywi and Ali ibn Sudan were questioned in the house of Dabbi Shutaywi in the presence of Hussein ibn Shalhoub and Zaydan ibn Turki and Saud ibn Said, the paternal uncle of Neyf ibn Murshid, and Haizum, the son of the brother of Dabbi, and in the presence of the old and young people of al Gomussa.

They were asked, "O Dabbi ibn Shutaywi and O Ali Sudan, you say that the horses of Ibn Sudan are Saqlawiyat Jedraniyat even though we have asked the descendants of al Jedran who were in charge of the stud of Saqlawiyat Jedraniyat and they have told us that as for the horses of Ibn Sudan, they have not heard about them from their grandfathers, nor from their elders that a mare from their stud has passed to you, O al Sab'a. But you, O Sab'a, said to the descendants of Jedran that the horses of Ibn Sudan are from your old stud, O Jedran. You told them during your conversation that if a mare passed to your friends they could claim her from your histories, O al Sab'a, and they have not investigated further. So you must tell us from where the horses of Ibn Sudan have been passed to you and how they passed to you.

Swear to us by your honor, good fortune, your religion and faith. Tell us the positive truth about what you have heard from your grandfathers' grandfathers concerning the history of the horses of Ibn Sudan."

They replied, "By Allah, O Ali Bey, we will tell you the history we have heard from our grandfathers and our elders and the white haired men of the tribe, that she originated from Hadhud ibn Jedran and passed to Beni Khaled on the day of al Arban when Abdul Aziz abu Saud ruled Nejd. It was on the day that water was sold that a shaqra 'awdah mare was disabled from Hadhud ibn Jedran. Hadhud ibn Jedran with his sick shaqra mare passed by al Khaledi, a shepherd of Shuwai of Beni Khaled who was tending his sheep. And Hadhud ibn Jedran went to him with the mare and implored him, 'Look at this mare, al 'awdah, she is disabled. You can see she is a Saqlawiya. Save her and as you save her, I will sell her to you for the price of your saving her life'.

And so the shepherd brought her a gourd of water and he began to let her drink and refreshed her by dampening her nostrils with cool water. And Hadhud went back to his people.

And when the mare had drunk the water she revived and she again became beautiful. And an Arab of Beni Hussein bought her from al Khaledi, who had given her water to drink and the mare went to Beni Hussein. Al Husseini went as a guest to the house of Hadhud ibn Jedran and al Husseini said to ibn Jedran, 'O Hadhud, that shaqra mare which had been disabled and was separated from you, O Hadhud, in the region of Beni Khaled, from which stud is she?'

And Hadhud ibn Jedran replied, 'By Allah, O Husseini, she is my mare, Saqlawiya Jedraniya.'

from the Abbas Pasha Manuscript

Matriarchs of the Roga/Moniet El Nefous Family: Mabrouka and Mona

Forbis photos

		Rabdan El Azrak	Dahman El Azrak
	Ibn Rabdan		Rabda
		Bint Gamila	Ibn Nadra
Shahloul			Gamila
		Mabrouk Manial	Saklawi II
	Bint Radia		Tarfa
		Radia	Feysul
Moniet El Nefous			*Ghazala
		Mansour	Gamil Manial
	Sheikh El Arab		Nafaa El Saghira
		Bint Sabah	Kazmeyn
Wanisa			Sabah
		Awad	Mabrouk Manial
	Medallela		Bint Obeya
		Khafifa	Ibn Samhan
			Dalal Al Zarka

The Saklawi strain has long been cherished for its refinement, elegance, pride, and courage. Part of its colorful history, as related in the *Abbas Pasha Manuscript*, introduces and sets the tone for this section. The Saklawi strain is listed second to the Dahman in the manuscript, and coincidentally the Saklawi Jedrans are also second in total number (3846) of straight Egyptians registered in America up to the millennium. Having such a collection of prominent horses within its family, makes it difficult to select a few from the many who have given the Saklawi strain such celebratory status. However, the Mahiba family has shared in the dreams of countless breeders, initially in Europe, and subsequently around the globe. Therefore it is appropriate to honor its contribution to the Saklawi Jedran strain in general, and to the Moniet El Nefous line in particular.

Queen of the Nile, she was affectionately called, and while Moniet El Nefous was still alive and gracing the pastures of El Zahraa, visitors dared not enter her paddock without bringing sugar cubes as tribute. She has been eulogized in magazine articles, and in *Authentic Arabian Bloodstock*, as well as in a charming little book, *Moniet el Nefous, Mouna, Mahiba, Eine edle Fuchsfamilie aus Äegypten* by Dr. H. J. Nagel and W. Eberhardt. Dam of 14 foals, Moniet's Sid Abouhom daughter, Mabrouka, generated an imposing male line (the stallion *Morafic, and through his full sister, *Ansata Bint Mabrouka, evolved the stallions Ansata Ibn Sudan, Ansata Abbas Pasha and Ansata Shah Zaman).

Mabrouka's full sister, Mouna became the most influential in production of females to carry on Moniet's line. Significant among those was the enchanting grey, *Bint Mouna (by Nazeer), who impacted worldwide programs through her daughters and particularly through her inbred son, The

Mabrouka (Sid Abouhom x Moniet El Nefous)

Mona (Sid Abouhom x Moniet El Nefous)

*Morafic (Nazeer x Mabrouka)

*Bint Mona (Nazeer x Mona)

Mahiba (Alaa El Din x Mona). Nagel collection.

Mona II (Mahomed x Mahiba x Mona). Nagel collection.

Egyptian Prince (by *Morafic – her full blood brother). *Bint Mouna had been sold to the Minister of Agriculture, Sayed Marei, but when Douglas Marshall wanted to buy her for his Gleannloch program, Dr. Marei graciously let her go – a gesture for which he will be favorably remembered. While *Bint Mouna sowed fertile seeds in America, her attractive but less comely chestnut half-sister, Mahiba (by Alaa El Din), was methodically going about her work in Germany. Imported in 1968 by Dr. Nagel, she passed on to greener pastures at the age of nine, having produced three influential daughters: 1) Sabah (by Ibn Galal), 2) Mona II (by Mahomed) and 3) Kis Mahiba (by Ibn Galal).

The chestnut Sabah, when bred to Ansata Abbas Pasha, gave birth to the good-looking chestnut stallion, Sherif Pasha, who became World Champion in Paris at the Salon du Cheval. She also produced the distinguished chestnut broodmare, Bint Sabah (by Ibn Nazeema), influential at Al Rayyan Farm in Qatar.

Mona II, when serviced by Ibn Galal, brought forth the handsome bay mare, Madinah. Had Madinah done no more than deliver Madheen (nee *Imperial Madheen), her famous son by *Messaoud, her name in history would have been assured. Here was the combination of bloodlines that was also experiencing phenomenal success in America: the Halima line through *Ansata Ibn Halima (combined with *Morafic, *Ansata Bint Mabrouka and *Bint Mouna), versus Messaoud in Europe through Moheba/Halima combined with Mouna through Mahiba. These combinations of bloodlines in particular were consistent in producing a certain look that became appreciated on both continents and eventually in the Arab world during its renaissance of the 90's.

*Imperial Madheen deserves special mention as a world traveler and distinguished sire. Bred in Germany, he was sold to Imperial Stud in the U.S., who sold a shared interest in

Madinah (Ibn Galal x Mona II). Nagel collection.

Kis Mahiba (Ibn Galal x Mahiba). Nagel collection.

Maysouna (Ibn Galal x Kis Mahiba). Nagel collection.

*Imperial Madheen (*Messaoud x Madinah)

him to Montebello Farm in Canada, who eventually brought him back to Texas where was eventually purchased by Count Federico Zichy-Thyssen and sent to Argentina, South America. Omar Sakr then purchased him, brought him back to America and then shipped him after a brief interlude to Germany, and then on to Egypt, the land of his ancestors. By now he had many miles behind him and few left in front of him. With little time remaining in his life, he was still able to sire a special filly for Omar Sakr before he passed away in 2002 at the Sakr farm in Cairo. An international jet-setter, he left a legacy that Moniet El Nefous can be proud of. In the meantime, the Madinah line remains busily performing its broodmare duties through such mares as *Imperial Madheen's full sister, Mesoudah M, and also producing exceptional show horses like Maydan Madheen, and others.

Also of prominence was Mona II's daughter, Maheera (by Nizam), a foundation mare for C. Tauschke's El Thayeba Stud where Maheera's Ansata Halim Shah sons, El Thay Maher and El Thay Mansour have played a distinguishing role.

Mahiba's third daughter, Kis Mahiba, became the dam of Maysouna, a foundation mare for the Maiworm Stud, who when bred to Ansata Halim Shah, presented them with a priceless gift – the ever popular first class champion and champion sire, Maysoun, whose get carry on a special legacy worldwide.

Other descendants of Moniet are equally deserving of mention, as will be noted throughout this book. Their venerated assets may be summed up by the opening paragraph to Moniet's story found on page 319 in *Authentic Arabian Bloodstock*:

> There are two ways of spreading light:
> to be the candle,
> or the mirror that reflects it.

Sharing the Dream - Saklawi

Descending from Moniet through Mouna

Van Lent photo

*Imperial Madheen
1984 Grey Stallion - Deceased 2002
Sire: *Messaoud
(Madkour x Maymoonah x Malikah)
Dam: Madinah
(Ibn Galal x Mona II x Mahiba x Mouna)
Bred by Madheen Arabians, Dr. S. Paufler
Multi- Champion stallion with progeny
on all continents including
Imperial Madaar, Australia;
Imperial Madori, Egypt;
Imperial Mahzeer, Qatar;
Imperial Mashhar, Morocco.

Escher photo

Mesoudah-M
1985 Grey Mare
Full sister to Imperial Madheen
Bred by Madheen Arabians, Dr. S. Paufler
A Junior Champion, Mesoudah -M is the dam
of International Champions **Maydan Madheen**,
Midoun Madheen, & **Mutair HP.**

Maydan Madheen 1993 Grey Stallion
Sire: Maysoun (Ansata Halim Shah x Maysouna x Kis Mahiba)
Dam: Mesoudah-M
Bred by Madheen Arabians, Dr. S. Paufler
1998 Supreme Champion - Best Egyptian of the Show, Egyptian
Event, Kaub; 2000 Supreme Champion Asil Cup; 2002 German
National Senior Champion Neustadt/Dosse; Multi- Champion

MADHEEN ARABIANS
Dr. Siegfried & Ruth Paufler • Zum Ortloh 15 • Goettingen, Germany D37077

Descending from Moniet through Mouna

Filsinger photo

Kyro KA
2000 Grey Stallion
Sire: Ansata Iemhotep
(Prince Fa Moniet x
Ansata Nefara)
Dam: Desire RB
Bred by Karija Arabians
Top 5, 2002 Nation's Cup, Germany;
2002 Class Winner at Babolna,
Hungary

Famira KA
2002 Bay Mare
Sire: Farres
(Anaza El Farid x Shameerah)
Dam: Desire RB
Bred by Karija Arabians

Filsinger photo

Desire RB
1995 Bay Mare
Sire: Thee Desperado
(The Ministril x AK Amiri Asmarr)
Dam: RA Nahlima
(Nabiel x Halima El Hamra x
Bint Bint Hoyeda)
Bred by R&B&S Melton
Top 5 Egyptian Event Futurity 1996;
Egyptian Event 1997 Res. Jr. Champion Filly &
Futurity Champion Filly.

KARIJA ARABIANS
Michael & Valerie Resch • P.O. Box 525 • Emerson, Manitoba, Canada R0A 0L0

Descending from Moniet through Mouna

Makhnificent KA
1995 Grey Stallion
Sire: Makhsous (*Sultann x *Nabda)
Dam: Star Bint Hafiza)
Bred by Jean Rogers of Kehilan Bloodstock
U.S. Top Ten Yearling Sweepstakes & U.S. Top Ten Futurity Colt, Region 9 Res. Champion Stallion & Champion Yearling Sweepstakes,
Egyptian Event Top Ten. Sire of Multi-champions.

Gigi Grasso photo

Star Bint Hafiza
1986 Bay Mare
Sire: Ibn Dahmahn (*Ibn Hafiza x Dahmah Shahwaniah)
Dam: AK Amiri Fayrouz (The Egyptian Prince x *Maather x Fayrooz x Mouna)
Bred by Florence Lacy Goudeau
Dam of **Makhnificent KA**; **Bodacious KA**, Region 6 Top 5 Sweepstakes Colt, Jr. Champion Stallion, Egyptian Event Top Ten, U.S. Top Ten Hunter Futurity; and **Dahmahn Becca**, Class A Champion, Egyptian Event Top Ten.

Scott Trees photo

KEHILAN ARABIANS
Jean & Herb Rogers, Becky Rogers, Nancy Blankenship and Janice Bush
8059 FM 1187W • Fort Worth, Texas 76126

Descending from Moniet through Mouna

RN Dananeer
1995 Grey Mare
Sire: Ansata Halim Shah
(*Ansata Ibn Halima x Ansata Rosetta)
Dam: Bint Sabah
(Ibn Nazeema x Sabah x Mahiba)
Bred by Sheikh Abdul Aziz Bin Khaled Al Thani,
Al Rayyan Farm

Serene Dalila
2002 Grey Mare
Sire: Salaa El Dine
(Ansata Halim Shah x Ansata Rosetta)
Dam: RN Dananeer
Bred by Serene Egyptian Stud

Gigi Grasso photos

Serene Carima
2001 Chestnut Mare
Full sister to Serene Dalila
Bred by Serene Egyptian Stud

SERENE EGYPTIAN STUD
Anna & Stefano Galber • Loc. SERENE, Via Coletto 2 • Affi (VR), Italy 37010

Sharing the Dream - Saklawi

Descending from Moniet through Mouna

Zahara Keela
1984 Bay Mare
Bred by Jarrell McCracken

*The best treasure of
a man is a fruitful mare.
Allah bade them multiply,
and they have multiplied.*
    Amir Abd el Kader

	*Morafic	Nazeer
The Egyptian Prince		Mabrouka
	*Bint Mona	Nazeer
ZAHARA KEELA		Mouna
	*Ansata Ibn Halima	Nazeer
Nehaya		Halima
	*Hoyeda	*Morafic
		Mouna

QADAR ARABIANS, LLC
Jeff & Judy Barth • N677 Highway 49 • Weyauwega, Wisconsin 54983

Descending from Moniet through Mouna

Faarraah
1994 Grey Mare
Sire: Imperial Immarouf
(Ansata Imperial x
Glorieta Maarqesa)
Dam: Faaraah (Al Fattah x
Foula x *Hoyeda)
Bred by Montebello Farms, Inc.

Bikr NA
2001 Grey Stallion
Sire: Ansata Sirius
(Ansata Iemhotep x Ansata Sekhmet)
Dam: Ansata Fantazi
Bred by Ansata Arabian Stud

Ansata Fantazi
1997 Grey Mare
Sire: Ansata Hejazi
(Ansata Halim Shah x Ansata Sudarra)
Dam: G Fantazia
(Maar Ibn Amaal x *Fantasia x Nafteta x Moneera)
Bred by Ansata Arabian Stud

NAJOUBA ARABIANS
Jessie & Oren MacLean • 1292 Post Road • Sussex Corner, New Brunswick, Canada E4E 2V5

Sharing the Dream - Saklawi

Descending from Om Dalal through Rayana (Inshass)

Rababa
1983 Grey Mare
Bred by The Egyptian Agricultural Organization
Dam of **Turkiya Al Rayyan**,
2nd Place 1999 Qatar International Show, 2nd Place Qatar National Show 1999
and 4th Place Qatar International Show 2000

Van Lent Jr. photo

Horses are your safeguard:
Give them the good things you yourselves like best;
Love horses, and take care of them;
In them alone lie honor and beauty.
　　　　Ben-Sassa of the Beni Aamer

	Ikhnatoon	*Farazdac	Alaa El Din
			Farasha
RABABA		Bint Om El Saad	Nazeer
			Om El Saad
	Roba	Wahag	Anter
			Kamar
		Wahida	El Araby
			Rayana

AL-RAYAH
Ala'a H. Al-Roumi • Kuwait

Descending from Moniet through Mouna and Manaya and Om Dalal through Rayana (Inshass)

Maharan
1999 Grey Stallion
Sire: *Messaoud (Madkour x Maymoonah)
Dam: Hamasa Najiha (Farag x Nafteta x Moneera x Mouna)
Bred by Walter Martin, Germany

Malameh
1996 Grey Mare
Sire: Ibn Akhtal (Akhtal x Looza)
Dam: Mabkhoota (Wahag x Maali x Manaya x Moniet El Nefous)
Bred by The Egyptian Agricultural Organization

Escher photo

Anas El Wogoud
1996 Dark Brown Stallion
Sire: Anas (Alaa El Din x Ramza)
Dam: Sahlalah (Sawab x Soha x Thouraya x Rayana)
Bred by The Egyptian Agricultural Organization
1999 Egyptian Nationals 3rd Place

Escher photo

BEBO STUD
Miloslava Khamis • 4 Sesorstris Street • Korba- Heliopolis, Egypt

Sharing the Dream - Saklawi

Descending from Om Dalal through Wedad (Inshass)

Egyptian Anjali Amanda
1993 Grey Mare
Sire: Alidaar (Shaikh Al Badi x Bint Magidaa)
Dam: Wasoona Bint Kaisoon
(Kaisoon x Waseema x Bint Wedad)
Bred by Veruska Arabians, Switzerland

Egyptian Asinah
1997 Grey Mare
Sire: Ansata Sinan
(Prince Fa Moniet x
Ansata Nefara)
Dam: Egyptian Anjali Amanda
Bred by Sylke Schuhmacher,
Germany
1999 Jr. Champion
Giboldehausen, Germany;
2000 Junior Reserve
Champion D'Alsace, France;
Junior Champion Velgen,
Germany

Escher photo

EICHHOF-ARABER
Sylke Schuhmacher • Heckenweg 20 • 21220 Seevetal, Germany

Sharing the Dream - Saklawi

Descending from Moniet through *Bint Moniet, Om Dalal through Wedad (Inshass) and Radia through Zamzam

MB Jacinta
1993 Grey Mare
Sire: Imperial Immarouf
(Ansata Imperial x Glorieta Maarqesa)
Dam: Bint Jamilla (*Jamill x Monisa Halima x
AK Monisa Moniet x Monisa RSI)
Bred by Montebello Egyptian Bloodstock

photos by Deanna Boyd

Jabarut
1989 Grey Mare
Sire: Dalul (*Morafic x *Dawlat)
Dam: UP Murassa (Maddah x Mahrousa
x Bint Wedad)
Bred by Stonebridge Select
Egyptian Arabian Ltd. Partnership

Bint Tahia
1987 Grey Mare
Sire: Prince Fa Moniet
(The Egyptian Prince x Fa Moniet)
Dam: Gasaara (*Faleh x *Hayam x
Tahia x Kawsar x Zam Zam)
Bred by G. Clark Harrison

Susan McAdoo photo

ROCKY TOP RANCH
Kimberly Blythe & Joseph Murgola • 404 Polk 47 • Mena, Arkansas 71953

Sharing the Dream - Saklawi

Descending from Radia through Saklawia

Shahil
1989 Grey Stallion
Bred by Maiworm Stud, Germany
Champion Stallion:
Four times B-Shows Europe,
Two times National Championship,
and Top Ten Paris (2nd in Class).
Sire of winners:
Frasera Futura, Frasera Shahilla, Frasera Shahin, etc.

Gigi Grasso photo

*His neigh is like the bidding of a monarch,
and his countenance enforces homage....*
       Shakespeare

SHAHIL	Maysoun	Ansata Halim Shah	*Ansata Ibn Halima
			Ansata Rosetta
		Maysouna	Ibn Galal
			Kis Mahiba
	Shahila	Ibn Galal	Galal
			Mohga
		Salima	Kaisoon
			Salha (Sameh x Saklawia)

LA FRASERA
Francesco & Serenella Santoro • Via Borneo 25 • Roma, Italy 00144

Sharing the Dream - Saklawi

Descending from Moniet
through Mouna,
and Radia through Zaafarana

Mouna Al Rayyan
1999 Grey Mare
Sire: Alidaar
(Shaikh Al Badi x Bint Magidaa)
Dam: Bint Sabah
(Ibn Nazeema x Sabah x Mahiba)
Bred by
Sheikh Abdul Aziz Bin Khaled Al Thani,
Al Rayyan Farm, Qatar

Gigi Grasso photo

Amira Al Rayyan
2000 Bay Mare
Sire: Al Aadeed Al Shaqab
(Ansata Halim Shah x Sundar Alisayyah)
Dam: Bint Bint El Araby
(Madkour I x Bint El Araby x
El Amira x Zaafarana)
Bred by
Sheikh Abdul Aziz Bin Khaled Al Thani,
Al Rayyan Farm, Qatar

Filsinger photo

AL-SHARG FARM
Talal Al Nisf • P.O. Box 24989 • Safat 13110, Kuwait

Matriarchs of the Ghazieh/Radia Family: Zaafarana

		Samhan	Rabdan El Azrak
	Ibn Samhan		Om Dalal
		Nafaa El Saghira	Meanagi Sebeli
Balance			Nafaa El Kebira
		Saklawi II	Saklawi I
	Farida		El Dahma
Zaafarana		Nadra El Saghira	Samhan
			Nadra El Kebira
		Rabdan El Azrak	Dahman El Azrak
	Ibn Rabdan		Rabda
		Bint Gamila	Ibn Nadra
Samira			Gamila
		Mabrouk Manial	Saklawi II
	Bint Radia		Tarfa
		Radia (Ghadia)	Feysul
			*Ghazala

Zaafarana (Balance x Samira)

Bint Radia (Mabrouk Manial x Radia)

The Zaafarana line is one of the few remaining lineal descendants that trace to Ghazieh, an original grey Saklawiya Jedraniya Ibn Sudan mare of the Ruala Anazah and imported to Egypt for Abbas Pasha I somewhat prior to 1855. This was a family Lady Anne Blunt much heralded, her famed chestnut stallion, Mesaoud, and the white mares Bint Helwa and *Ghazala "El Beida" being admirable representatives of it. Universally celebrated for producing "elite" refinement, this family is personified by such mares as Serra, Radia "Ghadia" and Samira (full sister to the illustrious sires Shahloul and Hamdan). Zaafarana inherited celebrity status as a daughter of her beautful prize-winning dam, Samira, and her exceptionally athletic sire, Balance, Egypt's most famous race horse and holder of the mile record at 1:45. A mare of good quality, although not as beautiful as her dam, Zaafarana's brilliant trot and regal presence set her apart as something

Samira (Ibn Rabdan x Bint Radia)

Amrulla "Ziada" (Sid Abouhom x Zaafarana)

*Talal (Nazeer x Zaafarana)

Ansata El Nisr (*Ansata Ibn Halima x *Ansata Bint Zaafarana)

*Ansata Bint Zaafarana (Nazeer x Zaafarana). Sparagowski photo.

Ansata Abu Nazeer (*Ansata Ibn Halima x *Ansata Bint Zaafarana) Polly Knoll photo.

special. The mother of ten offspring (six mares, four stallions), she was a mare who could produce beautiful individuals that were also exceptional athletes.

Prominent among her racing sons were Farfour (by Sid Abouhom), who raced under the name of El Barrak, and his full brother Ziada (Amrulla), who was used as a breeding stallion at the E.A.O. when they deviated from the use of the Nazeer blood. A full sister to them was Nahid, also successful on the track, a mare resembling her dam, and whose line has continued to breed on.

Most notable of Zaafarana's sons was *Talal (by Nazeer), who raced successfully in Egypt under the name "Johnny Boy," completing 41 races and retiring sound. Inheriting his dam's type overlaid by his sire's quality, he was a handsome, spirited and regal white stallion with a spectacular trot. A crowd pleaser wherever he was shown, *Talal achieved U.S. National Top Ten Stallion at halter in 1969 for his owner, James F. Kline. Holding court as the senior stallion at Kline Arabians in California until his final days, *Talal's blood is found in many programs through several sons, numerous daughters and their progeny.

Sharing the Dream - Saklawi

Ansata Aziza (*Ansata Ibn Halima x *Ansata Bint Zaafarana) Sparagowski photo

Farid Nile Moon (Anaza El Farid x GA Moon Taj Halima x Ansata Aziza) Javan photo

Emeraldd Moon (Anaza El Farid x GA Moon Tajhalima x Ansata Aziza). Filsinger photo at Arabian Horse Center, Kuwait.

Most special among Zaafarana's daughters was *Ansata Bint Zaafarana (by Nazeer), a mare of exceptional beauty, who produced ten progeny of consistent type, all of them sired by *Ansata Ibn Halima. Most influential among the six of her handsome, high-quality sons were: Ansata El Nisr, a white stallion and good sire, who was shown to U.S. Top Ten Stallion by Don Forbis for owners Apple Hill Arabians, and Ansata Abu Nazeer, foundation sire for Glorieta Ranch, who unfortunately died young, but still left a legacy of marvelous daughters whose bloodlines are wide-spread.

*Ansata Bint Zaafarana's daughters bred on well, particularly the captivating Ansata Aziza, a multi-champion who became the foundation mare for Gibson Arabians. Bred to their stallion, Imperial Imperor (by Ansata Imperial), Ansata Aziza produced the celebrated grey mare, GA Moon Taj Mahal, a multi-champion and dam of champions. Shown at the U.S. Nationals, she barely missed winning Top Ten honors, but as her owner, Dr. Gibson, left the stands, he was heard to remark: "She's still Top Ten to me," a comment other breeders having been in that position well appreciate. Her captivating full sister, the radiant chestnut, GA Moon Taj Halima, is also the dam of champions, having produced the distinguished 2001 Egyptian Event Champion Stallion, Farid Nile Moon, whose full sister, Emeraldd Moon, now represents this family as a broodmare at the Arabian Horse Center in Kuwait.

Sharing the Dream - Saklawi

*Representing both the Ghazieh and Roga female Saklawi lines, left is *Talal (Nazeer x Zaafarana) right is *Bint Mona (Nazeer x Mouna). Patti photo, E. Kline collection.*

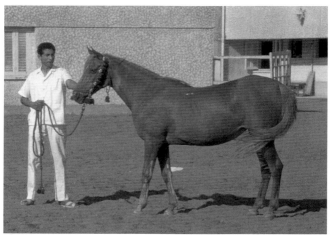

El Amira (Nazeer x Zaafarana). Forbis photo.

Saklawia II (Mashhour x Zamzam x Bint Radia). She produced Salha by Sameh, dam of Shahil. Forbis photo.

Another daughter of Zaafarana was the 1952 chestnut mare, El Amira (by Nazeer), who was sold from the E.A.O. and eventually came to the Marei's Al Badeia Stud. Quite unlike her grey full sister in type, she established herself as one of their foundation mares by producing one son and nine daughters, out of which several have made an international impact.

This Ghazieh line is also represented by a significant, but relatively small group of individuals descending from the elegant mare, Zamzam (Gamil III x Bint Radia "Ghadia), through her Mashhour daughter Saklawia II, the dam of Salha (by Sameh), who founded a popular line in Europe, and from whom descends the well known Italian stallion, Shahil (by Maysoun). In addition, a few Ghazieh lineal descendants can also be found through the Babson *Bint Serra line, thus contributing to the ongoing importance of this much revered Abbas Pasha root mare.

Where are those noble steeds
Whose dam never knew any but a noble sire?
The stirrup is their life; inaction is death to them.
O Father of cavaliers, the ignorant find them everywhere,
But they are as rare as true friends,
And when they die the very saddle sheds tears.

                                    Amir Abd El Kader

Sharing the Dream - Saklawi

Descending from Radia through Saklawia, and through Zaafarana

HF Amneris
2002 Bay Mare
Full sister to HF Aida
Bred by
Veruska Arabians

HF Aida 2001 Grey Mare
Sire: Immell
Dam: UP Sharaftana
(Nizam x Sadeedah x
Saady x Salha x Saklawia II)
Bred by Veruska Arabians

Immell
1997 Grey Stallion
Sire: Imtaarif
(Imperial Imdal x Taarifa)
Dam: Aniya
(*Ibn Moniet El Nefous x *Haniya x
Ferial x El Amira x Zaafarana)
Bred by David Houseknecht
Top 5 Regions 9 & 11;
Eight Class A Halter Championships,
Most Classic Arabian Wins;
Seven Egyptian Event Top 5's.
Sire of **H.F. Panache,**
Class A Res. Jr. Champion &
Egyptian Event Top 10

Polly Knoll photo

HOPE FARM
Joan Skeels or Sue Burnham • 194 Polk 21 • Cove, Arkansas 71937

Sharing the Dream - Saklawi

Randi Clark photo

Descending from Radia through Zaafarana

Zandai Jadallah
1995 Bay Mare
Bred by Dr. William M. Hudson

*Whoso keepeth and tendeth a horse
for the service of Allah
shall be recompensed
as one who fasts during the day
and passes the night in prayer.*
  Hadithes of the Prophet

	Anaza El Farid	Ruminaja Ali	Shaikh Al Badi
			Bint Magidaa
ZANDAI JADALLAH		Bint Deenaa	*Ansata Ibn Halima
			*Deenaa
	Ansata Zaafina	Ansata Mourad Bey	Ansata Ibn Sudan
			Ansata Regina
		Ansata Azalia	Maard
			Ansata Zareifa

ZANDAI ARABIANS
Dr. William Hudson • 6010 Riley Road • Cumming, Georgia 30040

Sharing the Dream - Saklawi

Descending from Radia through Zaafarana

Farid Nile Moon
1997 Bay Stallion
Sire: Anaza El Farid
(Ruminaja Ali x Bint Deenaa)
Dam: GA Moon Tajhalima
Bred by Karen Henwood
2001 Egyptian Event Unanimous
Supreme Champion Stallion,
2002 Grand Champion Old Dominion Show;
Grand Champion Fingerlakes;
Region 16 Res. Grand Champion.
Sire of **Farid Nile Dream**, 1st Egyptian Event
World Class Mares of 2001

Javan photo

GA Moon Tajhalima
1987 Chestnut Mare
Sire: Imperial Imperor
(Ansata Imperial x
Imperial Mistry)
Dam: Ansata Aziza
(*Ansata Ibn Halima x
*Ansata Bint Zaafarana)
Bred by Gibson Arabians

*This page is dedicated to the memory of *Ansata Ibn Halima and Anaza El Farid.*

Johnny Johnston photo

SANDYBROOK EGYPTIAN ARABIAN FARM
Karen Henwood • 13885 N.W. Highway 27 • Ocala, Florida 34482

Descending from Mabrouka (Inshass) through Mariam

Maria Halima
1995 Black Mare
Sire: Haziz Halim (*Ansata Ibn Halima x Serenity Shahrabi)
Dam: Chelsea (The Minstril x Muneera x *Musannah)
Bred by Gradin Arabians, U.S.A.
A Futurity Champion in the U.S.A.

GR Marianah
2000 Black Mare
Sire: Madallan-Madheen
Dam: Maria Halima
Jr. Champion Filly, Exter 2002
Class Winner, Borgloon 2002
Bred by Rothenberg Stud,
Annette & Erwin Escher

GR Marietta
1999 Black Mare
Sire: Madallan-Madheen
(Ansata El Salaam x Madinah)
Dam: Maria Halima
Bred by Rothenberg Stud, Annette & Erwin Escher
2000 European Res. Champion Jr. Fillies - Egyptian Event Europe;
2001 Liberty Championships
& Jr. Champion Filly Egyptian Classic Show

Escher photos

ROTHENBERG STUD
Annette & Erwin Escher • 86653 Monheim • Germany

Descending from El Samraa (Inshass) through Sameera

Escher photo

Classic Shadwan
1992 Grey Stallion
Bred by Jürgen Wölk, France
Res. Jr. Champion, Wels, 1994;
Res. Sr. Champion, Kaub 1997;
European Res. Sr. Champion,
Egyptian Event Europe 1997;
Sr. Champion International
Show, Kaub 2000;
Sr. Champion & Best in Show,
Egyptian Classic Show;
Platin Cup Champion 2000.
Sire of Multi- Champions and Gold-Medal
Stallions at German Stallion shows, Aachen.

CLASSIC SHADWAN	Alidaar	Shaikh Al Badi	*Morafic
			*Bint Maisa El Saghira
		Bint Magidaa	*Khofo
			*Magidaa
	Shagia Bint Shadwan	Shadwan	Shaarawi
			Kamar
		Samar	Shaarawi
			Sameera

ROTHENBERG STUD
Annette & Erwin Escher • 86653 Monheim • Germany

Sharing the Dream - Saklawi

Descending from Moniet through Mouna

Madallan-Madheen
1994 Black Stallion
Sire: Ansata El Salaam (Ansata Ibn Shah x Ansata Samantha)
Dam: Madinah (Ibn Galal x Mona II x Mahiba x Mouna)
Bred by Dr. Seigfried & Ruth Paufler
Gold Medal Winner Stallion Licensing Show Germany, Aachen 1997. Sire of Multi-Champion **GR Marietta,** and champion **GR Marianah**

Escher photos

GR Marja
2001 Black Mare
Sire: Prince Shetan
(Ansata El Salaam x Sphinx Bint Moheba)
Dam: Malinah (Ansata El Salaam x Madinah-M)
Bred by Rothenberg Stud, Annette & Erwin Escher

Maheeb
2001 Black Stallion
Sire: Classic Shadwan (Alidaar x Shagia Bint Shadwan)
Dam: Ken Bint Bint Mahiba
(Ansata Halim Shah x Kis Mahiba)
Bred by Erwin Dirnhofer
Owned by Maheeb Partnership

ROTHENBERG STUD
Annette & Erwin Escher • 86653 Monheim • Germany

Sharing the Dream - Saklawi

Descending from Moniet through Mouna

Mangalee Ibn Shadwan
1997 Grey Stallion
Sire: Classic Shadwan (Alidaar x Shagiah)
Dam: Ken Bint Bint Mahiba
(Ansata Halim Shah x Kis Mahiba x Mahiba x Mouna)
Bred by Dirnhofer Family
2000: Gold Medal Winner
Stallion Licensing Show Aachen &
National Championship Winner, Neustadt/Dosse

Escher photo

Toischel photo

Sameer
1983 Chestnut Stallion
Sire: Mohafez
(*Ibn Moniet El Nefous x Ahroufa)
Dam: Sabah
(Ibn Galal x Mahiba x Mouna)
Bred by Dr. Hans Nagel
An International Champion stallion.
1988 Gold Medal Winner Stallion
Licensing Show
Darmstadt-Kranichstein;
Performance-tested - Medingen.

BIRKHOF STUD
Dirnhofer Family • Birkhof 3 • Burglengenfeld, Germany 93133

Sharing the Dream - Saklawi

Descending from Moniet through Mouna, from Medallela through El Bataa, and from Om Dalal through Thouraya (Inshass)

Ghowleh Al Rayyan
1997 Grey Mare
Sire: Safir (Salaa El Dine x Aisha)
Dam: Bint Sabah (Ibn Nazeema x Sabah x Mahiba x Mouna)
Bred by Sheikh Abdul Aziz Bin Khaled Al Thani, Al Rayyan Farm, Qatar
1999 Qatar International Top Five

Gigi Grasso photos

Alenah
1987 Grey Mare
Sire: Amaal (*Morafic x *Bint Maisa El Saghira)
Dam: Adliah (*Morafic x Sooma x Thouraya I)
Bred by Gleannloch Farms, U.S.A.
2001 Qatar International Top Five

Ghazallah
1999 Chestnut Mare
Sire: Orfan Bey (Ansata Halim Shah x Imperial Sayyah)
Dam: MB Taffeta (Imperial Madheen x AK Intafa x Farakaa Bataa)
Bred by Bratt Arabians, U.S.A.

AL JAZIRA ARABIAN STUD
Mr. Talal Abdullah Al Mehri • P.O. Box 16991 • Qadisiya 35860, Kuwait

Sharing the Dream - Saklawi

Descending from Hind (Inshass) through Anzar

RN Marwa
1995 Grey Mare
Sire: Ansata Halim Shah
(*Ansata Ibn Halima x
Ansata Rosetta)
Dam: Baheyat Albadeia
(Amir Albadeia x Meseda x
Nagwa x Anzar)
Bred by
Sheikh Abdul Aziz Bin Khaled Al Thani,
Al Rayyan Farm, Qatar
1996 Qatar International Top Five
1999 Qatar National Top Five

Maha Al Jazira
2000 Grey Mare
Sire: Ansata Hejazi
(Ansata Halim Shah x
Ansata Sudarra)
Dam: RN Marwa
Bred by Mr. Talal Abdullah Al Mehri

Gigi Grasso photos

AL JAZIRA ARABIAN STUD
Mr. Talal Abdullah Al Mehri • P.O. Box 16991 • Qadisiya 35860, Kuwait

"You must…keep up with the mare in strength and courage." Qatar desert. Gigi Grasso photo.

The Hadban Enzahi Strain

Mani's Poem About Al Hadba

She loos'd the reins of the mare so she could run faster,
But the numbers of Turks were still pursuing,
And the mares who are ridden by women
    are not afraid when they see a raid coming upon them.
    and this shows the courage of the women who are riding.
And when the Arabs knew they were going to be attacked,
    they would beat the drums as a warning.
And the eyes of the mares sparkled,
    because they were excited about the coming battle.
O Abu Ali, our eyes are upon you,
    and we see you are hesitant to lead us.
The tribe dislikes you as they would a corpse
    because of your cowardice.
O brothers, look up to me because I am brave
    and get the grave diggers to bury him,
    because he is not fit to live.
I wish that they had not sent Beni Lam
    because he is hesitant,
But that they had sent me instead on my Hadba,
    for during the battle she is seized with fervor
    which pushes her on to the fray.
I wish, too, that the arrow [Abu Ali] had not fought the arrow [enemy],
    because his aim is not true and he trembled with fear.
You must be like the Hadba.
And the saddle and ornaments on her back are to her
    as the wings of a wild desert dove,
    and they make her fly.
She is as a wild dove among other doves,
    and she is not aware that they had scattered and left her
    for her mind is solely on the battle.
The dove in its swiftness barely stops to drink
    before she races back to the battle.
She misses not her companions who have deserted her.
You and the mare must be equal to a battalion of soldiers,
    and you must not weaken,
    but keep up with the mare in strength and courage.

Matriarchs of the Hadba Family: Bint Samiha

Bint Samiha

Bint Samiha	Kazmeen	Sotamm	*Astraled	Mesaoud
				Queen of Sheba
			Selma II	Ahmar
				Sobha
		*Kasima	Narkise	Mesaoud
				Nefisa
			Kasida	Nasr I
				Makbula
	Samiha	Samhan	Rabdan El Azrak	Dahman El Azrak
				Rabda
			Om Dalal	Sabbah I
				Bint Roga Al Zarka
		Bint Hadba El Saghira	El Halabi	Saklawi I
				Halabia
			Hadba	Saklawi I
				Venus

The Arab poem introducing this section describes the esteem held by the Bedouins for war mares from this family. Since the importation of Venus "Shekra Zefra" in 1875 to Egypt, the Hadban Enzahi strain has contributed distinguished broodmares as well as predominant sires in the straight Egyptian breed. Listed as the Pasha's fourth choice of strains within the *Abbas Pasha Manuscript*, the Hadban Enzahi strain coincidentally stands in fourth place as listed by the Pyramid Society with a total number of 2675 straight Egyptian horses registered in America as the new millennium begins.

One of the most venerated broodmares was the beautiful bay Kazmeen daughter, Bint Samiha. Two female branches descending from her are through her daughters, Samha (by Baiyyad), and Shams (by Mashaan). Of the males, Bint Samiha's legendary son, Nazeer (by Mansour), made such a remarkable contribution to the breed that had she done no

Kamla as a young mare (Sheikh El Arab x Samha)

more than produce him, her legacy would have been assured.

Samha, a handsome fleabitten grey mare, resulted from doubling the Hadban family through tail female sire and tail female dam. Her memorable contributions were three daughters: 1) Kamla (by Sheikh El Arab), 2) *Mamdouha (by Kheir) who produced *Gamila (by Enzahi) when she arrived in America, and 3) Zahia (by Sid Abouhom), who generated the line at Al Badeia in Egypt through Nagdia (by Nazeer). Nagdia became the dam of *Malekat El Gamal (by

Samha as a young mare (Baiyyad x Bint Samiha)

*Gamila (Enzahi x *Mamdouha x Samha)

*Malekat El Gamal (Waseem x Nagdia x Zahia x Samha). J. Johnston photo.

Ibn Galal I-7 (Ibn Galal I x Lotfeia/Lutfia)

Waseem). An attractive strong-bodied mare, like her sire, *Malekat El Gamal started a prominent family of her own when imported to America, and whose progeny are now represented worldwide, particularly in Qatar through the Sundar Ali Sayyah family and the stallion Al Aadeed Al Shaqab. *Gamila, also established a productive line through her daughter, Anchor Hill Hamla (by Hadbah), a foundation mare for the Glorieta Arabians program and the dam of renowned full sisters, Glorieta Zaafira and Glorieta Zaarina (by Ansata Abu Nazeer).

The Kamla line came into world prominence through her son, Hadban Enzahi (by Nazeer), the head sire at Germany's Marbach Stud, and in the 80's and 90's through the mare Lotfeia, (Alaa El Din x Bint Kamla) a foundation mare in Hungary's Babolna Stud. A classic individual with an alluring face, she closely resembled her granddam, Kamla, in type and hindquarter structure. Bred to Ibn Galal I (Ibn Galal x Hanan) she produced three excellent chestnut daughters, each bearing the traditional Hungarian sire-line names. Most notable among them was *Ibn Galal I-7. Affectionately nicknamed "Gala," the words "star quality" best described her. Remindful in type of 1971 U.S. National Champion Mare, *Serenity Sonbolah, Gala pranced out of the show ring to thunderous applause in third place as a U.S. National Top Ten Mare, a public relations success story for the Egyptian breed. However, dreams of her giving birth to progeny by Ansata Halim Shah were thwarted when she died presenting the first one - a filly, despite the efforts of five skilled veterinarians to save her. Her replacement full sister died young from a respiratory infection and also without production. Eventually Nagel brought Gala's dam, Lotfeia from Babolna to Germany and bred her to his Ansata Halim Shah son, Salaa El Dine, resulting in the birth of Nashua, a very special chestnut filly indeed. Fortunately she produced several fine daughters, but Nashua regrettably

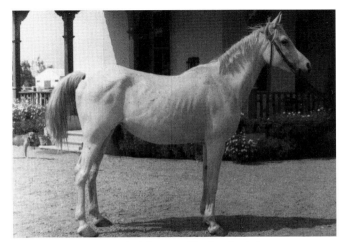

Dam of Lotfeia, Bint Kamla (El Sareei x Kamla x Samha)

Nashua (Salaa El Din x Lotfeia). Escher photo.

died young from laminitis. At this time her progeny are carrying on this somewhat fragile family's endearing traits through Nagel's program in Germany, at Ajmal, El Adiyat, and Ezzain stud farms in Kuwait, and in England with Al Nakeeb.

Bint Samiha's daughter, Shams, has been prominent in worldwide breeding programs through her daughters: 1) Maysouna (by Kheir), 2) Fathia (by Sid Abouhom) and 3) Sherifa (by Gassir). Maysouna's Nazeer daughter, Bint Maysouna (imported as *Ansata Bintmisuna whose story is featured in *Authentic Arabian Bloodstock*), and Hemmat (by Sid Abouhom), established excellent families. Bint Maysouna's daughter, Naama (by Antar), remained in Egypt to found a valued line, while in America her son, Ansata Abu Sudan (by Ansata Ibn Sudan), and his daughters carried her name forward. Hemmat produced the beautiful mare, *Nabda (by the Kamar son, Wahag), who was imported to America and played an important role at the Kehilan farm through her son, Makhsous. Hemmat produced the elegant *Morafic daughter, Neamat, who became the dam of *Nihal (by Antar), an exquisite multi-champion mare owned by Gleannloch. Full sister to Neamat was Afifa, quite opposite in type to her, but whose progeny blended well with Nagel's bloodlines in Germany, and with Ansata bloodlines in America. Hemmat also produced *Baheia (by Nasralla), from whose daughter Bakria (by Gharib), came the well-known multi-champion stallion and sire of champions, The Minstril, by Ruminaja Ali.

Shams second daughter, Fathia, produced the grey mare *Hoda (by Alaa El Din), who became instrumental in programs of Somerset and Kehilan farms. Fathia was also the dam of the captivating Nazeer daughter, Foze, sought after by many breeders when she was owned by the Hamdan Stud. Eventually Foze was sold to Germany where she appears in numerous pedigrees. Sherifa, Shams third daughter, remained in Egypt, also establishing an enduring family of which her grandson, Hafeed Antar, is well known as an influential sire at the E.A.O.

The broad panorama across which this family spreads resounds to the hoofbeats of many other outstanding individuals equally worthy of mention, because they too have made this important blending strain so valuable.

> *Her beauty surpasses the beauty of all other horses…*
> *Cavalry draw back before her.*
> *The base of her head is narrow,*
>   *her ears are small and pointed*
>   *like a finely-carved quill..*
> *She gallops like a greyhound…*
> *Her coat shines like a mirror.*
>
> Songs of the Beni Hilal

Shams (Mashaan x Bint Samiha)

*Neamat (*Morafic x Hemmat x Shams)*

Foze (Nazeer x Fathia x Shams)

Sharing the dream - Hadban

Gigi Grasso photo

Descending from Samiha through Kamla

NK Nadirah
2001 Grey Mare
Bred by
Dr. Hans Nagel

		Salaa El Dine	Ansata Halim Shah Hanan
	Adnan	Ghazala	Ghazal Hanan
NK NADIRAH		Salaa El Dine	Ansata Halim Shah Hanan
	Nashua	Lotfeia	Alaa El Din Bint Kamla

*Spare is her head and lean,
her ears pricked close together;
Her forelock is a net,
her forehead a lamp lighted...*
                Abu Zeyd

KATHARINENHOF
Dr. Hans-Joachim Nagel • Am Gräberfeld 13 • 26197 Grossenkneten, Germany

Sharing the dream - Hadban

Descending from Samiha through Kamla

Gigi Grasso photo

NK Nasrin
1995 Grey Mare
Bred by Dr. Hans Nagel

NK NASRIN	Adnan	Salaa El Dine	Ansata Halim Shah
			Hanan
		Ghazala	Ghazal
			Hanan
	Nashua	Salaa El Dine	Ansata Halim Shah
			Hanan
		Lotfeia	Alaa El Din
			Bint Kamla

KATHARINENHOF
Dr. Hans-Joachim Nagel • Am Gräberfeld 13 • 26197 Grossenkneten, Germany

Sharing the dream - Hadban

Descending from Samiha through Kamla

Van Lent, Jr. photo

NK Nabeelah
1995 Grey Mare
Bred by Dr. Hans Nagel

	Salaa El Dine	Ansata Halim Shah
Nahaman		Hanan
	Ameera	Madkour I
NK NABEELAH		Hanan
	Salaa El Dine	Ansata Halim Shah
Nashua		Hanan
	Lotfeia	Alaa El Din
		Bint Kamla

KATHARINENHOF
Dr. Hans-Joachim Nagel • Am Gräberfeld 13 • 26197 Grossenkneten, Germany

Sharing the dream - Hadban

Descending from Samiha through Kamla

NK Nada
1999 Grey Mare
Sire: Adnan (Salaa El Dine x Ghazala)
Dam: Nashua (Salaa El Dine x Lotfeia)
Bred by Dr. Hans Nagel, Katharinenhof

Allah preserve thee, my beloved;
thou art beautiful, thou art sweet,
thou art lovely!
Allah defend thee from envious eyes.
    Anonymous Arab Poet

Jaber Al Kazemi photos

EZZAIN ARABIANS
Usamah Zaid Al Kazemi • P.O. Box 30 • Safat 13001, Kuwait

Sharing the dream - Hadban

Descending from Samiha through Kamla

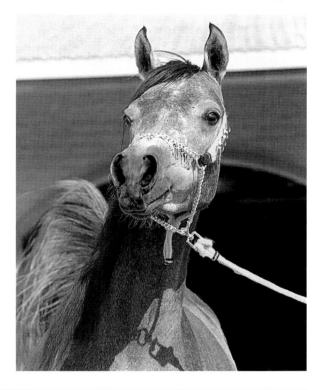

NK Nakeebya
2000 Grey Mare
Sire: NK Hafid Jamil (Ibn Nejdy x Helala)
Dam: NK Nabeelah (Nahaman x Nashua)
Bred by Dr. Hans Nagel, Katharinenhof

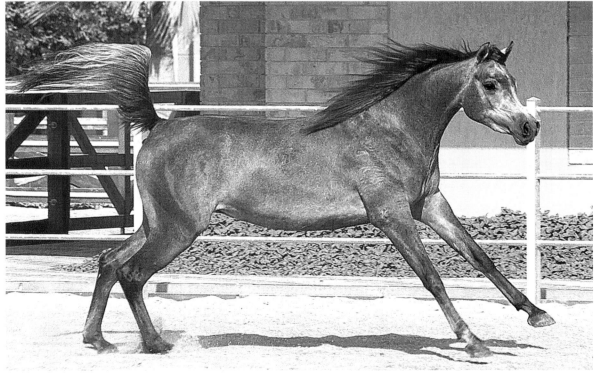

Jaber Al Kazemi photos

EZZAIN ARABIANS
Usamah Zaid Al Kazemi • P.O. Box 30 • Safat 13001, Kuwait

Sharing the dream - Hadban

Jaber Al Kazemi photos

Descending from Bint Rustem through Yosreia

Nayzak Ezzain
2001 Bay Stallion
Sire: Nabajah El Chamsin
(Nahaman x Fatima XVII)
Dam: S.E.A. Nazooka
(Beltagi x S.E.A. Inshallah x
Katr El Nada x Sara)

Some glory in their birth…
Some in their wealth,
Some in their body's force.
Some in their hounds and hawks,
Some in their horse.
                    Shakespeare

EZZAIN ARABIANS
Usamah Zaid Al Kazemi • P.O. Box 30 • Safat 13001, Kuwait

Sharing the dream - Hadban

Descending from Samiha through Kamla

Bint Nashua
1996 Bay Mare
Bred by Dr. Hans Nagel,
Katharinenhof

Gigi Grasso photos

		Salaa El Dine	Ansata Halim Shah
	Nahaman		Hanan
		Ameera	Madkour I
BINT NASHUA			Hanan
		Salaa El Dine	Ansata Halim Shah
	Nashua		Hanan
		Lotfeia	Alaa El Din
			Bint Kamla

EL ADIYAT ARABIANS
Mahmood A. Al Zubaid • P.O. Box 270 • Surra, Kuwait 45703

Sharing the dream - Hadban

Descending from Samiha through Samha and Bint Rustem through Salwa

Escher photos

Morning Glorii
1998 Grey Mare
Sire: Ansata Iemhotep
(Prince Fa Moniet x Ansata Nefara)
Dam: Glorieta Zaafira
(Ansata Abu Nazeer x Anchor Hill Hamla x
*Gamila x *Mamdouha x Samha)
Bred by Debbie Cooper and Tom Salome

Kholeif
1993 Black Stallion
Sire: Rahhal
(Aseel x Rawayeh)
Dam: El Khansaa
(Shaarawi x Mahbouba x Lateefa x Salwa)
Bred by the E.A.O., Cairo, Egypt

ABD-EL GAWAD GAWDAT
23 Kafr Abdou Street • Roushdy, Alexandria, Egypt

Matriarchs of the Hadba Family: Bint Rustem

			Aziz
		Mesaoud	Yemameh
	*Astraled		a Muniqi Hadruj
		Queen of Sheba	an Abayyah Sharrakia
Rustem			Wazir
		Merzuk	a Kuhaylah Jellabia
	Ridaa		Hadban (1878)
		*Rose of Sharon	Rodania
Bint Rustem			D.B.
		Saklawi I	D.B.
	El Halabi		D.B.
		Halabia	a Saqlawia Jidraniah
	Bint Hadba El Saghira		D.B.
		Saklawi I	D.B.
	Hadba		D.B.
		Venus	a Hadba Enzahia

Bint Rustem

Although somewhat less refined from the Samiha line, Bint Rustem, by the Blunt Rodan stallion, Rustem, became an effective matriarch through her two daughters by Ibn Rabdan: the bay Hind, and the black Salwa. They founded influential and important families in subsequent generations, particularly when the stallions Sheikh El Arab and Nazeer were incorporated into the family. The black Salwa contributed this color rather frequently to her line, particularly through her black son, Gharib (by Antar), who became a leading sire at the Marbach Stud in Germany.

Hind became the dam of Fasiha (by Awad), who in turn produced the exceptional broodmare, Yosreia (by Sheikh El Arab). Yosreia, when bred to Nazeer (doubling Mansour through the sire of the sire being the grandsire of the dam), gave such classic offspring as 1) Shahrzada, who had one of the most beautiful heads at El Zahraa during her time, and was the dam of the Gleannloch champion and foundation mare, *Dawlat (by Antar), 2) the elegant Bint Yosreia, who graced the Hamdan Stables in Egypt, and 3) the handsome white stallion, Aswan, a revolutionary sire who was presented as a gift to Russia by President Nasser. Aswan changed the face of the Russian breeding program and was subsequently influential in Poland and in America through the Russian-bred imports that were half-Egyptian, half-Russian, although at the time they were hyped as "straight Russian." Yosreia's stunning black daughter, 4) Mohga (by El Sarie) was also an excellent broodmare, and among other fine progeny, she became the dam of Gleannloch's memorable import, *Nahlah (by *Morafic), a U.S. Reserve National Champion Mare and a crowd favorite wherever she appeared.

Yosreia as a young mare (Sheikh El Arab x Hind)

Rouda (Sheikh El Arab x Fasiha x Hind)

Aswan (Nazeer x Yosreia), presented by President Nasser of Egypt to Russia in honor of the Aswan dam. He became influential in Polish and Russian programs, but he was not bred to straight Egyptian mares.

Bint Yosreia (Nazeer x Yosreia) full sister to Aswan and Shahrzada, was one of the important broodmares for Hamdan Stables in Egypt. JEF photo.

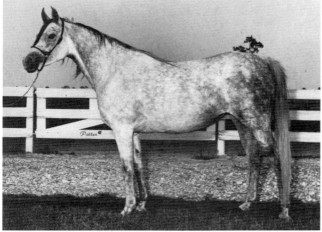

*Nabilahh (Antar x Frashah/Farasha x Yosreia), her name is found in countless important pedigrees. Potter photo.

Arabest Kalid (Ansata Ibn Sudan x Ruminaja Alia, tracing to Shahrzada x Yosreia), a champion stallion in the U.S. bred by the Trapps and exported to South Africa. Sparagowski photo.

*Faleh+ (Alaa El Din x Frashah/Farasha x Yosreia), half brother to *Nabilahh. Winner of multiple U.S. National Top Ten performance classes and a 100 mile Competitive Trail competition, he was one of Alaa El Din's best sons. Rhita McNair photo.

Richter MH (El Halimaar x Fasarra, tracing to Shahrzada x Yosreia). A U.S. and Canadian Top Ten Futurity Stallion and sire of numerous champions. Gigi Grasso photo.

Yosreia also produced 5) Farasha (by Sid Abouhom), dam of the chestnut stallion Galal (by Nazeer), who became an influential sire through his son, Ibn Galal "Magdi," (thus doubling again the Hadban line in tail female of sire and dam). Although not one of Nazeer's handsomest sons, Galal sired better individuals than himself. Farasha was also the dam of the two very refined and pretty-headed Alaa El Din sons, *Faleh and *Farazdac, both good contributors to American breeding programs. One of Farasha's special daughters was *Nabilahh (by Antar). From this line descends the internationally renowned Maar Bilahh (by El Halimaar) and the popular American-bred U.S. Reserve National Champion Stallion, Thee Desperado, whose dam's sire line is to Yosreia and whose sire's tail female line is to Shams.

The grey mare Rouda, Yosreia's full sister, founded her own dynasty through the E.A.O. mares Galila (by Sid Abouhom), dam of the stallion *Ghalii, and Rida (by El Sarie). Prominent descendants of Rouda directly from the E.A.O. can be found through Galila's daughters, Zebeda (by El Sarie), and Aziza (by Alaa El Din). Zebeda's pretty-headed daughter, Alifa (by Alaa El Din), was the valued combination of Alaa El Din on El Sarie daughters. Among Alifa's production was Kodwa (by Shadwan), a show champion and foundation mare for Nayla Hayek's Hanaya Stud in Switzerland.

Shadwanah (Classic Shadwan x Dalima Shah x AK Bint Dalia II)
Escher photo

Aziza produced the show winning mare, *AK Dalia (by *Ibn Hafiza), and the stallion, *Lancers Sahm (by Sabeel), both influential in American programs. *AK Dalia was imported by Bentwood. She became the dam of AK Bint Dalia II (by *Ibn Moniet El Nefous) who was bred to Ansata Halim Shah and produced the prominent show winners and breeding stock, Dalia Halim, Ansata Shahlia and Dalima Shah, imported by the Eschers to Germany.

The descendants of Rida are prominent through the Marei's Al Badeia Stud whose mare, Hosna (by Ibn Maisa), was exported to Babolna in Hungary where she produced a family of Ibn Galal progeny, many of whom are found at Simeon Stud in Australia and subsequently through its exports worldwide. From this line also descends the famed show horse, Simeon Shai (by Raadin Royal Star), who in 1991 came upon the scene at just the right time to capture Scottsdale Champion Stallion, both Canadian and U.S. National Stallion Championships, and World Champion Stallion at the Salon du Cheval in Paris, a compilation of wins unequalled by any show horse to this day.

References to both "the Samiha line" and "the Yosreia line," as they have become known, among others of this Hadban Enzahi strain, provide only a random sampling of the many excellent individuals who have contributed to this strain since the importation of the foundation mare, Venus (Shekra Zefra), who arrived in the Valley of the Nile over a century ago. While certain families differ in some points, as will be seen by careful study, each has its place in discriminating breeding programs.

Maar Bilahh (El Halimaar x Bint Nabilahh), *Stuart Vesty photo*

> *We play the fool in life if we adore*
> *A pretty face while merit we ignore;*
> *The beauty that enchants is often naught;*
> *'Tis dark within the light that lights the shore.*
>
> Rubiyat of Abu-Tayb-Al Mutanabi

Sharing the dream - Hadban

Descending from Bint Rustem through Yosreia

Filsinger photo

Om El Fotooh Hamdan
1982 Grey Mare
Bred by Hamdan Stables
Egyptian National Show E.A.O. 2001, 2nd Place Mares.
Dam of **El Sakb Hamdan**, Supreme Champion Stallion, E.A.O. National Show 1994;
Wadood Hamdan, Supreme Champion Colt, Sharkia National Show 1995;
Mofaget Hamdan, Winner of 3 races; **Al Sahlab Hamdan**, Winner of 3 races: Abu Dhabi Cup, E.A.O. Cup, Emirates Cup;
Motawag Hamdan, Winner of 2 races; **Ikram Allah Hamdan**, Winner of 9 races.

Love for Arabians is a magic addiction,
a world of sweet and caring sensation.
They are the Blessed creation,
whose magnificence
makes you live in
Heavenly fascination.
      – Quote by Fatma Hamza

In memory of my late father,
Ahmed Pasha Hamza,
the founder of Hamdan Stables,
whose sincere love and devotion
for the Arabian is a living legend.

	OM EL FOTOOH HAMDAN		
	Fateh	Shadwan	Shaarawi / Kamar
		Alifa	Alaa El Din / Zebeda
	Yosr Hamdan	Korayem	Nazeer / Helwa
		Bint Yosreia	Nazeer / Yosreia

HAMDAN STABLES
Fatma Ahmad Hamza • 1 Mohammad Mazhar Street, Zamalek • Cairo, Egypt 11211

Sharing the dream - Hadban

Descending from Bint Rustem through Yosreia

Al Sarawi photo

Taira
1982 Bay Mare
Sire: *Soufian
(Alaa El Din x Moniet El Nefous)
Dam: Nagliah
(*Ibn Hafiza x *Nabilahh x
Farasha x Yosreia)
Bred by Gleannloch Farms, U.S.A.

Ajeeb Al Jazira
2001 Grey Stallion
Sire: Safir (Salaa El Dine x Aisha)
Dam: Taira
Bred by Mr. Talal Abdullah Al Mehri

Gigi Grasso photo

AL JAZIRA ARABIAN STUD
Mr. Talal Abdullah Al Mehri • P.O. Box 16991 • Qadisiya 35860, Kuwait

Sharing the dream - Hadban

Gigi Grasso photo

Descending from Bint Rustem through Yosreia

Hilalia Al Shaqab
1995 Grey Mare
Bred by Al Shaqab Stud

	*Ansata Ibn Halima	Nazeer
Ansata Halim Shah		Halima
	Ansata Rosetta	Ansata Shah Zaman
HILALIA AL SHAQAB		*Ansata Bint Bukra
	Saheel	*Sultann
Wagda		Jehan
	Wagida	Galal
		Sara

*...if the sages ask thee why
This charm is wasted on the earth and sky,
Tell them, dear, that if eyes were made for seeing
Then Beauty is its own excuse for being.*

           Emerson

AL SHAQAB STUD
P.O. Box 90055 • Doha, Qatar

Descending from Samiha through Shams

Farida Al Shaqab
1998 Chestnut Mare
Sire: Imperial Pharouk (Imperial Al Kamar
x Imperial Phanadah)
Dam: Naajah (Gadallah x Fakhria)
Bred by Al Shaqab Stud

Gigi Grasso photos

Naajah
1988 Chestnut Mare
Sire: Gadallah
(Adeeb x Omnia)
Dam: Fakhria
(Farazdac x Sharara x Sherifa x Shams)
Bred by
The Egyptian
Agricultural Organization, Egypt

AL SHAQAB STUD
P.O. Box 90055 • Doha, Qatar

Sharing the dream - Hadban

Descending from Samiha through *Malekat El Gamal

Wahag Al Shaqab
1999 Grey Stallion
Sire: Safir (Salaa El Dine x Aisha)
Dam: Kamasayyah
Bred by Al Shaqab Stud
Third Place Yearling Colts Qatar National Show,
Qatar International Show, 1st Yearling Colts
& Champion Colt 2000

Filsinger photo

Kamasayyah
1992 Grey Mare
Sire: Imperial Al Kamar
(El Hilal x Imperial Sonbesjul)
Dam: Sundar Alisayyah
(Ruminaja Ali x Imperial Sayyah
x *Malekat El Gamal)
Bred by Hamilton Schmidt, U.S.A.
Qatar Res. Champion Filly 1993,
Middle East Champion Filly 1993,
Qatar Champion Filly 1994,
Top Ten Filly World
Championships Paris 1994,
Middle East Champion Mare 1998,
Qatar Champion Mare 1996 & 2001

Gigi Grasso photo

AL SHAQAB STUD
P.O. Box 90055 • Doha, Qatar

Sharing the dream - Hadban

Descending from Samiha through *Malekat El Gamal

Rik Van Lent Photo

Al Aadeed Al Shaqab
1995 Grey Stallion
Sire: Ansata Halim Shah
(*Ansata Ibn Halima x Ansata Rosetta)
Dam: Sundar Alisayyah
(Ruminaja Ali x Imperial Sayyah)
Bred by Al Shaqab Stud
Al Shaqab Herd Sire.
Winner of Qatar National &
International Junior Championships
& Qatar International Stallion Championships.
Champion Stallion - Nation's Cup, Aachen,
World Champion Stallion - Salon du Cheval, Paris, 2002
Sire of Champions.

If beauty came to be compared with him,
it would hang down its head in shame:
Or if it were said, O beauty,
hast thou seen the like? - it would answer,
The equal of this I have not.
      Tales of the Arabian Nights

AL SHAQAB STUD
P.O. Box 90055 • Doha, Qatar

Sharing the dream - Hadban

Little photo

Descending from Bint Rustem through Yosreia

EL Shaliah
1997 Black Mare
Bred by Peggy Wojchik & Steve Haeuser

We play the fool in life if we adore
A pretty face while merit we ignore;
The beauty that enchants is often naught;
'Tis dark within the light that lights the shore.

The hand of Allah turns the wheel of fate,
And we must act our part without debate.
Useless it is to question his decree—
What is to be was written on the slate.
                    Rubiyat of Abu-Tayb-Al-Mutanabi

		The Minstril	Ruminaja Ali
	Ali Azimm		*Bahila
		AliaShahm RA	Ruminaja Ali
EL Shaliah			Glorieta Zaarina
		Moniet El Sharaf	*Ibn Moniet El Nefous
	Monieta Sharie		Bint Bint Moniet
		*JK Blue Hasmeh	Gharib
			JKB Hamda x Bint Yosreia

In loving memory of my son, Weston

ROYAL SHAHARA ARABIANS
Michael & Leslie Nord • 13417 S. Greyhawk Lane • Spokane, Washington 99224

Sharing the dream - Hadban

Descending from Bint Rustem through Yosreia

Gigi Grasso photo

Thee Desperado
1989 Bay Stallion
Sire: The Minstril
(Ruminaja Ali x *Bahila)
Dam: AK Amiri Asmarr
(The Egyptian Prince x
Asmarr x Nagliah x *Nabilahh)
Bred by Tom Salome, Texas
Egyptian Event Jr. Champion,
Region 9 Champion Stallion,
Scottsdale Champion Stallion,
U.S. Res. National Champion Stallion,
Egyptian Event Overall Leading Sire
 - four consecutive years.
Sire of many notable champions including:
Thee Infidel, U.S National Top Ten Futurity
 and U.S. National Top Ten Stallion;
Sweet Eloise, U.S. National Top Ten
Futurity Filly, Regional Champion Mare;
BJ Thee Mustafa, World Champion Stallion, Paris;
PR Desert Desert Rose, U.S. National Champion;
BB Thee Renegade, U.S. National Top Ten Futurity
 Stallion & Egyptian Event Jr. Champion,
Thee Cyclone, South African National
Champion Stallion, **Botswana,** Scottsdale Top Ten,
and **Thee Asil,** Regional Res. Champion Stallion.

"Horse of a Lifetime"
        Judy Sirbasku

This page celebrates the memory of The Egyptian Prince

ARABIANS LTD.
Judy & Jim Sirbasku • 8459 Rock Creek Road • Waco, Texas 76708

Sharing the dream - Hadban

Gigi Grasso photo

Descending from Samiha through Shams

Ansata Najdi
1998 Grey Stallion
Bred by Ansata Arabian Stud

```
                                          The Egyptian Prince
                    Prince Fa Moniet       Fa Moniet
       Ansata Iemhotep                     Ansata Halim Shah
                    Ansata Nefara          Ansata Sudarra
ANSATA NAJDI
                    Ansata Halim Shah      *Ansata Ibn Halima
       PWA Ashera                          Ansata Rosetta
                    *Assiadeh              Mahomed
                                           Afifa x Hemmat x Maysouna
```

SERENE EGYPTIAN STUD
Anna & Stefano Galber • Via Coletto 2, Loc. SERENE • Affi (VR) Italy 37010

Sharing the dream - Hadban

Descending from Samiha
through Samha

MB Kariita
1997 Grey Mare
Sire: PVA Karim
(Imperial Imdal x BKA Rakiisah)
Dam: Riata Ali Sharafa
(Ruminaja Ali x Sharel Dorr)

Tuscan Keboush
2002 Grey Mare
Sire: Imperial Al Kamar
(El Hilal x Imperial Sonbesjul)
Dam: MB Kariita
Bred by Stuart & Brenda Schuettpelz

TUSCANI
Stuart & Brenda Schuettpelz • 1921 Niles-Buchanan Road • Niles, Michigan 49120

Sharing the dream - Hadban

Descending from Samiha through Samha

Ali Zafir
1986 Grey Stallion
Bred by PFG Trust, U.S.A.

Escher photo

	Shaikh Al Badi	*Morafic
Ruminaja Ali		*Bint Maisa El Saghira
	Bint Magidaa	*Khofo
ALI ZAFIR		*Magidaa
	Ansata Abu Nazeer	*Ansata Ibn Halima
Glorieta Zaafira		*Ansata Bint Zaafarana
	Anchor Hill Hamla	Hadbah
		*Gamila

….a horse is a thing of such beauty…
none will tire of looking at him
as long as he displays himself in his splendor.
Xenophon

BEBO STUD
Miloslava Khamis • 10th Ramadan City • Cairo, Egypt

Sharing the dream - Hadban

Descending from Bint Rustem through Salwa and Samiha through Shams

Mayhoub
1992 Grey Stallion
Sire: Adl (Ikhnatoon x Enayah)
Dam: Sobha (Shaarawi x Mahbouba x Lateefa x Salwa)
Bred by The Egyptian Agricultural Organization
Winner of the 100 km Endurance Competition in 2000

Escher photo

Horoob
1995 Grey Stallion
Sire: Harras (Kisra x Hebat El Nil)
Dam: Hadeel (Seef x Naama x
Bint Maysouna x Maysouna x Shams)
Bred by The Egyptian
Agricultural Organization
Third place Egyptian Nationals 1998

Amaar
1999 Grey Stallion
Sire: Rashdan (Aybac x Randa)
Dam: Orbah (Adl x Aleyat x Ola x
Naama x Bint Maysouna)
Bred by Bebo Stud

BEBO STUD
Miloslava Khamis • 10th Ramadan City • Cairo, Egypt

Sharing the dream - Hadban

Descending from Samiha through Samha

Thee Revolution
1996 Grey Stallion
Sire: Thee Desperado
(The Minstril x AK Amiri Asmaar)
Dam: Bint Zaarina
Bred by Marilyn Thomas

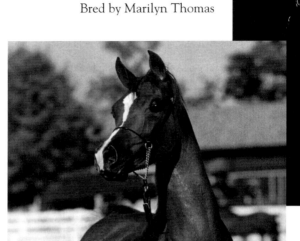

Gigi Grasso photo

FW Zaariya
2001 Chestnut Mare
Sire: Thee Gambler
(Thee Desperado x Aliadaarra)
Dam: Bint Zaarina
Bred by Marilyn Thomas
Egyptian Event Top Ten
World Class Yearling Fillies

Javan photo

Bint Zaarina
1988 Grey Mare
Sire: Ruminaja Ali
(Shaikh Al Badi x Bint Magidaa)
Dam: Glorieta Zaarina
(Ansata Abu Nazeer x
Anchor Hill Hamla x *Gamila)
Bred by James and Judy Sirbasku
Owned by Marilyn Thomas
Egyptian Event Res. Senior Champion Mare
U.S. National Top Ten Mare Halter
Canadian National Top Ten Mare Halter

FALCONWOOD LEGACY BLOODSTOCK
Marilyn Thomas • P.O. Box 840 • Texarkana, Texas 75504

Sharing the dream - Hadban

Descending from Samiha through Shams

Antigua Dance
1987 Grey Mare
Sire: Ruminaja Bahjat
(Shaikh Al Badi x Bint Magidaa)
Dam: Talyla (Ansata Shah Zaman x
Talya x Hoda x Fathia x Shams)
Bred by Leonor Romney
Dam of champions including multi-champions
Simply Seductive, Dance Diva, Marquis I, and
Calypso Dance

Gigi Grasso photo

Marquis I
1993 Grey Stallion
Sire: Makhsous
Dam: Antigua Dance
Bred by Janice Bush
Region 6 Top 5 Stallion, Class A Champion,
Egyptian Event Res. Champion World Class
Stallion. Sire of Champions, Regional and
International Winners.

Makhsous
1979 Grey Stallion
Sire: *Sultann (Sameh x Lubna)
Dam: *Nabda (Wahag x Neamat x
Hemmat x Maysouna x Shams)
Bred by Stanley G. White
Sire of Halter & Performance U.S. National Champions
&Top Tens & multi- Regional & International Winners
and Champions.

"I have come from the desert with its closeness to the spirit of Nature, which you do not understand. I was born of the wind. Mine is a warrior spirit. I cannot be humiliated in punishment or defeated even in death. For my spirit lives on in my children's children."

Javan photo

KEHILAN ARABIANS
Jean & Herb Rogers, Becky Rogers, Nancy Blankenship and Janice Bush
8059 FM 1187W • Fort Worth, Texas 76126

Sharing the dream - Hadban

Descending from Bint Rustem through Rouda

Dalima Shah
1993 Grey Mare
Sire: Ansata Halim Shah (*Ansata Ibn Halima x Ansata Rosetta)
Dam: AK Bint Dalia II (*Ibn Moniet El Nefous x
AK Dalia x Aziza x Galila)
Bred by Magness Arabians
1995: Jr. Champion Mare & Best in Show;
1999: Champion Mare Res. Egyptian Classic Show; Liberty
Champion; European Supreme Champion Mare Egyptian Event
Europe; Egyptian of the Year/Europe 2000: Platin Cup Champion

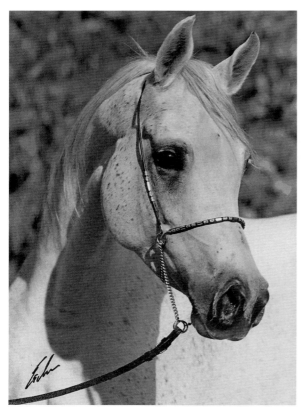
Escher photos

GR Shadiva
2001 Grey Mare
Full sister to Shadwanah
Bred by Rothenberg Stud, Annette & Erwin Escher

Shadwanah
1997 Grey Mare
Sire: Classic Shadwan (Alidaar x Shagia Bint Shadwan)
Dam: Dalima Shah
Bred by Rothenberg Stud, Annette & Erwin Escher
1999 Jr. Champion Egyptian Classic Show, European
Champion Jr. Fillies Egyptian Event Europe, National Champion
Res. Jr. Fillies, Neustadt; 1998-99 Multi-Liberty Championships

ROTHENBERG STUD
Annette & Erwin Escher • 86653 Monheim, Germany

Sharing the dream - Hadban

Descending from Bint Rustem through Rouda

Escher photo

Dalia Halim
1991 Grey Stallion
Bred by Magness Arabians, U.S.A.
1995 Champion of Stallion Licensing, Darmstadt, Germany; Class Champion Wells, Austria; Placed in Ridden Class Level E & A, Meissenheim, Germany.
Senior Res. Champion National Show, Winterlingen 2002
Senior Champion National Show, Ströhen 2002
Sire of Liberty and Show Champions

DALIA HALIM	Ansata Halim Shah	*Ansata Ibn Halima	Nazeer
			Halima
		Ansata Rosetta	Ansata Shah Zaman
			*Ansata Bint Bukra
	AK Bint Dalia II	*Ibn Moniet El Nefous	*Morafic
			Moniet El Nefous
		*AK Dalia	*Ibn Hafiza
			Aziza x Galila x Rouda

AL MUNTAHA
Christian Kesseböhmer • Waldstrasse 56 • Bad Essen 49152 Germany

Sharing the dream - Hadban

Descending from Bint Rustem through Hosna

Simeon Sehavi
1999 Chestnut Mare
Sire: Asfour (Malik x Hanan)
Dam: Simeon Sheba
(Ra'adin Royal Star x 27 Ibn Galal V)
Bred by Simeon Stud
2001 National Champion Filly,
Supreme Exhibit A Class Show
2003 triple crown: Australian Champion Filly,
National Stud Show Champion,
and East Coast Champion

Simeon Sadran
2001 Chestnut Stallion
Sire: Asfour (Malik x Hanan)
Dam: Simeon Simona (Asfour x 27 Ibn Galal V)
Bred by Simeon Stud

Simeon Sukari
1987 Chestnut Mare
Sire: Asfour (Malik x Hanan)
Dam: 27 Ibn Galal V
(Ibn Galal x Hosna x Rida x Rouda)
Bred by Simeon Stud
1991 Supreme Champion Mare Victorian Classic;
East Coast Jr. Champion Filly
1992 Australian Champion Mare and
Supreme Champion Qld. Challenge.

Already created her own dynasty.

Rob Jerrard photo

SIMEON STUD
Marion Richmond & Peter & Ruth Simon • 44 Bulkara Road • Bellevue Hill, • Sydney, Australia, NSW 2023

Sharing the dream - Hadban

Descending from Bint Rustem through Rouda

Van Lent photo

Kodwa
1977 Grey Mare
Bred by The Egyptian
Agricultural Organization
Kodwa is a Multi-Champion Mare
and is the dam of **Hanaya Rafek**,
Ridden and In Hand Champion.

KODWA	Shadwan	Shaarawi	*Morafic
			Bint Kamla
		Kamar	Nazeer
			Komeira
	Alifa	Alaa El Din	Nazeer
			Kateefa
		Zebeda	El Sareei
			Galila x Rouda

HANAYA ARABIANS
Nayla Hayek • Expohof • Schleinikon, Switzerland CH -8165

"From where did they come to you…" Van Lent Jr. photo, Al Rayyan mares, Qatar.

The 'Abeyyan Strain

Ramih ibn Fuhayd ibn Sultan ibn Moabhel of al Ruwala was questioned in the house of Sheikh Sha'lan and in the presence of Sheikh Faysal and Huran ibn Moabhel and Faris al Zayd of al Sha'lan,

"O Ramih, tell us about the 'Abeyyat al Sho'ayri because it is said that they are yours. From where did they come to you, and to whom did they pass from you, and what kind of 'Abeyyat are they? By your honor and good fortune and religion and faith, tell us truthfully at this gathering what you have heard from your grandfathers."

Ramih ibn Moabhel replied, "The way she passed to my uncle, Mushwit ibn Sha'lan, from al Jowayhani of al Suwayt of al Dafeer, is that my uncle, Mushwit ibn Sha'lan, captured her during a battle between al Dafeer and Abdullah ibn Sha'lan, the grandfather of Sheikh Faysal Sha'lan. On the day of al Arban at Nejd, Mushwit unhorsed Jowayhani and took her, and she was a hamra mare. The mare went to Mushwit. He mated her and she gave birth to a safra filly.

On that day we and Shammar battled, the day of al Mokhrim, we, O al Ruwala, attacked al Qasim. And we carried the filly on the side of a camel and went with our people to Hojaylan, the Sheikh of al Qasim, in the direction of Burayda at al Qasim. And al Qasim and we were friends at the time. And Ibn Safeer, the Ra'i of al Khabrah of al Qasim, got up and saw the filly and he said, 'This is my methnawi because on the day that the hamra mare was at al Dafeer, I had made the methnawi agreement.' And at that time, because al Qasim and al Ruwala were friends, we handed over the safra filly to Ibn al Safeer as methnawi, and at that time Ibn al Safeer was asked about the hamra mare and what she was.

Ibn al Safeer replied, 'She is 'Abeyya Sherrakiyah who passed to us from Ibn Sakhya of Beni Khaled.'

**A Poem of Hussein ibn Safeer,
the Ra'i of the Stud**

> O my filly, do not shy away from me, al Rahimi.
> I know that you belong to al Lokhaymi,
> but I love you as much as he did.
> She has beautiful legs and a beautiful body
> and a Lokhaymi face.
> And we thank Allah that there were good people
> to take care of you until you grew up
> to be such a fine mare.
> And I thank Allah that he has granted me this mare
> as my destiny,
> And I thank Allah that he has given me this mare who
> is without blemish.

Sharing the Dream - Abayyan

Safir, a grandson of Hanan, and rider Houssein Issa in the Qatar desert. Gigi Grasso photo.

RN Ajeeba, a Hanan great granddaughter, and rider Houssein Issa, Qatar. Gigi Grasso photo.

Matriarchs of the El Shahbaa Family: Hanan and Bint Magidaa

The Abeyyan Om Jurays family was discussed in Chapter III and also in *Authentic Arabian Bloodstock*. This once rare Abeyyan family in Egypt has become more prolific globally, and it is timely to pay tribute in this section to two highly influential mares of this strain: Hanan, who was bred in Egypt and exported to Germany, and Bint Magidaa, who was born in America and whose sons in particular, like Hanan, brought her world acclaim. Although there are more descendants of Hanan represented here, primarily due to her major incorporation with Ansata bloodlines in the Arab world and Europe, the influence of Bint Magidaa is no less significant, and many of her progeny have blended exceptionally well with the Ansata stock. Their pedigrees are noted for comparison of bloodlines. Hanan is featured here and Bint Magidaa further on in this section.

Hanan as a mature mare. Van Lent photo

The mare Inshass Mona, who descended from the root mare El Shahbaa, founded a valuable family through her daughter, Hanan, an elegant bay mare sired by Alaa El Din. Hanan was foaled in 1967 and was purchased by Dr. Hans J. Nagel of Germany. She was shipped to the Babolna State Stud in Hungary along with five other horses that Nagel had selected to establish a new Egyptian breeding program for this historically important stud. Other relatives were her very refined and elegant grey half-brother, Bilal (by *Morafic), who remained in Egypt and became a well-known sire for the Shams El Asil Stud. Her bay half-sister, *Serenity Montaha (by Galal), was imported to America where she became a good broodmare, and a chestnut half-brother, Maher (by Galal) remained in Egypt but was unfortunately not used to the degree he could have been.

Hanan	Alaa El Din	Nazeer	Mansour
			Gamil Manial
			Nafaa El Saghira
		Bint Samiha	Kazmeyn
			Samiha
	Kateefa	Shahloul	Ibn Rabdan
			Bint Radia
		Bint Rissala	Ibn Yashmak
			Risala
	Badr	Besheir El Ashgar	a Saqlawi
			a Kahila Ajouz
		Badria	an Abayyan
			a Hamdaniah
Mona (Inshass)	Mahdia	Hamdan	Ibn Rabdan
			Bint Radia
		El Mahrousa	El Zafir
			El Shahbaa

When Hanan arrived at Dr. Nagel's newly purchased farm in northern Germany near Bremen, she was accompanied by Mahiba and Marah, two other foundation mares he had selected. However, Hanan overshadowed any horse that came near her. "At Babolna, she was the regular show star whether turned loose or shown in hand, and at her new home, it was at her stall that visitors stopped the longest," Nagel reflects. "A dark bay with black mane and tail, with no white other than a tiny white fleck on her forehead, and four dark legs and hooves, she was an object of admiration, but by temperament could be irritable; she was friendly and sweet to people she knew, but extremely unsociable towards strangers." Those who had the pleasure of seeing her and being in her presence over a period of time would agree with her owner's comment that Hanan was a type then unknown in Europe. Expounding candidly on both her positive and negative traits in his book, *Hanan, The Story of An Arabian Mare*, which he dedicated to her, Dr. Nagel lovingly points out that: "Life in its fullness always shone in her eyes, which were set prominently in her well-defined sockets. Such expressive eyes can only be the mirror of a character of a noble Arabian horse, a means of communication, and a window to the soul of such an animal. A quiet, gentle eye has always been the attribute of a reliable horse."

For those who knew her, Hanan was also a pleasure to watch in action, and, as Nagel reminisces: "She moved with a playful lightness, flying along at the trot as if weightless. Although her hocks were a bit straight, she used them well and powerfully. It was no spectacular show trot with a stiff front leg but rather an appealing movement forward with a slight knee action; nothing was angular or uncouth but at all phases an example of balance and harmony. Her way of going emphasized the self-confidence which dominated her personality."

*Jamilll (Madkour I x Hanan), prior to his lease to Ansata. Van Lent photo

Ameera (Madkour I x Hanan). Van Lent photo

Among Hanan's admirers was Dr. Rudofsky, one of the renowned traditional horseman and authorities in Germany and former director of Radowce Stud. "You have found an exceptional dam for producing stallions. Take good care of her; she is a rarity," he told Nagel. And, as mentioned in Chapter IV, she was indeed an "incubator" mare - producing sons (*Jamilll, Asfour, Ibn Galal I) and daughters (Ghazala, Ameera, Amal and Ashraff) that took after their sires to a great extent, yet also leaving some of her own impression upon them.

After delivering Salaa El Din at age 18, Hanan went into retirement and enjoyed her final days as undisputed queen of Katharinenhof. When at last she was peacefully laid to rest, her name had achieved celebrity status throughout the Arabian horse world as a mare of legendary significance.

Ibn Galal I (Ibn Galal x Hanan), at Babolna.

There is not a creature on the earth
But God provides its sustenance.
He knows its dwelling, and its resting place.

The Holy Quran: *Surah - The Bee*

Salaa El Dine (Ansata Halim Shah x Hanan) as a young horse

Aisha (Ansata Halim Shah x Ghazala x Hanan). Van Lent photo

Asfour (Malik x Hanan)

Shahin (Salaa El Din x Ameera). Filsinger photo

Ghazala (Ghazal x Hanan). Van Lent photo

RN Ajeeba (Ruminaja Ali x Bint Amal x Amal x Hanan). Filsinger photo

Sharing the Dream - Abayyan

Gigi Grasso photo

Descending from El Shahbaa through Mona (Inshass)

Adnan
1989 Grey Stallion
Bred by Dr. Hans Nagel

ADNAN	Salaa El Dine	Ansata Halim Shah	*Ansata Ibn Halima
			Ansata Rosetta
		Hanan	Alaa El Din
			Mona
	Ghazala	Ghazal	Nazeer
			Bukra
		Hanan	Alaa El Din
			Mona

AJMAL ARABIAN STUD
Mohammed Jassim Al Marzouk • Area 11, Farm No. 1 & 2 • Wafra, Kuwait 3131

Descending from El Shahbaa through Mona (Inshass)

NK Aischa
1996 Grey Mare
Sire: Salaa El Dine
(Ansata Halim Shah x Hanan)
Dam: Amarilla (Jamil x Ghazala x Hanan)
Bred by Dr. Hans Nagel

Ajmal Obbeyah 2001 Grey Mare
Sire: Ansata Hejazi
(Ansata Halim Shah x Ansata Sudarra)
Dam: NK Aischa
Bred by Mohammed J. Al Marzouk

Gigi Grasso photo

Gigi Grasso photo

Bint Amal 1985 Grey Mare
Sire: Ansata Halim Shah
(*Ansata Ibn Halima x Ansata Rosetta)
Dam: KEN Amal (Mohafez x Hanan x Mona)
Multiple Class A Championships in U.S.A.
Dam of **RN Ajeeba**, 1997 Qatar
Int'l Res. Jr. Champion Filly

Van Lent photo

AJMAL ARABIAN STUD
Mohammed Jassim Al Marzouk • Area 11, Farm No. 1 & 2 • Wafra, Kuwait 3131

Sharing the Dream - Abayyan

Descending from El Shahbaa through Mona (Inshass)

NK Jurie
1999 Grey Mare
Sire: NK Hafid Jamil
(Ibn Nejdy x Helala x Ansata Gloriana)
Dam: NK Nariman
(Salaa El Dine x Amarilla x Ghazala x Hanan)

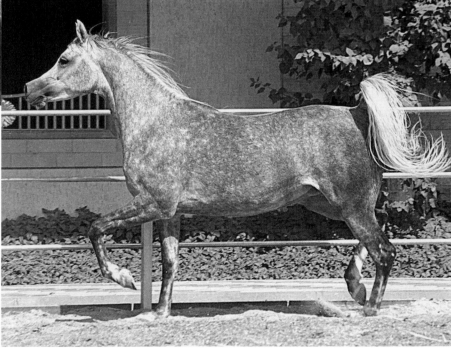

A silken coat of silver light,
No star outshines her in the night.
    Abdul Hafiz

Jaber Al Kazemi photos

EZZAIN ARABIANS
Usamah Zaid Al Kazemi • P.O. Box 30 • Safat 13001, Kuwait

Sharing the Dream - Abayyan

Descending from El Shahbaa through Mona (Inshass)

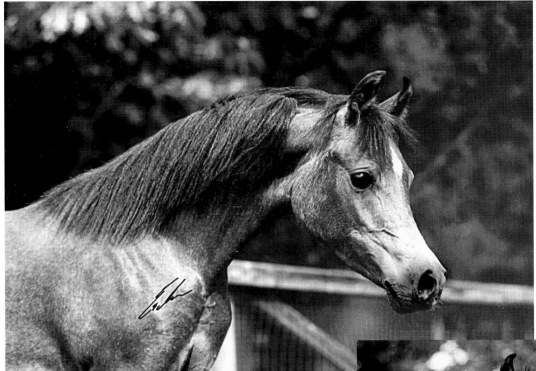

Escher photos

NK Al Amirah
2001 Grey Mare
Sire: NK Hafid Jamil
(Ibn Nejdy x Helala x Ansata Gloriana)
Dam: NK Nariman
(Salaa El Dine x Amarilla x Ghazala x Hanan)
Full sister to NK Jurie
Bred by Dr. Hans J. Nagel

Quality is a treasure that cannot be touched,
An inspiration that cannot be lost,
A source of pride that never loses its lustre.
Anonymous poet

EZZAIN ARABIANS
Usamah Zaid Al Kazemi • P.O. Box 30 • Safat 13001, Kuwait

Sharing the Dream - Abayyan

Descending from El Shahbaa through Mona (Inshass)

Suror Ezzain
2001 Grey Stallion
Sire: Ansata Al Murtajiz
(Ansata Hejazi x Ansata Samsara)
Dam: Zahra
(KP Nameed x Assal x Ameera x Hanan)
Bred by Ezzain Arabians

…Through his mane and tail the high wind sings,
Fanning the hairs, who wave like feathered wings.
Shakespeare

Jaber Al Kazemi photos

EZZAIN ARABIANS
Usamah Zaid Al Kazemi • P.O. Box 30 • Safat 13001, Kuwait

Sharing the Dream - Abayyan

Descending from El Shahbaa through Mona (Inshass)

Nadeema
1991 Grey Mare
Sire: Salaa El Dine (Ansata Halim Shah x Hanan)
Dam: Amarilla (*Jamilll x Ghazala x Hanan)
Bred by Dr. Hans Nagel, Katharinenhof

Gigi Grasso photos

NK Halla
1998 Grey Mare
Sire: Adnan (Salaa El Dine x Ghazala)
Dam: Asfoura (Ansata Halim Shah x Ameera x Hanan)
Bred by Dr. Hans Nagel, Katharinenhof
NK Halla is co-owned by Dr. Hans Nagel
and Mahmood Al Zubaid.

EL ADIYAT ARABIANS
Mahmood A. Al Zubaid • P.O. Box 270 • Surra, Kuwait 45703

Sharing the Dream - Abayyan

Pamela Amberson photo

Descending from El Shahbaa through Mona (Inshass)

RAS Amala
1993 Grey Mare
Sire: Prince Fa Moniet
(The Egyptian Prince x Fa Moniet)
Dam: Bint Amal
(Ansata Halim Shah x Ken Amal x Hanan)
Bred by Robert or Erika Brunson
1994-95 Egyptian Event
Top Ten Futurity Winner

The wild element condensed in Jibrail's hand,
and by the majesty of the Living God
emerged as the steed of the desert - the Drinker of the Wind
                Carl Raswan

NF Safiyyah
2001 Grey Mare
Sire: Ansata Iemhotep
(Prince Fa Moniet x Ansata Nefara)
Dam: RAS Amala
Bred by Ray & Jamie Roberts

NIRVANA FARMS
Ray & Jamie Roberts • 13913 Cameron Road • Excelsior Springs, Missouri 64024

Descending from El Shahbaa through Mona (Inshass)

Gigi Grasso photo

Mansoura Al Rayyan
2000 Grey Mare
Sire: Alidaar
(Shaikh Al Badi x Bint Magidaa)
Dam: Bint Amal
(Ansata Halim Shah x Ken Amal
x Hanan)
Bred by Al Rayyan Farm
Class Winner Qatar International Show 2001

Gigi Grasso photo

RN Sultana
1995 Grey Mare
Sire: Ansata Halim Shah (*Ansata Ibn Halima x Ansata Rosetta)
Dam: Ameena (*Jamilll x Ken Amal x Hanan)
Bred by Al Rayyan Farm
Class Winner 1998 Qatar National Show

Gigi Grasso photo

Jaflah Al Rayyan
2002 Grey Mare
Sire: Al Aadeed Al Shaqab
(Ansata Halim Shah x Sundar Alisayyah)
Dam: RN Sultana

AL RAYYAN FARM
Sheikh Abdul Aziz Bin Khaled Bin Hamad Al Thani • P.O. 375 • Doha, Qatar

Sharing the Dream - Abayyan

Descending from El Shahbaa through Mona (Inshass)

Alfaya Kamar
2000 Chestnut Mare
Sire: Imperial Kamar (El Hilal x Imperial Sonbesjul)
Dam: Noble Imdalia
Bred by Sharon Litizzette

Asami Ruuh
1999 Grey Stallion
Sire: Hadaya El Tareef (Imperial Imdal x Hadaya NMerytaten)
Dam: Noble Imdalia
Bred by Sharon Litizzette

With flowing tail and flying mane, wide nostrils never stretched by pain, mouth bloodless to the bit or rein, and feet that iron never shod, and flanks unscarred by spur or rod.
   Lord Byron

For lovliness needs not the foreign aid of ornament, but is when unadorned adorned the most.
   James Thompson

Noble Imdalia
1994 Grey Mare
Sire: Imperial Imdal (Ansata Imperial x Dalia)
Dam: Anisah (Malik x Ashraff x Hanan)
Bred by Dr. Donald J. Flores
Dam of Halter Winners, Top Ten's &
Top Five's in Class A Shows and The Egyptian Event

She moves a goddess, and she looks a queen.
   Alexander Pope

Sally photo

WINDAMERE ARABIANS
Ed & Sharon Litizzette • 9121 Aspen Drive • Weed, California 96094

Sharing the Dream - Abayyan

Descending from El Shahbaa through Mahfouza

Sulifah
1996 Grey Mare
Sire: Alaa Al Din (Salaa El Dine x Ashraff x Hanan)
Dam: Matala Bint Marah
(Jamil x Marah x Maysa x Mahfouza)
Bred by Siegfried Manz, Germany

TB Qadifa
2002 Bay Mare
Sire: Bayfyrre (M A Bayhajt x Firih)
Dam: Sulifah
Bred by Monika Savier

Gigi Grasso photos

TRE-BALZANE STUD
Monika Savier • Colle del Marchese, 61 • Castel Ritaldi, Italy 06044

Descending from El Shahbaa through Mahfouza

Ward Shah
1986 Grey Mare
Bred by E.A.O.
Senior Champion Mare
European Champion Straight Egyptian Top Ten

Jenni May photos

WARD SHAH			
Adeeb	Shaarawi	*Morafic	Bint Kamla
	Nawal	*Tuhotmos	Nagda
Wasfia	Ibn Maysouna	Nazeer	Maysouna
	Shamaa	*Morafic	Maysa x Mahfouza

DHS Shaakirah
1998 Grey Mare
Sire: Alidaar (Shaikh Al Badi x Bint Magidaa)
Dam: Ward Shah
Bred by Mr. and Mrs. F.J. Schwestermann

DOWDSTOWN HOUSE STUD
Mr. & Mrs. F. J. Schwestermann-Lawless • Dowdstown • Maynooth, Co. Kildare, Ireland

Sharing the Dream - Abayyan

Descending from El Shahbaa through Mahfouza

Maysa
1983 Chestnut Mare
Sire: Moniet El Sharaf
(*Ibn Moniet El Nefous x Bint Bint Moniet)
Dam: *Magidaa (Alaa El Din x Maysa x Mahfouza)
Bred by Jarrell McCracken
Class A Halter Champion

Filsinger photo

Bint Maysa 1999 Chestnut Mare
Sire: BB Thee Renegade
(Thee Desperado x PH Safina)
Dam: Maysa
Bred by Karija Arabians
2002 Reserve Champion Futurity Filly,
Egyptian Event

Safir KA 2001 Bay Stallion
Sire: Farres (Anaza El Farid x Shameerah)
Dam: Maysa
Bred by Karija Arabians

Filsinger photo

KARIJA ARABIANS
Michael & Valerie Resch • P.O. Box 525 • Emerson, Manitoba, Canada R0A 0L0

Sharing the Dream - Abayyan

Descending from El Shahbaa through Mona (Inshass)

RN Ajlay
1996 Grey Mare

Filsinger photo

RN AJLAY	Safir	Salaa El Dine	Ansata Halim Shah
			Hanan
		Aisha	Ansata Halim Shah
			Ghazala
	Ameena	*Jamilll	Madkour I
			Hanan
		Amal	Mohafez
			Hanan

AL SHAQAB STUD
P.O. Box 90055 • Doha, Qatar

Sharing the Dream - Abayyan

Descending from El Shahbaa
through Mahfouza

Alidaar
1980 Grey Stallion
Bred by Russell & Mildred Jameson

U.S. & Canadian National
Top Ten Futurity Colt.
Qatar International Shows:
Res. Champion 1999,
2nd Place Aged Stallions 2000 & 2001.
Qatar National Shows 2000 & 2001:
First Place Aged Stallions &
Res. Champion Stallion

Gigi Grasso photo

	Shaikh Al Badi	*Morafic	Nazeer
			Mabrouka
		*Bint Maisa El Saghira	Nazeer
ALIDAAR			Maisa
	Bint Magidaa	*Khofo	*Morafic
			*Nabilahh
		*Magidaa	Alaa El Din
			Maysa

AL SHAQAB STUD
P.O. Box 90055 • Doha, Qatar

555

Sharing the Dream - Abayyan

Descending from El Shahbaa through Mahfouza

Al Shahhaa Hamdan
1992 Grey Mare
Bred by Hamdan Stables
Champion Mare,
Egyptian National E.A.O. Show 2000
Dam of **Al Ghalya Hamdan,**
1st Place, Yearling Fillies,
Egyptian National E.A.O. Show 1997;
Ridwan Allah Hamdan,
Winner of 3 races.

Arabians are the crystallization of pride,
embodiment of serenity,
fountain of youth in which we dwell.
The inner beauty of their souls
reflects on their expression
and in their eyes there is more to tell.
— Quote by Fatma Hamza

Escher photo

	Fateh	Shadwan	Shaarawi / Kamar
		Alifa	Alaa El Din / Zebeda
AL SHAHHAA HAMDAN			
	Wardshah	Adeeb	Shaarawi / Nawal
		Wasfia	Ibn Maysouna / Shamaa

In memory of my ancestor, our Holy Prophet Mohammed (PBUH), our Ideal for his great love for the Arabian.
And to my grandfather, Sayed Pasha Khashaba, one of the founders of the E.A.O.,
a dedicated breeder and an exceptional horseman.

HAMDAN STABLES
Fatma Ahmad Hamza • 1 Mohammed Mazhar Street, Zamalek • Cairo, Egypt 11211

Descending from El Shahbaa
through Mona (Inshass) and Mahfouza

Nefertiti
1999 Grey Mare
Sire: Alidaar (Shaikh Al Badi x Bint Magidaa)
Dam: Nessma (Salaa El Dine x Amarilla x Ghazala x Hanan)
Bred by Al-Nakeeb Arabians
Middle East Champion Foal 1999
and Jordan National Champion 2001

Badawia
1997 Grey Mare
Sire: Nahaman
(Salaa El Dine x Ameera
x Hanan)
Dam: Beshira
(Jamil x Bushra x Marah x
Maysa x Mahfouza)
Bred by Dr. Hans Nagel

Nagda
1986 Grey Mare
Sire: Nabiel (*Sakr x *Magidaa)
Dam: Bint Magidaa (*Khofo x *Magidaa x
Maysa x Mahfouza)
Bred by Russell & Mildred Jameson

Gigi Grasso photos

AL-NAKEEB ARABIANS
Mr. Hassanain Al-Nakeeb • Gillingham Hall • Gillingham, Norfolk, United Kingdom NR34OED

Sharing the Dream - Abayyan

Matriarchs of the El Shahbaa Family: Bint Magidaa

		Nazeer	Mansour
	*Morafic		Bint Samiha
		Mabrouka	Sid Abouhom
*Khofo			Moniet El Nefous
		Antar	Hamdan
	*Nabilahh		Obeya
		Farasha	Sid Abouhom
Bint Magidaa			Yosreia
		Nazeer	Mansour
	Alaa El Din		Bint Samiha
		Kateefa	Shahloul
*Magidaa			Bint Rissala
		Antar	Hamdan
	Maysa		Obeya
		Mahfouza	Hamdan
			El Mahrousa

Polly Knoll photo

Magidaa (Alaa El Din x Maysa) Johnny Johnston photo.

Dreams do become realities for many, and the conception of Bint Magidaa was to prove a dream come true for everyone whose life she touched. Her dam, by the same sire as Hanan, was the elegant chestnut mare, *Magidaa, imported from Egypt by Gleannloch. Bred to the handsome well-bred *Morafic son, *Khofo, Bint Magidaa resulted and was eventually sold to become a foundation mare for the Jamesons of Ranch Ruminaja. There she produced outstanding colts and daughters when bred to their *Morafic son, Shaikh El Badi (*Morafic son to *Morafic granddaughter - the sire of the sire being the grandsire of the dam). Many of her progeny bearing the Ruminaja prefix became show champions, and Ruminaja Ali became Res. U.S. National Champion Stallion. One of

Bint Magidaa's closest associates was David Gardner, a most talented trainer, farm manager, and marketing genius. David and his wife, Marion, survived a terrible car accident, in which they were nearly killed after he retired from his Gardner Bloodstock business. Many of his fulfilled dreams flashed before him during his miraculous recovery. This story is one which he wrote especially for this chapter.

Ruminaja Ali (Shaikh Al Badi x Bint Magidaa) Polly Knoll photo.

Sharing the Dream - Abayyan

Ruminaja Bahjat (Shaikh Al Badi x Bint Magidaa)

Ruminaja Majed (Shaikh Al Badi x Bint Magidaa). Polly Knoll photo

Bint Bint Magidaa (Shaikh Al Badi x Bint Magidaa). Forbis photo

Richteous (Richter MH x Bint Bint Magidaa). Scott Trees photo

Karen Kasper sculpture of Bint Magidaa. Javan photo

Alidaar
(Shaikh Al Badi x
Bint Magidaa)
Escher photo

559

Sharing the Dream - Abayyan

Left is Bint Bint Magidaa (x Shaikh Al Badi) with her dam Bint Magidaa at 26 years old. Van Lent photo.

"I guess my long love affair with Bint Magidaa began before I ever saw her. In 1971, as a young trainer working for Bentwood Farm, I was sent to Egypt to learn the various bloodlines. I was struck by the beauty of an old chestnut Nazeer son named Alaa El Din. He had the most perfect head and neck of any of the El Zahraa stallions at that time. Although his sons were uninteresting, his daughters were the most beautiful mares I had seen.

I returned to America with the image of Alaa El Din and his beautiful daughters in my imagination. By this time the best of the Egyptian stallions were in America. I felt if those fabulous Egyptian stallions were bred to mares of the Alaa El Din line, the result could be a generation of superstars. I dreamed of being part of that adventure.

A decade after my first trip to Egypt, a young stallion named Ruminaja Ali burst onto the scene. Along with the rest of the world, I saw him as one of the most beautiful stallions alive. He was the very combination of bloodlines I had been so smitten with. His sire was Shaikh Al Badi, a modern Nazeer-bred stallion. Ruminaja Ali's dam was the Alaa El Din granddaughter Bint Magidaa. I went to see Ruminaja Ali's granddam, *Magidaa, and his dam Bint Magidaa. They were superb. After seeing his first foals, I vowed I must own him. It took 3 years to convince his owners to sell him, but in 1982, I cashed a life insurance policy and acquired a 30 day option to buy Ruminaja Ali. On the 30th day, with a syndicate made up of 10 of our best clients, I hand delivered a check to Ruminaja Ali's owners for over $1,000,000.

Just before Ruminaja Ali became U.S. Reserve National Champion Stallion in 1983, I purchased Bint Magidaa in partnership with my friend Joe Krushinski. We owned her for five wonderful years. During those years Bint Magidaa gave us one beautiful foal after another including Bint Bint Magidaa and a handsome grey full brother to Ruminaja Ali who we named Alidaar. As Alidaar began to develop it was apparent having two full brothers of such class in one breeding farm was a waste, and I agreed to sell him to the Poth family in Europe where he was became one of Europe's prominent stars. Eventually Alidaar was sold to Qatar where, like Ruminaja Ali, he has seeded the earth with beautiful sons and daughters.

Later, my good friend, Jim Sirbasku, of Rock Creek Arabians, bought Bint Magidaa as a birthday present for his wife, Judy. Bint Magidaa lived to be 28 years old and had her 18th foal at 26 years old. She was the queen of Rock Creek and cherished by the Sirbaskus, their staff, and the scores of visitors who came to see her each year.

All of the *Magidaa and Bint Magidaa family had interesting characters. They were kind and very intelligent. The females were wonderful mothers and raised healthy, confident foals. Like many of the great mares I have observed, they were slightly distant, and royally aloof. When mares of this family were raising a foal, they kept to themselves. The males although courageous, were tractable, kind and of course exceedingly beautiful.

It is no exaggeration to say Bint Magidaa and her dam, *Magidaa, founded a breeding legacy. Nearly thirty years after their first foals were born, one can see their characteristics in their relatives. They were prolific and prepotent. They have crossed successfully with a variety of bloodlines. The stallions breed as true as do the mares of this line. Today, many of the great champions trace to this mare line. If the ideal Arabian horses are those depicted in the art of Carl Vernet, *Magidaa and Bint Magidaa moved modern Arabian horses closer to that ideal."

Time is Too Slow for those who Wait,
Too Swift for those who Fear,
Too Long for those who Grieve,
Too Short for those who Rejoice;
But for those who Love, Time is eternity.

Henry Van Dyke

Sharing the Dream - Abayyan

Descending from El Shahbaa through Mahfouza

Bint Bint Magidaa
1987 Grey Mare
Sire: Shaikh Al Badi
(*Morafic x *Bint Maisa El Saghira)
Dam: Bint Magidaa
Bred by
Russell & Mildred Jameson, Michigan
Egyptian Event Res. Jr. Champion Filly;
Region 9 Top Five.
Scottsdale Top Ten Mare

"More beautiful than her mother!"
    Judy Sirbasku

Sparagowski photo

"Our Future"
Judy Sirbasku

Thee Infidel
1995 Bay Stallion
Sire: Thee Desperado
(The Minstril x AK Amiri Asmarr)
Dam: Bint Magidaa
Bred by James & Judy Sirbasku
Egyptian Event EBC Champion Colt;
Res. Jr. Champion; U.S. National Top Ten Futurity,
Top Ten Stallion; Region 9 Top Five Stallion.
Sire of Champions.
2002 Egyptian Event 2nd Leading Sire.

Jean Pierre photo

ARABIANS LTD.
Judy & Jim Sirbasku • 8459 Rock Creek Road • Waco, Texas 76708

Sharing the Dream - Abayyan

Scott Trees photo

Bint Magidaa
1970 - 1998

"My beloved Bint Magidaa died November 19, 1998, and I still miss her."
Judy Sirbasku

The Kuhaylan Jellabi Strain

The Poem of Ibn Mesaoud al Qahtani about his Mare

I feel worried that the Kuhayla will not perform
as she always has,

And because of my great fear, she appeared to
be yellow in my eyes.
And because of my apprehension
we gathered around her the best horses and horsemen.

And when praying for her victory
it was as if we asked for the stars.

We strike our banner amidst the enemy,
And when we approached our enemy
our arrows were like a forest of iron
and we were able to defeat all the enemy's position.

And it was late in the day on Thursday
and the sight of our army was splendid to behold.

And we accomplished all that we had desired,
and more
in our battle against the enemy
astride our slight horses
with our weapons aimed before us.

And when the enemy heard the sound made by our
advancing horses,
they knew at once we were strong and powerful.

And with the sound we made, and at our sight
they were frightened.

And many prey have seen the foreheads of the horses
belonging to the powerful hunters who are
our enemies,

For they have slain the wild cow with their spears.

And we have entered the enemy country
each rider and his horse,

And the horses were like brides
garbed in fine raiment.

And everyone admired them
and they walked with the people in a very calm
and sedate manner,
surrounded by the old men
and followed by the children.

Ibn Saud says we must protect her
because every man would covet her.

She is a shaqra,
and she has a long tail
and flies like the wild dove.

When she is standing, she appears calm,
but when she runs, it is as if she has wings.

And there is a string of prayer beads
adorning her neck
to bring good fortune.

And the blessings of our Lord Mohamed
are upon her because of those beads.

Sharing the dream - Kuhaylan Jellabi

Matriarchs of the Yamama/*Maaroufa Family: Maar-Ree and RDM Maar Hala

Much has been written about this family, particularly about the grand matriarch, *Maaroufa. She was purchased from her breeder, H.H. Prince Mohamed Aly in Egypt and imported to Illinois in 1932 by Henry Babson, whose royally-bred horses are the subject of the book, *The Royal Arabians of Egypt and the Stud of Henry B. Babson*. In the Forward, Mr. Babson's daughter, Elizabeth Tieken, pays tribute to this grand mare and to *Maaroufa's full brother, *Fadl, remarking that: "My father liked *Maaroufa better than any of his other horses. She was typey and beautiful, but he wouldn't have been as pleased with her had she not been a good broodmare." She went on to reflect: "*Fadl was his second favorite; he too did his job. My father's task, as he saw it, was to use his human knowledge to get the best out of his horses, and he worked at this endeavor to the best of his ability. He never foaled what he believed was the 'perfect horse' but he had many splendid near misses, and very real pleasure in his quest for perfection."

This family requires a book in itself, as there are countless valuable descendants from it. Two matriarchs among them stand out, and they, along with other familial ancestors, are worthy of special notice in this section where their bloodlines are significantly featured: RDM Maar Hala descending from Maar Ree, and Fa Moniet descending from Fada.

"I took one look, and that was it," Dr. Felino Cruz remembers about the first time he saw the two year old Maar Hala at Bob and Jean Middleton's place. He purchased her, taking great pleasure in the beauty she brought into their lives, and from then on he devoted himself to breeding straight Egyptian bloodlines. Maar Hala spent some days visiting at Ansata

Maar-Ree	Fasaab	*Fadl	Ibn Rabdan	Rabdan El Azrak
				Bint Gamila
			Mahroussa	Mabrouk Manial
				Negma
		*Bint Bint Sabbah	Baiyyad	Mabrouk Manial
				Bint Gamila
			Bint Sabah	Kazmeyn
				Sabah
	Maarou	Fay El Dine	*Fadl	Ibn Rabdan
				Mahroussa
			*Bint Serra	Sotamm
				Serra
		*Maaroufa	Ibn Rabdan	Rabdan El Azrak
				Bint Gamila
			Mahroussa	Mabrouk Manial
				Negma

Maar-Ree (Fasaab x Maarou). J. Johnston photo

RDM Maar Hala (El Hilal x Maar Jumana x Maar-Ree) as she appeared on the cover of Arabian Horse World *magazine. Javan photo*

Sharing the dream - Kuhaylan Jellabi

Far left RDM Maar Hala
with two of her daughters:
Bint Maar Hala (x Shaikh Al Badi)
and Maar Halaa (x Abenhetep)
J. Little photo.

Ansata Mahroussa (Ansata Hejazi x Shahlikah to Maar-Ree)
Jaber Al Kazemi photo.

Halim El Mansour (*Ansata Ibn Halima x RDM Maar Hala)
J. Little photo.

MFA Bint Monien (*Farazdac x MFA Monien x Maar-Juahara x Maar-Ree). K. Kasper photo.

Sharing the dream - Kuhaylan Jellabi

Arabian Stud in Texas, where she was scheduled to be bred to *Ansata Ibn Halima after her first foal, ET Crown Prince, by the Egyptian Prince, was born there. During the time she graced Ansata with her presence, she was lovingly appreciated by all the staff and visitors for her special nobility, serenity and beauty. When she arrived the first thing one noticed was her beautifully shaped head and big dark black eyes with soulful expression. Her overall harmony, wonderful balance, and correctness perfectly reflected her pedigree that combined the classic lineage of Nazeer, Halima, and Mahroussa. An excellent blending of these individuals, Maar Hala was also an ideal representative of her important tail female family. Originally iron grey, her color eventually turned fleabitten, like her paternal grandsire, giving her an even more regal desert appearance. Her handsome son, El Halimaar, resulting from her mating to *Ansata Ibn Halima, brought her prominence by winning halter and "most classic" honors, as well as U.S. Top Ten Futurity Stallion in 1983, an honor that he achieved along with his paternal half-brother, Ansata Halim Shah, at a time when "classic" ruled the show ring. Another worthy son was Halimaar's full brother, the handsome bay multi-champion, Halim El Mansour.

Over the years Maar Hala presented the Cruz family with 11 foals sired by top straight Egyptian stallions, including those mentioned above, as well as Shaikh Al Badi (sire of Prince Ibn Shaikh), Ruminaja Ali (sire of Maar Ibn Ali), and Al Metrabbi (sire of Maartrabbi). Of her offspring, eight became champions, thus ranking her as the leading champion-producing Egyptian mare, and she was on *Arabian Horse World's* elite list of "Aristocrat Mares." Of course, Maar Hala knew "who" she was. She would selectively nibble only special carrots when offered a bunch, remembers Jody Cruz. Although she aged gracefully, one day after she had reached age 27, Jody sensed her time was at hand and took her out of her stall into the soft California sunlight where they could commune peacefully with nature together. Although she hadn't eaten much for awhile, she took the carrots he offered as a gesture to their loving friendship before Jody returned her to her spacious stall where they silently bade each other final farewells. An hour later her spirit soared to a higher plane. Today her blood enriches the breed worldwide. Many who saw her called her phenomenal, extraordinary, and royal. "To me," said Dr. Felino Cruz, "she was the most beautiful horse in the world. It was as simple as that."

> *Green be the turf above thee,*
> *Friend of my better days!*
> *None knew thee but to love thee,*
> *Nor named thee but to praise.*
>
> Fitz-Greene Halleck

ET Crown Prince (The Egyptian Prince x Maar Hala). Jeff Little photo

*El Halimaar (*Ansata Ibn Halima x Maar Hala). Jeff Little photo*

Sharing the dream - Kuhaylan Jellabi

Jeff Little photo

Descending from *Maaroufa through Maar Ree

RDM Maar Hala
1973 - 2000 Grey Mare
Sire: El Hilal
(*Ansata Ibn Halima x *Bint Nefisaa)
Dam: Maar Jumana (Disaan x Maar Ree)
Bred by Dr. & Mrs. R.L. Weaver
All time leading Egyptian
Dam of Champions.
Dam of Champions
El Halimaar & **Prince Ibn Shaikh.**

Long Live the Queen!

Javan photo

El Halimaar
1980 - 2002 Grey Stallion
Sire: *Ansata Ibn Halima (Nazeer x Halima)
Dam: RDM Maar Hala
Bred by Dr. Felino Cruz
U.S. Top Ten Futurity Stallion 1983; Sire of Multi-Champions.

Vesty photo

*Maarauder MH
1996 Grey Stallion
Sire: Crusader (Salaa El Dine x AK Kastana)
Dam: Maartrabbi (Al Metrabbi x RDM Maar Hala)

RANCHO BULAKENYO
Felino, Jody, Karen & Ryan Cruz • 2755 Los Osos Valley Road • Los Osos, California 93402

Sharing the dream - Kuhaylan Jellabi

Descending from *Maaroufa through Maar-Ree

Photo Randi Clark

Zandai Rubayana
1994 Grey Mare
Bred by Dr. William Hudson

	*Ansata Ibn Halima	Nazeer
Ansata Omar Halim		Halima
	Ansata Rosetta	Ansata Shah Zaman
ZANDAI RUBAYANA		*Ansata Bint Bukra
	Alcibiades	*Rashad Ibn Nazeer
MFA Tassiadeh		*Bint Moniet El Nefous
	MFA Tashery	The Egyptian Prince
		Maar-Juahara x Maar Ree

*Though art thy mother's glass and she in thee
Calls back the lovely April of her prime.*
  Shakespeare

ZANDAI ARABIANS
Dr. William Hudson • 6010 Riley Road • Cumming, Georgia 30040

Descending from *Maaroufa through Maar-Ree

Sharing the dream - Kuhaylan Jellabi

Abraxas Moonbeam
1999 Bay Mare
Sire: Abraxas-Moonstruk (ET Crown Prince x SF Moon Maiden)
Dam: Abraxas-Maar Hala (ET Crown Prince x Bint Maar Hala)
Bred by Abraxas Arabians; owned by Desert Straights

J. Little photo

Abraxas- Maar-Hala
1991 Bay Mare
Sire: ET Crown Prince (The Egyptian Prince x RDM Maar Hala)
Dam: Bint Maar Hala (Shaikh Al Badi x RDM Maar Hala)
Bred by Dr. Felino Cruz

J. Little photo

Abraxas Moon Glow
1996 Bay Mare
Full sister to Abraxas Moonbeam
Bred by Abraxas Arabians

Randall photo

ABRAXAS ARABIANS
Emil & Debra Nowak • 32415 Yates Road • Winchester, California 92596

Sharing the dream - Kuhaylan Jellabi

Descending from Negma through Ghazala I (Inshass)

Escher Photo

218 Elf Layla Walayla 1991 Grey Mare
Bred by Babolna State Stud, Hungary
Asil Cup Winner; Supreme Champion Omtende; Champion Marbach; Champion Res. Nation's Cup; Champion Res. Europe Championship; Champion Res. World Championship, Salon du Cheval

			*Ansata Ibn Halima
	Assad	Ansata Halim Shah	Ansata Rosetta
		Arussa	Madkour I
218 ELF LAYLA WALAYLA			Hanan
		Ibn Galal I	Ibn Galal
	223 Ibn Galal I-13		Hanan
		23 Ghalion-2	Ghalion
			6 El Aziza

*She walks in beauty like the night
Of cloudless climes and starry skies
And all that's best of dark and bright
Meet in her aspect and her eyes.*

                Lord Byron

HANAYA ARABIANS
Nayla Hayek • Expohof • Schleinikon, Switzerland CH-8165

Sharing the dream - Kuhaylan Jellabi

Descending from *Maaroufa through Maar-Ree

MFA Bint Monien
1989 Grey Mare

	Alaa El Din	Nazeer
*Farazdac		Kateefa
	Farasha	Sid Abouhom
MFA BINT MONIEN		Yosreia
	The Egyptian Prince	*Morafic
MFA Monien		*Bint Mona
	Maar-Juahara	*Ibn Moniet El Nefous
		Maar-Ree

Seeds spring from seeds, beauty breedeth beauty…
By law of Nature thou art bound to breed,
That thine may live…

        Shakespeare

HANAYA ARABIANS
Nayla Hayek • Expohof • Schleinikon, Switzerland CH-8165

Sharing the dream - Kuhaylan Jellabi

Descending from *Maaroufa through Maar-Ree

Ansata Mahroussa
1997 Grey Mare
Sire: Ansata Hejazi
(Ansata Halim Shah x Ansata Sudarra)
Dam: Shahlika (Ansata Halim Shah x
Bint Hassema x Maar-Malika x
Maar Jumana x Maar-Ree)
Bred by Ansata Arabian Stud

Jaber Al Kazemi photos

Massarrah Ezzain
2002 Grey Mare
Sire: Ansata Al Murtajiz
(Ansata Hejazi x Ansata Samsara)
Dam: Ansata Mahroussa
Bred by Ezzain Arabians

EZZAIN ARABIANS
Usamah Zaid Al Kazemi • P.O. Box 30 • Safat 13001, Kuwait

Sharing the dream - Kuhaylan Jellabi

Descending from *Maaroufa through Fada

Nabajah El Chamsin
1998 Grey Stallion

	Salaa El Dine	Ansata Halim Shah
Nahaman		Hanan
	Ameera	Madkour I
NABAJAH EL CHAMSIN		Hanan
	Nejdy	Salaa El Dine
Fatima XVII		Lotfeia
	Maarshafa	Moniet El Sharaf
		Bint Bint Maarena

Jaber Al Kazemi photos

EZZAIN ARABIANS
Usamah Zaid Al Kazemi • P.O. Box 30 • Safat 13001, Kuwait

Matriarchs of the Yamama/*Maaroufa Family: Fada and Fa Moniet

Fada (Faddan x Aaroufa x *Maaroufa). Judith Forbis photo.

Fa Moniet (*Ibn Moniet El Nefous x Fada). Johnny Johnston photo.

Another exceptional broodmare of the *Maaroufa line was Fada, bred and owned by the Babson Farm. Unlike most of the *Maaroufa line, which was grey, Fada was different in type. Foaled in March 1956, she was chestnut with a lollipop blaze and three white socks. A noble, kindly mare, very tall (over 15 hands), she was rather plain-headed with high-placed dark expressive eyes of average size, deep jowls, and a long neck, traits that she transmitted to her foals. She was high-withered, relatively long in body, good topline, with a slightly peaky, but strong croup, well set and good tail carriage, wide and deep, and good movement. She was not a beauty one would get excited about in comparison to some of her more classic grey relatives. Nevertheless, she distinguished herself as an important broodmare of this family. Her handsome, bold-moving sire, Faddan (by *Fadl), was out of the Saklawi Jedran mare, *Bint Saada, thus giving Fada strong paternal tail-female lines to the Saklawi family through Faddan, and through her dam's sire, Fay El Dine.

				Nazeer	Mansour
		*Morafic			Bint Samiha
				Mabrouka	Sid Abouhom
	*Ibn Moniet El Nefous				Moniet El Nefous
				Shahloul	Ibn Rabdan
		Moniet El Nefous			Bint Radia
				Wanisa	Sheikh El Arab
Fa Moniet					Medallela
				*Fadl	Ibn Rabdan
		Faddan			Mahroussa
				*Bint Saada	Ibn Samhan
Fada					Saada
				Fay El Dine	*Fadl
		Aaroufa			*Bint Serra
				*Maaroufa	Ibn Rabdan
					Mahroussa

Faddan (*Fadl x *Bint Saada) sire of Fada. Babson Farm photo.

Aaroufa (Fay El Dine x *Maaroufa) dam of Fada. Jerry Sparagowski photo.

Sharing the dream - Kuhaylan Jellabi

Maarena (Fabah x Fada), an excellent broodmare. Polly Knoll photo.

Bint Atallah (Ruminaja Ali x AK Atallah x Al Nahr Mon Ami x Bint Fada). J. Guess photo.

Among her progeny who have had immense impact were her daughter, Bint Fada (by Fa-Serr), the good broodmares Maarena and Idaa, and the stallion, El Ibn Fabah (by Fabah). One of Fada's most beautiful daughters was Fa Moniet, the result of mating her to *Ibn Moniet El Nefous at the time he was in Illinois with his importer/owner, Jay Stream. A very elegant mare from the day she was born, Fa Moniet was one of the early crosses of the new Egyptian imports on the old Egyptian bloodlines. She stood out in the Babson pastures from all the rest as someone unique. When Jarrell McCracken of Bentwood Farms saw her as a young mare, he persuaded the Babson farm manager, Homer Watson, into selling her, and she was delivered to Texas and put in the Bentwood show string. Taken to Scottsdale, she won a coveted Top Ten placement in strong halter competition.

Retired from the show ring, Fa Moniet was bred to The Egyptian Prince and produced the elegant mare Akid Fa Mona, but tragically Fa Moniet died after foaling a beautiful colt who was named Prince Fa Moniet in her honor. When Bentwood was dispersed several breeders tried to purchase him, but Ansata obtained him for use as a senior sire.

Although Fa Moniet did not have a long life, she did her job, as Mr. Babson required of her granddam, *Maaroufa, and served her family line with distinction.

Akid Fa Mona (The Egyptian Prince x Fa Moniet) a full sister to Prince Fa Moniet

Prince Fa Moniet (The Egyptian Prince x Fa Moniet), sire of international champions. Sparagowski photo.

> The flowers, still faithful to the stems,
> Their fellowship renew;
> The stems are faithful to the root,
> That worketh out of vie;
> And to the rock the root adheres
> In every fibre true.
>
> William Wordsworth

Sharing the dream - Kuhaylan Jellabi

Sparagowski photo

Descending from *Maaroufa through Fada

Akid Geshan
1986 Grey Stallion
Bred by Alyce Burgess
Class A Grand Champion
Most Classic Arabian wins in U.S.A.;
Supreme Champion Exhibit
in Australia.
Sire of Multi-Champions
in U.S.A. and Australia.

Nabiel	*Sakr	*Sultann
		Enayat
	*Magidaa	Alaa El Din
AKID GESHAN		Maysa
	AK El Zahra Moniet	*Ibn Moniet El Nefous
Akid Hanalei		Maarena
	Akid Serra Moniet	Serr Al Sahih
		Fa Moniet x Fada

DARA MEADOWS FARM
Debra & David Geiser • 449 E. Nelson Street • Lexington, Virginia 24450

Sharing the dream - Kuhaylan Jellabi

Descending from *Maaroufa through Fada

Phaaros
2000 Grey Stallion
Bred by Judy Guess, U.S.A.

Gigi Grasso photos

	Anaza El Farid	Ruminaja Ali
ZT Faa'iq		Bint Deenaa
	ZT Jamdusah	*Jamilll
PHAAROS		IES Sondusah
	Ruminaja Ali	Shaikh Al Badi
Bint Atallah		Bint Magidaa
	AK Atallah	*Ansata Ibn Halima
		Al Nahr Mon Ami x Bint Fada

He neighs as the lion roars in the mountain;
He is an eagle who soars through the skies…
A poem recited by the Emir Abd el Kader

ALFABIA STUD
Gigi Grasso & Paolo Damilano • Via Pittamiglio 1 • Cherasco, Italy 12062

Sharing the dream - Kuhaylan Jellabi

Descending from *Maaroufa through Aaroufa

Javan photo

Imperial Baarez
1996 Grey Stallion
Bred by Imperial Egyptian Stud
Egyptian Event Supreme
Champion Stallion;
U.S. Top Ten Futurity Stallion;
East Coast Champion Stallion;
Region 15 Champion Stallion.
Undefeated in Futurity classes.
Sire of Champions including
Imperial Baarakah and
Imperial Baaraah

Where in this wide world can (a person) find nobility without pride, friendship without envy or beauty without vanity? Here, where grace is laced with muscle and strength by gentleness confined. He serves without servility, he has fought without enmity. There is nothing so powerful, nothing less violent; there is nothing so quick, nothing more patient.
                    Ronald Duncan

	Imerial Imdal	Ansata Imperial
PVA Kariim		Dalia
	BKA Rakiisah	The Egyptian Prince
IMPERIAL BAAREZ		AK Radia
	*Orashan	Messaoud
BB Ora Kalilah		Ora
	PH Safina	*Lancers Sahm
		Noufina

IMPERIAL EGYPTIAN STUD
Barbara A. Griffith • 2642 Mt. Carmel Road • Parkton, Maryland 21120

The Kuhaylan Rodan Strain

The Poem of Hussein Ibn Farz about Al Harqa who reached The Stud of Abbas Pasha I

O you, who would try to outdistance me,
    O people who are unstable, following just anyone,
You have never seen a mare like her among the horses of the tribes.
She accepts not calm horses in her presence.
She lets not just anyone ride her.
Indeed there is nothing which compares with her
    save for the raging torrents
    which sweep the rocks before them.
O my mare, you give more than is your due,
You are like a generous giver of camels,
    like Abu Saud, whose generosity can reach everyone,
And if you were for anyone else but him,
I would not have given or sold you,
    not even for the highest prices.
O you who should want to have such a mare,
    it is futile even thinking about it,
    for she is unique.
My enemies who try to outdistance me
    ride their horses holding the reins,
And behind them are the servants who carry the weapons
    riding upon calm horses.
And when they stopped at the curve to rest their mounts,
    they were big and strong.
And they called out, in pure Shemi
    encouraging an attack.
And their swords they unsheathed
    which glittered in the sunlight
And because of their ignorance,
    they sang of victory which they envisioned to come.
But I raised my sharp sword high,
    and I cleared the way for the camels which came behind me,
    so we could plunder the enemy.
And those long dresses [of the women]
    rustled like the branches of the trees,
    and our prisoners came to us
    their appearance a joy to behold.
Even though there was bloodshed and killing between us,
    we beg of Allah
    that he should heal and compensate all those who were
    wounded and lost the battle.
And we thank Allah, who has given us all we would ever need,
He who is most generous.

Sharing the dream - Kuhaylan Rodan

Matriarchs of the Rodania- Riyala Family: Malaka

Malaka as a weanling (Kheir x Bint Bint Riyala).

Malaka	Kheir	Ibn Samhan	Samhan	Rabdan El Azrak
				Om Dalal
			Nafaa El Saghira	Meanagi Sebeli
				Nafaa El Kebira
		Badaouia	D.B.	D.B.
				D.B.
			a Saqlawi Shafiyah	D.B.
				Saqlawi Shafiyah
	Bint Bint Riyala	Gamil Manial	Saklawi II	Saklawi I
				El Dahma
			Dalal I	Rabdan El Azrak
				Om Dalal
		Bint Riyala	Nadir	Mesaoud
				Nefisa (1885)
			Riyala	*Astraled
				Ridaa

The Rodan strain tracing to the root desert-bred Rodania mare appears third after the Dahman and Saklawi strains with a total of 3122 straight Egyptian horses registered in the U.S. at the beginning of the new millennium. The Rodan has always produced popular and winning performance, endurance, racing, and halter horses since it appeared on the western scene after being imported and incorporated into the Crabbet program in England by the Blunts and their daughter, Lady Wentworth. Later it became successful in America through the Spencer Borden and W.R. Brown Crabbet imports of the early 1900's. Eventually these bloodlines spread worldwide and also became one of the foundation strains in Egypt with the royal famiy studs and at the R.A.S./E.A.O.

The Riyala branch of this family through Malaka's daughters, Samia and Nazeera, produced many of the top Gleannloch show horses, particularly in the 70's and 80's, who won fame for that stud, e.g., *Romanaa II (U.S. National Top Ten Mare) *Fawkia (U.S. National Top Ten Mare), Dalia (Scottsdale Champion and Legion of Merit winner, and dam of U.S. Res. National Champion and Salon du Cheval World Champion, Imperial Imdal). The 1975 U.S. National Champion Stallion and Legion of Merit winner, *Asadd, was also of this family.

The Rodan strain continues to produce popular performance, endurance, racing and halter horses on every continent. The Riyala branch has generally been deeper

Nazeera (Nazeer x Malaka)

Samia (Nazeer x Malaka)

bodied, shorter coupled and somewhat shorter headed compared to the Bint Rissala line, in the R.A.S./E.A.O. programs as well as in later years, again depending on the linebreeding and combinations of families used. The crosses with Sameh, *Ansata Ibn Halima and *Messaoud are particularly noteworthy.

Among the numerous matriarchs that stand out in this Riyala family are the three full sisters, Mamlouka, Nazeera, and Samia, all sired by Nazeer. On the patriarchal side stands their full brother, the powerful white full-bodied stallion, Waseem, sire of the Imperial import, *Malekat El Gamal, a mare much resembling him in type and who is found in numerous pedigrees.

One of the successful families emerging in the new millennium is from *Hekmat, (Antar x Samia). Prettier than her dam, she was a substantial strong-bodied, well-structured chestnut Antar daughter imported by Gleannloch Farms, who represented in the flesh what her pedigree indicated she could be. Dam of 7 mares, 6 geldings, and two stallions, her most notable champion son was Al Nimr (by *Morafic). Her prominently known daughters are Royal Gemm (by El Hilal), influential in the Abraxas Arabians program, Il Durra (by *Morafic), influential in the Silver Maple program through Hi-Fashion Hitesa, and Hamamaa (by Ibn Antar and, thus, double Antar), the dam of successful international bloodstock such as Peter Pond's Australian National Champion Stallion, Hansan (by El Hilal) and Bint Hamamaa (by Ansata Ibn Sudan), the dam of Al-Nakeeb's champion Nabiel daughter, Bint Bint Hamamaa, now in England. Bint Bint Hamamaa also produced B A Halamet (by El Halimaar), a valuable mare for Abraxas Arabians.

*Hekmat (Antar x Samia x Malaka)

Hansan (El Hilal x Hamamaa x *Hekmat). Pond photo

Waseem (Nazeer x Malaka), a powerful and handsome Malaka son. Forbis photo.

*Asadd (*Sultann x Amani x Nazeera x Malaka), a U.S. National Champion. Johnny Johnston photo.

Sharing the dream - Kuhaylan Rodan

The wheel of fortune turns, and the time came when Egypt would receive back some of the bloodlines they had exported so freely in the 60's, 70's and 80's, and which they now needed to restore certain classic characteristics. Thus when Omar Sakr imported from America Bint Bint Hamamaa's multi-champion son, Shaheen (by El Hadiyyah), and Imperial Madori (by *Imperial Madheen), who traces in tail female to *Fawkia, these stallions created a small revolution in the country by winning show championships and then going on to become sires that influenced Egyptian breeding in a most classic way. To date, the widely used Imperial Madori has sired more champions than any other stallion in Egypt, and Shaheen, who does not stand at public stud, has become a valued foundation sire of classic type and show-winning progeny for Sakr Arabians.

*Hekmat's full sister in blood, the iron-grey Kawmia (Antar x Nazeera), different in type, prettier and a very handsome strong-bodied mare, attracted much interest from the Russians when they were buying for Tersk during the late 60's. However, they did not take any stock directly from the E.A.O., preferring to buy lesser-quality horses off the race track. Kawmia produced the Lancer Arabians import, *Masria, by *Sultann, and this line became influential in the Abraxas program. The Nazeera daughter, Hebah (by *Ibn Hafiza), produced U.S. Top Ten Mare *Lancer's Asmara (by Seef) who crossed well with Ansata bloodlines as did the descendants of *Salomy (El Sarie x Malaka) and Amani (El Sarie x Nazeera) in America. Another successful combination of Ansata bloodlines with Malaka descended from the mare *Omayma (Sameh x Nazeera), imported by Gleannloch and later owned by the Pond's Forest Hill Arabians of Australia.

This Malaka family has a strong following worldwide, as will be noted from this chapter and through study of the Pyramid Society's *Reference Handbooks of Straight Egyptian Horses*, *Asil Arabians* books, and other reference material and magazines featuring Egyptian horses.

> *In the hands of a craftsman,*
> *The roughest diamond shines.*
> *The skill lies in recognizing the potential*
> *And encouraging it to glow.*
> Anonymous

Imperial Madori (*Imperial Madheen x Imperial Orianah x Imperial BtFawkia x *Fawkia x Mamlouka x Malaka). Escher photo.

The lovely Faras Azali (Fehris x Moradil, descending from *Salomy). J. Johnston photo.

Halims Asmara (Ansata Halim Shah x *Lancers Asmara). Escher photo.

Sharing the dream - Kuhaylan Rodan

Descending from Malaka through Nazeera

***HS Hero**
1998 Grey Stallion

	Asfour	Malik
Simeon Sadik		Hanan
*HS HERO		Simeon Safanad	Sankt Georg
		27 Ibn Galal-5
	El Hilal	*Ansata Ibn Halima
Helwa Lancer		*Bint Nefisaa
	Nahidd	*Nagid
		*Masria

J. Little photos

ABRAXAS ARABIANS
Emil & Debra Nowak • 32415 Yates Road • Winchester, California 92596

Sharing the dream - Kuhaylan Rodan

Descending from Malaka through Samia

Abraxas Halamaa
1994 Grey Mare
Sire: ET Crown Prince
(The Egyptian Prince x RDM Maar Hala)
Dam: B A Halamet
Bred by Abraxas Arabians
Top Ten Egyptian Event, 2nd Place

J. Little photos

B A Halamet
1990 Grey Mare
Sire: El Halimaar
(*Ansata Ibn Halima x RDM Maar Hala)
Dam: Bint Bint Hamamaa
(Nabiel x Bint Hamama x Hamamaa)
Bred by Dr. Felino Cruz

Abraxas Habielaa
1996 Grey Mare
Full sister to Abraxas Halamaa
Bred by Abraxas Arabians
Owned by Desert Straights
Top Ten Egyptian Event -Third place.

ABRAXAS ARABIANS
Emil & Debra Nowak • 32415 Yates Road • Winchester, California 92596

Descending from Malaka through Samia

Abraxas Hilals Gemm
1998 Grey Mare
Sire: ET Crown Prince
(The Egyptian Prince x RDM Maar Hala)
Dam: AH Abraxas
(El Hilal x Royal Gemm x *Hekmat x Samia)
Bred by Abraxas Arabians

J. Little photos

Abraxas Abu Hilal
1999 Bay Stallion
Full brother to Abraxas Hilals Gemm

Abraxas Amir
2000 Grey Stallion
Full brother to Abraxas Hilals Gemm

ABRAXAS ARABIANS
Emil & Debra Nowak • 32415 Yates Road • Winchester, California 92596

Sharing the dream - Kuhaylan Rodan

Darryl photo

Descending from Malaka through Samia

Hi-Fashion Hitesa
1983 Bay Mare
Sire: El Hilal (*Ansata Ibn Halima x *Bint Nefisaa)
Dam: Yasamin (Samim x Il Durra x *Hekmat x Samia)
Bred by Hi-Fashion Arabians
Pictured with her BB Thee Renegade colt,
Firestorm SMF, Region 12 Top Ten Yearling,
Egyptian Event Res. Champion Yearling
and 2 Year old Egyptian Champion.

Polly Knoll photo

HF Tesa Reekh
1989 Bay Mare
Sire: Ibn El Mareekh (*El Mareekh x Bint Deenaa)
Dam: Hi-Fashion Hitesa
Bred by Hi-Fashion Arabians

Polly Knoll photo

Martessa
1996 Bay Mare
Full sister to HF Tesa Reekh
Bred by Silver Maple Farm
Owned by Round Mountain Arabians, Inc.

SILVER MAPLE FARM, INC.
Christie & Henry Metz • 5125 Happy Canyon Road • Santa Ynez, California 93460

Sharing the dream - Kuhaylan Rodan

Descending from Malaka through Samia

Azure Moon
1998 - Deceased - Grey Stallion
Sire: Ali Saroukh
Dam: Hi-Fashion Hitesa
Bred by Silver Maple Farm, Inc.
Region 12 Top Ten

Born during a "Blue Moon"

Gigi Grasso photo

Darryl photo

Tempest Moon
2000 Grey Stallion. Full brother to Azure Moon
Bred by Silver Maple Farm, Inc.
Egyptian Event Top Five Yearling, Top Three Egyptian
Futurity; Third Place Egyptian Breeders Challenge

Scott Trees photo

Tranquility Moon
2001 Grey Mare. Full sister to Azure Moon
Bred by Silver Maple Farm, Inc.
Egyptian Event Futurity Yearling Filly Top Ten,
and Top Ten Egyptian Breeders Challenge

SILVER MAPLE FARM, INC.
Christie & Henry Metz • 1825 8th Street South • Naples, Florida 34102

Sharing the dream - Kuhaylan Rodan

Escher photo

Descending from Malaka through Samia

Shaheen
1993 Grey Stallion
Sire: El Hadiyyah
Dam: Bint Bint Hamamaa
Bred by Bryant Arabians
Two times Egyptian Event
West Coast Jr. Champion;
Reserve Jr. Champion,
Egyptian Event, Lexington, KY;
Junior Champion - Egypt Breeders Open;
Senior Champion - Egypt Breeders Open;
Most Classic - Egypt Breeders Open;
Get of Sire - Egypt Breeders Open;
Sire of numerous champions

		*Ansata Ibn Halima	Nazeer
	El Hadiyyah		Halima
		Ansata Jellabia	Ansata Ibn Sudan
SHAHEEN			Maarqada
		Nabiel	*Sakr
	Bint Bint Hamamaa		*Magidaa
		Bint Hamamaa	Ansata Ibn Sudan
			Hamamaa

SAKR ARABIANS
Sherifa & Omar Sakr • 6 Hassan Sabry Street • Zamalek, Cairo, Egypt

Sharing the dream - Kuhaylan Rodan

Descending from Malaka through Samia

Bint Bint Hamamaa
1985 Grey Mare
Bred by R.J. & J. Thorndike
A Champion Mare and Dam of Champions.
Dam of Shaheen, a Multi-Champion in Egypt.

	*Sakr	*Sultann	
Nabiel		Enayat	
	*Magidaa	Alaa El Din	
BINT BINT HAMAMAA		Maysa	
	Ansata Ibn Sudan	*Ansata Ibn Halima	
Bint Hamamaa		*Ansata Bint Mabrouka	
	Hamamaa	*Ibn Antar	
		*Hekmat x Samia	

Gigi Grasso photos

AL-NAKEEB ARABIANS
Mr. Hassanain Al-Nakeeb • Gillingham Hall • Gillingham, Norfolk, U.K. NR34 0ED

Sharing the dream - Kuhaylan Rodan

Descending from Malaka through Samia

MA Victoria
1999 Grey Mare
Bred by Karen Maisano

```
                                    The Egyptian Prince
              Prince Fa Moniet      Fa Moniet
   Ansata Iemhotep                  Ansata Halim Shah
              Ansata Nefara         Ansata Sudarra
MA VICTORIA                         The Minstril
              Thee Desperado        AK Amiri Asmarr
   Desperados Isybela               The Minstril
              Sybella               *Nibrias to Samia
```

Quality is a treasure that cannot be touched,
An inspiration that cannot be lost,
A source of pride that never loses its lustre.
                    Anonymous

Cameron photos

MA Belladonna
2002 Bay Mare
Bred by Karen Maisano
Sire: *HS Tadeusz
(Simeon Sadik x MA Tatia)
Dam: Desperados Isybela
(Thee Desperado x Sybella)

MAISANO ARABIANS
Karen Maisano • 9188 Eagle Run Drive • Brighton, Michigan 48116

Sharing the dream - Kuhaylan Rodan

Descending from Malaka through Salomy

Ansata Azali
1998 Grey Mare
Bred by Ansata Arabian Stud

J. Wich photo

		Prince Fa Moniet	The Egyptian Prince
	Ansata Iemhotep		Fa Moniet
		Ansata Nefara	Ansata Halim Shah
ANSATA AZALI			Ansata Sudarra
		Ansata El Naseri	Ansata Ibn Sudan
	Bint Faras Azali		*Ansata Bint Bukra
		Faras Azali	Fehris
			Moradil x Il Mandil x *Salomy

ORIENTA ARABIANS
Judith Wich • Eichenbuehl 26 • Wilhelmsthal, Germany 96352

Sharing the dream - Kuhaylan Rodan

Gigi Grasso photo

Descending from Malaka through Nazeera

Jamil Ezzain
2002 Grey Stallion
Bred by Ezzain Arabians

	Ibn Nejdy	Nejdy
NK Hafid Jamil		Ghazala
	Helala	Salaa El Dine
JAMIL EZZAIN		Ansata Gloriana
	The Minstril	Ruminaja Ali
TH Camille		*Bahila
	Amarela Al Badi	Shaikh Al Badi
		Amal Ali

*"…such as ancient sculptors modeled…
he was…the composite of all equestrian
statues of history."*

Felix Salten

EZZAIN ARABIANS
Usamah Zaid Al Kazemi • P.O. Box 30 • Safat 13001, Kuwait

Sharing the dream - Kuhaylan Rodan

Descending from Malaka through Nazeera

Halims Asmara
1992 Grey Mare
Sire: Ansata Halim Shah
(*Ansata Ibn Halima x Ansata Rosetta)
Dam: *Lancers Asmara (Seef x Hebah x Nazeera x Malaka)
Bred by Magness Arabians
European Champion Filly Egyptian Event Europe 1994
Jr. Champion Filly, Reserve, Kaub 1994;
European Res. Champion Mare Egyptian Event Europe 1997

Escher photos

GR Amaretto 1999 Grey Stallion
Sire: Classic Shadwan (Alidaar x Shagia Bint Shadwan)
Dam: Halims Asmara
Bred by Rothenberg Stud - Annette and Erwin Escher
Jr. Champion Egyptian Classic Show 2000
National Res. Champion, Jr. Colts 2001, Germany
Gold-Medal, German Stallion Show, Aachen 2002

GR Amaretta 2000 Grey Mare
Full sister to GR Amaretto
Bred by Rothenberg Stud - Annette and Erwin Escher
Class Winner Pyramid Cup 2001,
National Res. Champion Jr. Fillies, Germany 2002

ROTHENBERG STUD
Annette & Erwin Escher • 86653 Monheim • Germany

Matriarchs of the Rodania-Risala Family: Bint Rissala

Bint Rissala (Ibn Yashmak x Risala). Forbis collection

Bint Rissala	Ibn Yashmak	Feysul	Ibn Nura	Sottam (1860)
				Bint Nura
			El Argaa	Waziri Al Auwal
				Bint Jellabiet Feysul
		Yashmak (1893)	*Shahwan	Wazir
				Dahmeh Shahwaniyah
			Yamama (1885)	D.B.
				a Kuhaylah Jellabiyah
	Risala	Mesaoud	Aziz	Harkan
				Aziza (1868)
			Yemameh	Shueyman
				Bint Ghazieh
		Ridaa	Merzuk	Wazir
				a Kuhaylah Jellabiyah
			*Rose of Sharon	Hadban (1878)
				Rodania

No history book would be complete without mention of one of the most beloved and popular mares of her time, the remarkable *Serenity Sonbolah (Sameh x Bint Om El Saad). Bred at the E.A.O., she was beautiful as a young filly and remained so throughout her life. She was selected in Egypt for Serenity Egyptian Stud when the Marshalls, McNairs and Forbises were acquiring a large group of horses for themselves and for Serenity's owners, Hansi and Bradford Heck. Arriving in Canada, she was registered as *Serenity Sonbolah, and became Hansi Heck's pride and joy. It wasn't long before she grew into a brilliant show mare, achieving multiple championships and qualifying for the U.S. National Championships in 1971. Hansi, concerned over her welfare and maintenance prior to the show, brought her to the Ansata Stud, then located in Chickasha, Oklahoma, and left her there some days until the show commenced at the nearby Oklahoma City Fairgrounds. Her stablemate at the farm was the regal and very classic multi-champion stallion, Ansata Ibn Sudan, son of *Ansata Ibn Halima, who was also destined for the big event. Few who saw that U.S. National Show, with its huge entries in both the mare and stallion classes, will ever forget the ethereal chestnut mare who was crowned U.S. National Champion Mare, and her kingly escort, Ansata Ibn Sudan, who was enthroned as the U.S. National Champion Stallion. Together they made history, but unfortunately, they were never mated.

After her successful show career, *Serenity Sonbolah was retired to become a broodmare. Eventually she was acquired by Doug and Barbara Griffith of Imperial Egyptian Stud where she was also much loved and spent her remaining days as the queen of the farm and star of their film, To Fly Without Wings.

*Serenity Sonbolah a U.S. National Champion Mare (Sameh x Bint Om El Saad x Om El Saad x Yashmak x Bint Rissala). Johnny Johnston photo.

During her lifetime she produced 10 foals: 7 grey mares and one chestnut, and two grey stallions that were gelded. Like some superstar show horses, she was unique unto herself, and she produced mostly greys of a different type and quality. Her champion grey daughter, Imperial Sonboleen, by Moniet El Nafis, brought the high selling mare price at a Pyramid Society Breeder's Sale and was eventually exported to Australia. Another valuable daughter was the grey Imperial Sonbesjul, a Hossny daughter who, when bred back to El Hilal, doubled the *Ansata Ibn Halima blood (sire of the sire being the grandsire of the dam), producing the multiple halter and performance champion, Imperial Al Kamar, a handsome grey stallion and very prepotent sire of champions who favors many of his sire's and maternal grandsire's traits.

The mare Yashmak (by Sheikh El Arab x Bint Rissala), a tall bay with rather plain long head, huge expressive black eyes, long neck and rather long body, produced important mares, including Om El Saad (by Shahloul), *Sanaaa, (by Sid Abouhom), who was one of Gleannloch's early imports and the dam of multi-champion stallion, Hossny, and Rahma (by Mashhour), who became the dam of 11 Hanadi (by Alaa El Din) who was exported to the Babolna Stud in Hungary as a foundation mare. The Yashmak son, *Rashad Ibn Nazeer, was imported to America and used in the Pritzlaff program.

The grey mare Kateefa (by Shahloul x Bint Rissala), was an outstanding broodmare. She appeared to be more elegant than her half-sister, Yashmak, and generally produced somewhat finer bone structure and overall refinement. Her important contribution to Egyptian breeding was the stallion, Alaa El Din (by Nazeer), whose bloodlines permeate the global breeding programs, particularly through his daughters, but also through a few of his sons. Her chestnut daughter, Bint Kateefa, (by Sid Abouhom), made a significant contribution by producing the elegant Kaisoon (by Nazeer), and Farag (by *Morafic), both white stallions who originally were imported from Egypt to Germany where they made a mark in Europe and eventually worldwide. Their half-sister, the chestnut mare, *Bint Bint Kateefa (by Antar), was imported by Gleannloch to America, from whose line several well-known black horses descended.

Another noteworthy mare of this Rissala line was Ameena, an elegant grey Hamdan daughter, from whom descended some extremely beautiful grey mares: Enayet (by *Morafic), the dam of *Sakr, and Enayet's daughter *Daad (by Alaa El Din); the elegant Gleannloch import, *Omnia (by Alaa El Din), dam of the white stallion Zedann (by *Morafic) and the lovely AK Khattaara (by *Ibn Moniet El Nefous).

The Rodania family's influence can be very strong and it requires careful study if it is to be linebred or inbred, especially when chestnut is bred to chestnut whereby considerable white markings could result. Lady Anne Blunt in her dairy said that Rodania should be bred to Saklawis only (i.e., Rodania mares to Saklawi stallions, not the reverse). Had that mating occurred between *Serenity Sonbolah and Ansata Ibn Sudan, another masterpiece might have been achieved for this historical family.

Though nothing can bring back the hour
Of splendour in the grass, or glory in the flower;
We will grieve not, rather find
Strength in what remains behind.
    William Wordsworth

*Imperial Al Kamar (El Hilal x Imperial Sonbesjul x *Serenity Sonbolah)*

*Sanaaa, at Gleannloch (Sid Abouhom x Yashmak)

*Omnia (Alaa El Din x Ameena). J. Johnston photo

Sharing the dream - Kuhaylan Rodan

Gigi Grasso photo

Descending from Bint Rissala through Yashmak

Nahma Bint Ibn Halim Shah
1988 Grey Mare

The grey mare, the renowned,
in the world there is none like her...
Abu Zeyd

		Ansata Halim Shah	*Ansata Ibn Halima
	El Thay Ibn Halim Shah		Ansata Rosetta
			Ibrahim
		Mahameh	Mona III
NAHMA BINT IBN HALIM SHAH			Madkour
		Madmaymour	Maymoonah
	Nadra		Madkour
		Nagma	Nagha (to Yashmak)

ALFABIA STUD
Gigi Grasso & Paolo Damilano • Via Pittamiglio 1 • Cherasco (CN) Italy 12062

Sharing the dream - Kuhaylan Rodan

Descending from Bint Rissala through Yashmak

Asra Salaa
1995 Grey Mare
Sire: Salaa El Dine (Ansata Halim Shah x Hanan)
Dam: Asra Ganiya (Messaoud x Bint Garia x El Garia x Noosa x Saaida)

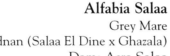

Alfabia Salaa
Grey Mare
Sire: Adnan (Salaa El Dine x Ghazala)
Dam: Asra Salaa
Bred by Alfabia Stud

Gigi Grasso photos

Serene Bint Asra Salaa
2000 Grey Mare
Sire: Teymur B
(Assad x 214 Ibn Galal I)
Dam: Asra Salaa
Bred and owned by Serene Egyptian Stud

ALFABIA STUD
Gigi Grasso & Paolo Damilano • Via Pittamiglio 1 • Cherasco (CN) Italy 12062

Sharing the dream - Kuhaylan Rodan

Descending from Bint Rissala through Ameena

Gawharah Al Shaqab
1998 Grey Mare
Bred by Al Shaqab Stud

Filsinger photo

GAWHARAH AL SHAQAB	Imperial Mahzeer	*Imperial Madheen	Messaoud
			Madinah
		Maar Bilahh	El Halimaar
			Bint Nabilahh
	Sahaba	Adeeb	Shaarawi
			Nawal
		Hasanat	Shaarawi
			Sarkha x Ameena

AL SHAQAB STUD
P.O. Box 90055 • Doha, Qatar

"We stand on the threshold of a new era." - Judith Forbis

Gigi Grasso photo, Qatar.

IN CONCLUSION

Why Breed Arab Horses

"As old as time itself and as fleet as its flying moments," the Arabian horse, and the Egyptian-Arabian in particular, represents the last vestiges of Nature's premier achievement of evolutionary beauty. Forged in the crucible of Arabia Deserta, these unique creatures evolved in the harshest conceivable conditions. This natural process honed and refined them, separating the wheat from the chaff, and brutally culling all but the most perfect. As a result of his graceful symmetry, speed, vigor, docility, and endurance, this horse we cherish as a friend and companion has played an important part in the history of mankind. Without his aid we probably would not have fulfilled our destiny to "multiply and replenish the earth, and subdue it, and have dominion over the fish of the sea, and the fowl of the air, and over every living thing that moveth upon the earth."

The resourceful pharaohs of Egypt were golden opportunists. They wasted no time in acquiring the choicest desert-bred horses and establishing planned breeding programs. The Egyptian horses quickly surpassed all competitors, and with the aid of their swift and courageous war steeds these dynasty-building kings wielded great power and built vast empires. The artists of ancient Egypt provide us with excellent representations of these fabulous horses existing more than 3000 years ago, capturing the turmoil of the battle or the hunt and bringing the animal world to life with great sensitivity and understanding. The Egyptians believed that something translated into art would become a reality in the spirit world and they strived to achieve contact with the absolute through sympathetic renderings.

Today the archetypical Arabian horse is referred to as the classic Arabian horse. The term "classic" immediately brings to mind the idea of something singularly pure and perfect that for centuries has been accepted as a model by which everything else of the kind is measured. The most frequent association of the word is with ancient Greek culture. When the ancient Greeks obtained the Arab horse they became conscious of its immortal and celestial derivation. In his honor they fashioned a god called Pegasus – a symbol of perfection, poetry raised to the heights, one of God's most beautiful and spiritual ideas. This horse, then, not surprisingly, is deeply imbedded in the bedrock of human subconscious – the dream world we visit daily. No wonder it intrigues us still.

As breeders, although we may age in years, each year a new and exciting life comes into our experience and refreshes our outlook, raises hope in our hearts, and is cause to continue in pursuit of that elusive goal – perfection – in all facets concerned with this new creature. Indeed we renew our lives through our horses. They give us "something to do, something to love and something to hope for."

William Beebe (1877-1962), the celebrated American naturalist, wrote: "When the last individual of a race of living things breathes no more, another Heaven and another Earth must pass before such a one can be again."

Our responsibility as stewards is thus made explicit.

Johara Al Naif, by Ansata Shalim (Ansata Halim Shah x Ansata Nefertari) x Al Johara (Prince Fa Moniet x Ansata Majesta), Jr. Champion Filly, Salon du Cheval, Paris 2002. Bred and owned by Al Naif Stud, Qatar. Judith Wich photo.

Art is the original manna in the desert.
It alone enables everyone to taste the true values of life.
All artists live in the future. They design future worlds.
We stand on the threshold of a new era, which is revealed to us
Through the prophetic quality of art —
The art of breeding Egyptian-Arabian horses.

    Judith Forbis
    Pyramid Society Breeders Sale I
    June 11-13, 1982

Jerry Sparagowski's granddaughter, Brooke

The world is your kaleidoscope, and the varying combinations of colors which at every succeeding moment it presents you are the exquisitely adjusted pictures of your evermoving thoughts.

You will be what you will be;
   Let failure find its false content
      In that poor word, 'environment,'
But spirit scorns it, and is free.

It masters time, it conquers space;
   It cows that boastful trickster, Chance,
      And bids the tyrant Circumstance
Uncrown, and fill a servant's place.

The human Will, that force unseen,
   The offspring of a deathless Soul,
      Can hew a way to any goal,
Though walls of granite intervene.

Selected quotes from James Allen - *As A Man Thinketh*

Major 2002 International Shows with Percentages of Winners Carrying the Blood of Ansata Horses (as of December 31, 2002)

Ansata bloodlines have made a major impact in worldwide competitions over the past four decades. In this book, emphasis has purposely not been placed on championships won by Ansata horses or their derivatives. Nevertheless, their continual success in the show ring and in various performance disciplines has been phenomenal for one private breeding farm. Therefore, to complete the picture, a select summary of wins by Ansata bloodlines ONLY in the year 2002 is included to give a general idea of their past, the present, and continuing influence.

2002 World Championships, Paris
100% of the Four World Champions
50% of the Reserve World Champions
50% of the Top Ten Fillies
30% of the Top Ten Mares
60% of the Top Ten Colts
50% of the Top Ten Stallions

Aachen (All Nations Cup)
50% of the Champions & Reserves
60% of all Class Winners

Qatar
100% of the Champions
86% (six of seven) of the female class winners
100% of the male class winners

Middle East Championships (Jordan)
50% of all Champions and Reserves

European Championships
37.5% of all Champions and Reserves

Merrist Wood (UK)
62.5% of all Champions and Reserves

British Nationals
25% of all Champions and Reserves

Towerlands (UK)
50% of all Champions and Reserves (including Geldings)

Elran Arabian Cup (Belgium)
75% of all Champions and Reserves

German Stallion Show
(equivalent to State Licensing Show in Past)
66.67% (6 of 9) Gold Medal Winners

German National Championships
37.5% (3 of 8) National Champions and Reserves

Norwegian Nationals
62.5% (5 of 8) National Champions and Reserves

2002 U.S. Nationals
37.5% (3 of 8) Senior and Junior National Champions (Mares & Stallions Only)

EGYPTIAN ONLY SHOWS

2002 Egyptian Event (Lexington, KY, USA)
87.5% (7 of 8) Halter Champions and Reserve
100% (4 of 4) Supreme Champion and Reserve Mares and Stallions
High Scoring Egyptian

2002 Egyptian Event Europe
62.5% of Champions and Reserves
70% of all Class Winners

2002 Egyptian Event Italy
100% (8 of 8) Champions and Reserves

ARABIAN BREEDERS ASSOCIATION

P.O. Box 43 • Millville, CA 96062 • (530) 547-4030

SUSAN MEYER
President

CORY SOLTAU, D.V.M.
Vice President

MARY SYPOLT
Secretary

DON SEVERA
Treasurer

BRUCE CLARK
Director

NIKKI CRISCOLO
Director

CLIFF McCURDY
Director

HENRY METZ
Director

HOWARD PIKE
Director

SHEILA VARIAN
Director

MICHELLE WATSON
Director

DIANE BUTLER
Executive Officer

September 1, 2002

Judith and Don Forbis
Ansata Arabian Stud
234 Polk 130
Mena, Arkansas 71953

Dear Judith and Don,

Each year the Directors of the Arabian Breeders Association take great pleasure in announcing the recipient of the ABA Lifetime Breeders Award. This year we are delighted to honor you with our 2002 ABA Lifetime Breeders Award which recognizes your outstanding achievements in breeding Arabian horses.

In 1999 the Arabian Breeders Association presented its first ABA Lifetime Breeders Award to Jeanne and Charles Craver for their work in preserving the Davenport line of horses. Leland and Esther Mae Mekeel received the 2000 ABA Lifetime Breeders Award for their program which continues to influence the Arabian horse world at the highest levels. The ABA honored Martha Ann and Joe Cassel with the 2001 award for the excellence of their pure Polish breeding program and for their outstanding achievements in Arabian racing.

The Ansata Arabian Stud breeding program is appreciated internationally for its straight Egyptian breeding program which you established more than 40 years ago. The Ansata horses have been recognized through the years for their great type and beauty with championships at the U.S Nationals, the Salon du Cheval, and at many competitions across the world. The entire Arabian horse community owes a debt of gratitude to you, Judith, for your many wonderful books on the heritage of the Arabian horse, including your newest, <u>Ansata Ibn Halima, The Gift</u>.

We are so pleased to have this opportunity to recognize your many accomplishments and to thank you for your contribution to the Arabian breed.

With our heartfelt congratulations,

The Board of Directors
Arabian Breeders Association

Benefactors

It is with deep appreciation that I thank the distinguished benefactors who went the extra mile to help make this reference book a reality.

THE ARABIAN GULF: A special thanks and much gratitude is extended to those breeders in the Arab world for their overwhelming support. Particular appreciation goes to Al Shaqab Stud, owned by The Emir of Qatar, **H.H. Sheikh Hamad Bin Khalifa Al Thani**, and to his manager, **Sheikh Hamad Bin Ali Al Thani**, for extra participation as benefactors and to whom Chapter V, in memory of Ansata Halim Shah, is dedicated. Al Shaqab is a Life Member of The Pyramid Society and its breeding program is primarily straight-Egyptian. The Qatari team, led by Al Shaqab's stallion, Al Aadeed Al Shaqab, won the prestigious 2003 Nation's Cup in Europe, and the Al Shaqab horses have won consistently at the Salon du Cheval and other European shows. Their hospitality to the worldwide Arabian horse community during their Qatar International Arabian Horse Shows, which H.H. the Emir enthusiastically supports, is legendary...... **Sheikh Abdulaziz Bin Khaled Bin Hamad Al Thani**, established Al Rayyan Farm in 1987 and made a solid commitment to Egyptian-Arabian bloodlines. He believes the Arabian horse is part of the Arab world's history and tradition, Qatar's in particular, and he considers it the Arab's duty and responsibility to preserve the breed and to further its importance worldwide. Chairman of the State of Qatar's Racing and Equestrian Club, he is admired as the leader of the Arabian horse renaissance in the Arab world during the 1990's and is respected worldwide for breeding and publicizing traditional classic Arabian horse type.

The Kuwaiti Arabian horse breeders have taken a keen interest in breeding Egyptian Arabian bloodlines in particular after the devastation of their country by Iraq. **Mr. Usamah Al Kazemi**, one of this book's major benefactors, was instrumental in guiding the Arabian Horse Center of Kuwait to acquire straight Egyptian bloodstock as a nucleus after the Gulf War. In 1994 he purchased his first straight Egyptian Arabians from Egypt and founded his own farm, Ezzain Arabians. He has gradually selected an admirable group of horses and is now developing a serious breeding program at his magnificent new facility that he designed to showcase his beautiful horses, including the very classic young sire, Ansata Al Murtajiz...... **Mr. Mohammed J. K. Al Marzouk** is a young breeder and Arabian horse art collector dedicated to the development of the Egyptian Arabian horse in his country, and to educating the Arab people about their Arabian horse cultural heritage through sponsorship of authoritative books and Arabian Horse Center activities. He created his attractive and functional Ajmal Stud to be a peaceful haven for his horses where Ansata Hejazi reigns as sire supreme, along with a fine collection of Egyptian mares..... **Mr. Talal Al Mehri** is the enthusiastic owner of Al Jazira Stud. He bases his program on straight Egyptian bloodlines, some of which were imported from U.S.A., others come from Al Rayyan Farm in Qatar, including the beautiful mare, Ansata Selma. He is a dedicated and enthusiastic public relations representative for the Egyptian Arabian breed as he continues to encourage everyone to learn more about their Arabian horse heritage by visiting the Kuwaiti breeders' farms and supporting the Arabian Horse Center activities.

SAUDI ARABIA: In the Kingdom of Saudi Arabia, **Mr. Khalid Al-Haddad**, owner of Al-Aadeyat Arabians, is acknowledged by his peers as a respected breeder. He first developed a straight Egyptian Arabian program at his farm in Egypt, but eventually moved his horses to Jeddah and established a stud farm there. He has shown his horses successfully at shows in Saudi Arabia and Egypt, and he has helped to popularize Egyptian Arabians in both countries.

EGYPT: Egypt is home to the Egyptian Arabian horse as we know it today. **Mr. Gawdat Abd-El Gawad** has built his attractive farm not far from the historical city of Alexandria. Having imported a number of straight Egyptian Arabians from the U.S.A. during the 90's as well as having purchased stock bred in Egypt, he is committed to developing a valued breeding program that will be worthy of Egypt's Arabian horse heritage...... **Mr. Omar Sakr**, of Sakr Arabians near Cairo, has made a major contribution to the Egyptian community in the Land of the Nile. His beautifully landscaped farm is sanctuary to a straight-Egyptian herd of superb quality. A Life Member of The Pyramid Society, and President of the Egyptian Breeders Association of Egypt, he has been responsible for educational projects and horse shows in Egypt. He also imported classic Egyptian Arabians from the U.S.A. and Europe in the 90's when new blood was needed to restore traits being lost in Egyptian programs after the mass exodus of important bloodstock from Egypt to America and Europe during the 70's and 80's.

EUROPE: European breeders have a long history of breeding Egyptian Arabian horses and traditional classic Arabian type. A Life Member and supporter of The Pyramid Society, **Mr. Hassanain Al-Nakeeb** is the owner of Al-Nakeeb Arabians. Formerly of Iraq, he moved to Jordan where he established a straight Egyptian breeding farm and showed his horses successfully in that country. Now headquartered at his magnificent Gillingham Hall estate in Norfolk, he brought with him his carefully bred Egyptian herd thus introducing a new dimension of high-quality Egyptian bloodlines to England...... **Ms. Nayla Hayek**, of Hanaya Arabians in Switzerland, is a long-time breeder of Egyptian horses and has been a leader in developing these bloodlines in her country for many years. She has served as President of the Swiss Arabian Horse Society, and her enthusiasm as a breeder, competitor in the show ring, and as judge is recognized around the world..... **Dr. Hans Nagel** of Katharinenhof Farm in Germany, has been our friend and associate since the 60's. A major force in the Egyptian Arabian horse community, he is a respected judge, author, and breeder. He has served the Arabian breed faithfully as President of the German Arabian Horse Society and Vice President of the World Arabian Horse Association, and he was instrumental in developing the first Egyptian Event held in Weisbaden, Germany..... **Ferdinand and Janette Schwestermann** enjoy breeding Egyptian Arabian horses in Kildare, Ireland, at their charming historical farm, Dowdstown Arabian Stud. Ferdinand is active within the international Egyptian-Arabian community, having become a Life Member of The Pyramid Society as well as an active participant in The Pyramid Society Europe functions. He is now involved in developing the new Egyptian Event Europe show to be held at Baden Baden in 2003.

NORTH AMERICA: In North America the Egyptian Arabian horse continues to retain its popularity and acquire new enthusiasts, with much credit given to The Pyramid Society through its membership services and ongoing Egyptian Event. **Steve and Mauri Chase**, owners of Chase Arabians in Texas, became committed to an Egyptian program after attending the 1991 Egyptian Event where they purchased *Authentic Arabian Bloodstock* as a guide. They favor *Ansata Ibn Halima and *Morafic bloodlines (through The Egyptian Prince and Ruminaja Ali) and as a small farm currently owning 10 mares and fillies, are working diligently to breed horses that capture one's attention and say, "Look at me... I'm beautiful, I'm confident, I'm an Arabian!"....... **Dr. Chess Hudson** of Zandai Arabians in Florida, has been a dedicated breeder of Egyptian Arabians and a friend to Ansata for many years. Despite an overwhelming medical practice, he remains personally active in his farm where he stands at stud the handsome sire, Zandai Ibn Omar. The featured speaker at several Ansata seminars, his unique lectures were based on an artistic Arabian horse video he had created for which he wrote the music. The talented doctor gave a presentation seminar guests will never forget. ..... **Henry and Christie Metz** of Silver Maple Farm, were formerly located in Iowa and Florida, and are now settled with their Egyptian horses in California. Life Members of The Pyramid Society, they have a prominent program and their dedication and generosity within the Arabian horse community is legendary. Henry Metz serves on various boards within the Arabian horse community, including the Arabian Horse Registry of America. Christie Metz has served on The Pyramid Society Board of Directors and is President of The Pyramid Foundation..... **Michael and Leslie Nord** of Royal Shahara Arabians in Washington, are relatively new on the Egyptian scene. Very enthusiastic about their future with this breed, they have already become Life Members of The Pyramid Society and have dedicated their breeding program of straight Egyptians to Leslie's late son, Wesley.... **Emil and Debra Nowak** of Abraxas Arabians in California have been committed to Egyptian Arabian bloodlines since 1974 after receiving valued guidance and assistance from the respected pioneer breeders, Mr. and Mrs. Roy Jackson. The Nowaks base their carefully planned program around certain family lines and Debra is developing a Breeders Workshop to help educate breeders about genetics, structure, movement and the art of breeding .... Our Canadian participants, **Michael and Valerie Resch**, own the beautiful farm, Karija Arabians in Emerson, Manitoba. They are devoting much energy to popularizing the Egyptian Arabian in Canada, and to successfully supporting these bloodlines in the show ring both at home and abroad. They take great pride in their developing herd of straight Egyptians, and in their beautiful stallion, Farres, who was awarded a gold medal at the 2002 licensing show in Germany.

SOUTH AMERICA: **Claudia Quentin**, owner of Haras Las Cortaderas, Argentina, has been an admirer and breeder of Ansata horses since the 70's, and was co-owner with Ansata of the famed mare, Ansata Selket, who now graces Al Rayyan Farm in Qatar. A horsewoman and connoisseur of the arts, she has graciously served the breed both as President of the Arabian Horse Society of Argentina, and as a W.A.H.O. Executive Committee member.

To each one of you, thank you, and good luck on your quest for the perfect Arabian horse.

Index to BENEFACTORS

Abraxas Arabians - 466, 467, 569, 583, 584, 585
Emil & Debra Nowak
32415 Yates Road
Winchester, California 92596 USA
Res. 909-926-3728
Fax 909-676-5592
Email: abraxasarb@aol.com

Ajmal Arabian Stud - 422, 423, 424, 425, 542, 543
Mohammed Jassim Al Marzouk
Area 11, Farm No. 1 & 2
Wafra, KUWAIT
Bus. 965-245-0055
Fax 965-242-4132
Email: ajmalarabianstud@hotmail.com

Al-Aadeyat Arabian Stud - 432, 433, 434
Khalid S. Al-Haddad
PO Box 54110
Jeddah 21514 KINGDOM OF SAUDI ARABIA
Res. (966) 2 6066110
Bus. (966) 2 6612285
Fax (966) 2 6635275

Al Jazira Arabian Stud - 429, 500, 501, 519
Mr. Talal Abdullah Al Mehri
PO Box 16991, Qadisiya – Code 35860
KUWAIT
Res. 965 9683022
Bus. 965 2528350
Fax 965 2523255
Email: aljazira_stud@hotmail.com

Al-Nakeeb Arabians - 374, 557, 589
Mr. Hassanain Al-Nakeeb
Gillingham Hall
Gillingham Norfolk
NR340ED UNITED KINGDOM
Res. (+44) 1 502 714141
Fax: (+44) 1 502 719634
Email: Arabians@Al-Nakeeb.com

Al Rayyan Farm - 406, 407, 408, 409, 410, 549
Sheikh Abdulaziz Bin Khaled Bin Hamad Al Thani
PO Box 375
Doha, QATAR
Bus. + 974 4805150
Fax + 974 4805150

Al Shaqab Stud - 413, 414, 415, 416, 417, 418, 419, 520, 521, 522, 523, 554, 555, 598
Sheikh Hamad Bin Ali Al-Thani, Mgr.
PO Box 90055
Doha, QATAR
Bus. 974 4800 348
Fax 974 4800 347

Chase Arabians - 459, 460, 461
Steve & Mauri Chase
4641 New Hope Road
Aubrey, Texas 76227 USA
Res. (940) 440-3123
Bus. (940) 440-3123
Fax (940) 365-7407
Email: mauri@chasearabians.com

Dowdstown House Stud - 468, 469, 552
Mr. & Mrs. F.J. Schwestermann-Lawless
Dowdstown, Maynooth
Co. Kildare IRELAND
Res. (+41) 1 780 0120
Bus. (+353) 1 628 6227
Fax (+41) 1 780 0129
Email: mail@dowdstownarabians.com

Ezzain Arabians - 380, 381, 510, 511, 512, 544, 545, 546, 572, 573, 592
Usamah Zaid Al Kazemi
PO Box 30
Safat 13001 KUWAIT
Res. (965) 5343004/5
Bus. (965) 2404067/8
Fax (965) 2404035
Email: uzalkazemi@hotmail.com

Abd-el Gawad Gawdat - 430, 431, 514
23 Kafr Abdou St.
Roushdy
Alexandria EGYPT
Mobile 010 1075107
Fax 010 5463868
Email: GawdatHG@yahoo.com

Hanaya Arabians - 444, 445, 535, 570, 571
Nayla Hayek
Expohof
Schleinikon 8165 SWITZERLAND
Res. (41) 1 856 0995
Fax (41) 875 0255
Email: info@hanaya.ch

Karija Arabians - 447, 478, 553
Michael & Valerie Resch
PO Box 525
Emerson, Manitoba R0A 0L0 CANADA
Res. (204) 373-2168
Bus. (204) 373-2905
Fax (204) 373-2203
Email: karijafarm@hotmail.com

Katharinenhof - 507, 508, 509
Dr. Hans-Joachim Nagel
Am Graberfeld 13
26197 Grossenkneten GERMANY
Res. (0421) 323013
Bus. (04433) 1535
Fax (04433) 1564
Email: office @nagels-arabianstud.de

Claudia Quentin – 382, 383, 384
Buenos Aires, ARGENTINA

Royal Shahara Arabians – 457, 458, 524
Michael & Leslie Nord
13517 S. Greyhawk Lane
Spokane, Washington 99224 USA
Res. (509) 448-1577
Bus. (509) 979-9312
Fax (509) 448-1650
Email: RoyalShaharaArab@.com

Sakr Arabians – 390, 391, 588
Sherifa & Omar Sakr
6 Hassan Sabry Street
Zamalek
Cairo, EGYPT
Bus.: 202 8900318/9
Res.: 202 7382892
Mobile 2012 2137239
Email: osakr@intouch.com

Silver Maple Farm, Inc. – 394, 395, 396, 586, 587
Christie & Henry Metz
5125 Happy Canyon Road
Santa Ynez, California 93460 USA
Res. (805) 688-9873
Bus. (805) 688-9873
Email: info@smfarabs.com

Zandai Arabians – 386, 494, 568
Dr. William Hudson
6010 Riley Road
Cumming, Georgia 30040 USA
Bus. (770) 887-0472
Fax (770) 887-1140

Patrons

The list of distinguished patrons of this book reads like a Who's Who in the Arabian horse world. Many have participated in previous Ansata publications; others are supporting this one for the first time. It is with sincere gratitude that I thank each and every one for their generous patronage and belief in the educational value of this book. As a result, future generations will have historical reference material about Egyptian horses born in the 19th, 20th and early 21st century, thus giving them better educational tools than many of us had when we began our quest to learn about this breed.

Two patrons did not submit pages of their own. Therefore, the following special pages are dedicated to them for their generosity and steadfast support of Ansata publications and admiration of Ansata horses over the years:

> Page 59 featuring Ansata Ibn Sudan, to long-time friends, **Robert** and **Erika Brunson** of Royal Arabian Stud in California, because of their special affection for this magnificent stallion, and because one of his daughters was a beloved companion and riding horse at their ranch for many years.

> Page 96 picturing *Ansata Ibn Halima, to my friend, **Walter Mishek** of *Arabian Horse Times* magazine, where this beloved stallion and many other Ansata-bred horses have been featured since this world famous magazine began as a newspaper in Walter's garage thirty-three years ago.

Once again, my heartfelt appreciation to everyone for your support, and may all your dreams come true!

Index to PATRONS

AA Arabians – 448, 449
Brooke & Candi Weeks
14528 Lime Kiln Road
Grass Valley, California 95949 USA
Res.   530-268-0468
Bus.   702-236-8888
Email: Oceanwst@aol.com

Alfabia Stud – 454, 455, 577, 596, 597
Gigi Grasso & Paolo Damilano
Via Pittamiglio 1
Cherasco 12062 ITALY
Res.   (39) 338-7320391
Bus.   (39) 0124 629533
Fax   (39) 0171411223
Email: alfabia.stud@libero.it

Al Muntaha Stud – 446, 533
Christian Kessebőhmer
Waldstrasse 56
D-49152 Bad Essen GERMANY
Res.   49(0)5742-2006
Bus.   49(0)173-5242734
Fax   49(0)5741-297020
Email: almuntaha@t-online.de

Al-Rayah - 483
Ala'a H. Al-Roumi
KUWAIT
Res.   (965) 9033465
Fax   (965) 2402085

Al Safinat Farm – 426, 427, 428
Khaled Ben Shokr
PO Box 158
Sura, KUWAIT 45702
Fax   (965) 532 5971
Email: wafrall@yahoo.com

Al-Sharg Farm - 488
Talal Al-Nisf
PO Box 24989
Safat 13110 KUWAIT
Res.   (965) 2515367
Bus.   (965) 2415488
Fax   (965) 2451960
Mobile (965) 6642000

Al-Zamet - 453
Susanne and Ingo Schreibvogel
Zum Jugendheim 4
D-21220 Seevetal GERMANY
Res.   (0049)-4185-2986
Fax   (0049)-4185-2986
Email: Farm@al-zamet.de

Antara Egyptian Arabians - 403
Jerry and Nancy Gates
550 Gavin Lane
Rison, Arkansas 71665 USA
Res.   (870) 357-2698
Bus.   (870) 357-2698
Email: antaraegyptians@yahoo.com

Arabian Horse Center – 420, 421
PO Box 22436
Safat 13085 KUWAIT
Bus.   (965) 476-0632
Fax   (965) 476-0898

Arabian Horse Times
Walter Mishek
1050 8th Street N.E.
Waseca, Minnesota 56093 USA
Bus.   (507) 835-3204
Fax   (507) 835-5138
Email: wrm@ahtimes.com

Arabian Horse World - 471
Denise P. Hearst, Publisher
1316 Tamson Dr., Suite 101
Cambria, California 93428 USA
Phone (800) 955-9423 or (805) 771-2300
Fax   (805) 927-6522
Email: info@ahwmagazine.com

Arabians Ltd. – 525, 561
Judy and Jim Sirbasku
8459 Rock Creek Road
Waco, Texas 76708 USA
Res.   (254) 772-0597
Bus.   (254) 753-3199
Fax   (254) 752-6056
Email: arabiansltd@arabiansltd.com

Azda Arabians - 398
Kimberly McGill
PO Box 522
Hobe Sound, Florida 33475 USA
Res.   + 954 610-9072
Bus.   + 41 79 2043339
Email: mcgillresearch@msn.com

Baraka Farm - 385
Barbara & Tyrone Lewis
5223 Hwy 71 S.
Cove, Arkansas 71937 USA
Res.   870-387-8439
Email: bsl@barakafarm.com

Bebo Stud – 484, 528, 529
Miloslava Khamis
10th Ramadan City
Cairo, EGYPT
Res.   (202) 419-1122
Bus.   (2010) 0166 00 42 / (2012) 21 329 22
Fax   (2015) 370 777 / 02 267 22 41
Email: bebostud@yahoo.com

Birkhof Stud – 375, 499
Dirnhofer Family
Birkhof 3
93133 Burglengenfeld GERMANY
Res.   (0049)9473 8702
Bus.   (0049)9473 8702
Fax   (0049)9473 910175
Email: birkhofstud@t-online.de

Isabell Boersing & Reinhold Szabo - 440
Obergruenthal 6
83064 Raubling GERMANY
Res.   +49-(0)8035-1510
Fax   +49-(0)8035-6652
Email: IsaBoersing@web.de

Chapel Farms Arabians - 465
Robert & Christine Fauls
3129 Smokey Road
Newnan, Georgia 30263 USA
Bus.   (770) 253-0355
Fax   (770) 253-0354
Email: chapelga@aol.com

Dara Meadows Farm - 576
Debra & David Geiser
449 E. Nelson Street
Lexington, Virginia 24450 USA
Res.   (540) 464-5588
Fax   (540) 463-1073
Email: darameadowsfarm@nexet.net

DeShazer Arabians – 442, 443
Hank & Sandra DeShazer
17025 Shaw Road
Cypress, Texas 77429 USA
Res.   (281) 351-7829
Bus.   (281) 290 9585
Fax   (281) 255 9475
Email: sandy@deshazer.com

Eichhof-Araber - 485
Sylke Schuhmacher
Heckenweg 20
21220 Seevetal GERMANY
Res.   (49) 4185 3171
Fax   (49) 4185 3173
Email: Sylke.Schuhmacher@web.de

Index to PATRONS (*continued*)

El Adiyat Arabians – 378, 513, 547
Mahmood A. Al Zubaid
PO Box 270
Surra, KUWAIT 45703
Tel.   (965) 4311685
Fax   (965) 4312295

El Miladi Arabian Stud - 387
Cynthia Culbertson
PO Box 928
Carrizozo, New Mexico 88301 USA
Res.   (505) 648-2612
Fax   (505) 648-2612
Email: elmiladi@tularosa.net

El Thayeba Arabian Stud - 439
Cornelia Tauschke
Kneter Sand 1
D-26197 Grossenkneten GERMANY
Res.   49(0)44 35-5552
Fax   49(0)44 35-973803
Email: tauschke@el-thayeba.de

Falconwood Legacy Bloodstock - 530
Marilyn Thomas
PO Box 840
Texarkana, Texas 75504 USA
Bus.   (870) 542-7009
Fax   (870) 542-6110
Email: info@falconwoodarabians.com

Halypa-Al Duhaym - 397
Annalisa & Pasquale Monticelli Berloco
Via Luca Spaziante No. 20
Altamura – Bari - 70022 ITALY
Res.   (39) 80 310 53 47
Bus.   (39) 333 68 54 027
Fax   (39) 80 310 53 47
Email: halypa@jumpy.it

Hamdan Stables – 518, 556
Fatma Ahmad Hamza
1 Mohammad Mazhar St. Zamalek
11211 Cairo, EGYPT
Res.   (+202) 738-4082
Bus.   (+20) 010-100-3880
Fax   (+202) 736-4835
Email: hamdanstables@yahoo.com

Hope Farm - 493
Joan Skeels or Sue Burnham
194 Polk 21
Cove, Arkansas 71937 USA
Res.   870-387-8231
Bus.   870-387-8231
Fax   870-387-8231

Ibriz Arabians - 376
Wilfred & Nancy Bourque
2271 Route 890
Cornhill, N.B. E4Z 1B5 CANADA
Res.   (506) 756-3875
Email: ibriz@nb.sympatico.ca

Imperial Egyptian Stud - 578
Barbara A. Griffith
2642 Mt. Carmel Road
Parkton, Maryland 21120 USA
Bus.   (410) 329-6380
Fax   (410) 329-8039
Email: IESArabs@aol.com

Kehilan Arabians – 479, 531
8059 FM 1187 West
Ft. Worth, Texas 76126 USA
Res.   (817) 443-6124
Bus.   (817) 443-6124
Fax   (817)443-0288
Email: kehilan@aol.com

La Frasera – 388, 389, 487
Dr. Francesco & Serenella Santoro
Via Borneo 25
Roma 00144 ITALY
Res.   0039065913526
Bus.   0039065913526
Fax   0039065913526
Email: frasera@lafresera.com

Madheen Arabians - 477
Dr. Siegfried & Ruth Paufler
Zum Ortloh 15
Goettingen D37077 GERMANY
Res.   (049) 551-22530
Fax   (049) 551-22530
Email: drpaufler@surfeu.de

Maisano Arabians – 405, 590
Karen Maisano
9188 Eagle Run Drive
Brighton, MI 48116 USA
Res.   (810) 231-4043
Bus.   (810) 231-4043
Fax   (810) 231-6733
Email: kmaisano@rovin.net

Najouba Arabians - 482
Oren & Jessie MacLean
1292 Post Road, Sussex Corner
New Brunswick E4E 2V5 CANADA
Res.   (506) 433-6098
Fax   (506) 433-2565 or 433-6098

Nedschd Arab - 377
Gabriele Schuster & Dr. Ferdinand Denzinger
Hans-Watzlik-Strasse 21
D-92539 Schoensee GERMANY
Bus.   (+49) 9674500
Fax   (+49) 967491156
Email: mail@nedschd-arab.com

Nile Bay Arabians - 379
Gary & Tracy Davis
1241 Crabapple Road
Ozark, Missouri 65721 USA
Res.   (417) 443-6199
Bus.   (417) 861-2021
Fax   (417) 443-2073
Email: GDavis9755@aol.com

Nirvana Farms - 548
Ray & Jamie Roberts
13913 Cameron Road
Excelsior Springs, Missouri 64024 USA
Res.   (816) 628-4306
Email: nirvanaf@birch.net

Orienta Arabians – 441, 591
Judith Wich
Eichenbuehl 26
96352 Wilhelmsthal GERMANY
Res.   (+49) 9260265
Bus.   (+49) 92609639022
Fax   (+49) 9260962380
Email: JudithWich@t-online.de

Qadar Arabians LLC - 481
Jeff & Judy Barth
N677 Hwy. 49
Weyauwega, Wisconsin 54983 USA
Res.   (920) 867-4599
Bus.   (920) 867-4599
Fax   (920) 867-2286
Email: qadar@centurytel.net

Rahim Arabians - 402
Dr. Aly Abdel Rahim
439 Pyramids Street
Giza 11511 EGYPT
Res.   (202) 3838656
Bus.   (202) 3884351
Fax   (202) 3884350
Email: palma_pyramids@hotmail.com

Rancho Bulakenyo - 567
Felino, Jody, Karen & Ryan Cruz
2755 Los Osos Valley Road
Los Osos, California 93402 USA
Res.   (805)534-1391
Fax   (805) 534-1392
Email: richtermh@aol.com

Index to PATRONS (continued)

Rocky Top Ranch - 486
Kimberly Blythe & Joseph Murgola
404 Polk 47
Mena, Arkansas 71953 USA
Res.   (479) 243-0416
Bus.   (479) 243-0416
Email: joe@rockytopranch.com

Rothenberg Stud – 452, 496, 497, 498, 532, 593
Annette & Erwin Escher
86653 Monheim GERMANY
Res.   (+49) (0) 9091 3707
Fax   (+49) (0) 9091 3708
Email: rothenberg@straightegyptians.com

Royal Arabian Stud
Erika & Bob Brunson
1300 Dawnridge Drive
Beverly Hills, California 90210 USA
Bus.   (310) 203-8404
Fax   (310) 203-8247

Sandybrook Egyptian Arabian Farm – 470, 495
Karen Henwood
13885 N.W. Hwy. 27
Ocala, Florida 34482 USA
Res.   (352)732-7430
Fax   (352)732-7431
Email: sandybrookarabs@aol.com

Second Wind - 450
M. Kent Mayfield / John R. Ford, Jr.
5653 State Road 130
Dodgeville, Wisconsin 53533 USA
Res.   (608)935-3540
Email: swind@mhtc.net

Vollblutaraber-Gestut
Familie Gunter W. Seidlitz - 451
Gut Muggenbach
Sesslach D-96145 GERMANY
Res.   (0049) 9567-230
Bus.   (0049) 9567-981030
Fax   (0049) 9567-1576

Serene Egyptian Stud – 404, 480, 526
Anna & Stefano Galber
Via Coletto 2 Loc. Serene
Affi (VR) 37010 ITALY
Res.   0039 045 7200929
Bus.   0039 335 8373255
Fax   0039 0456206553
Email: Serenestud@tin.it

Simeon Stud - 534
Marion Richmond & Peter & Ruth Simon
44 Bulkara Road
Bellevue Hill, Sydney, New South Wales
AUSTRALIA 2023
Res.   (61) 293271649
Mobile   (61) 418268749
Fax   (61) 292451929
Email: simeonstud@hpm.com.au

Tre-Balzane Stud - 551
Monika Savier
Colle del Marchese, 61
Castel Ritaldi 06044 ITALY
Res.   +39 0743 252 004
Bus.   +39 335 5235135
Fax   +39 0743252305
Email: savier@arabi–egiciani.it

Tuscani - 527
Stuart and Brenda Schuettpelz
1921 Niles-Buchanan Road
Niles, Michigan 49120 USA
Res.   (269) 683-5449
Bus.   (269) 683-5449
Fax   (269) 683-5449
Email: info@tuscaniarabians.com

Windamere Arabians – 456, 550
Ed & Sharon Litizzette
9121 Aspen Drive
Weed, California 96094 USA
Res.   (530) 938-3558
Bus.   (530) 938-3558
Fax   (530) 938-1143
Email: windamer@inreach.com

The Pilgrimage, Carl Göbel, 1855 lithograph, Dr. Karin Thieme Archives.

PICTORIAL ACKNOWLEDGMENTS

Many of the photographs in this book were taken by the author over more than 45 years, or were gifts from private collections. Grateful acknowledgment is extended to contributors of photographs and to all those photographers whose names are not listed on the photos, especially to professional photographers of our era who have done so much to enhance and artistically represent the breed. Without them, reference books like this one could not exist.

The photographs in this book are intended to be historical and educational and to give the reader a picture of what a specific horse may look like, or have looked like, bearing in mind the angles used in photographing that horse, and the state of the art of equine photography at the time the photo was taken. Again, it must be emphasized that all kinds of visual media can be deceptive, and there is no substitute for seeing and evaluating a horse in the flesh.

Without dedicated managers, trainers and grooms, many of the horses depicted in this book would not have looked as good. A perfect example of this care in posing and highlighting is illustrated by Bart Van Buggenhout, Manager-Trainer of Al Rayyan Farm, Qatar, who sought out the unique locations and organized the horses for the artistic photographs taken by Gigi Grasso and Rik Van Lent Jr. that grace this and other publications.

Most sincere gratitude is extended to Jerry Sparagowski, the official Ansata farm photographer, who began his career photographing horses at Ansata in the early 70's and who has chronicled the evolution of Ansata Arabian Stud ever since.

Special thanks go to Erwin Escher, Irina Filsinger, Gigi Grasso, Rik Van Lent Sr., Rik Van Lent Jr., and Bert Van Lent, for their contributions of additional reference photos vital in illustrating certain texts. Thanks also to Carol Lyons for photographs relating to the W.R. Brown importation, to Charles Craver, Julie Pagliaro, and John Vogel for those of several early Babson horses, and to Karen Thieme for providing photographs of Arabian horse lithographs from her unique collection.

Every effort has been made to show the photographers' names as written on their photograph and to also credit them beside their photographs wherever possible, especially when the photographer's name does not appear on the photograph itself, and provided the photographer is known. However, there are some exceptions to this precedent in Chapter IV.

Among the professional and amateur photographers whose work appears in this volume are:
Jaber Al Kazemi, Alexander Photography, Pamela Amberson, George Axt,
Bickle Photography, Jean-Pierre Bissilliat, Deanna Boyd
Cathy Childs, Randi Clark,
Mike Gorecki, Gotcha Photography, Melanie Groger, Judy Guess
Rob Hess
Rob Jerrard, Johnny Johnston,
Polly Knoll,
Darryl Larson, Barbara Lewis,
Jenni May, Susan McAdoo, Rhita McNair,
Patti, Judy Perry, Peter Pond, Alan Potter,
Randall Photography, Carl Raswan,
Nicole Sachs, Sally Photography, Moshe Sandbank, Javan Schaller, Erika Schiele, Ron Shimer, Don Stine, Lois St. Clair
Carola Toischel, Scott Trees
Yvette Van Natta, Stuart Vesty,
Judith Wagner, Judith Wich, Wright Photography

Index of HORSE PHOTOS

A

Aana	21, 183, 207
Aarak Al Safinat	426
Aaroufa	318, 574
Abbas Pasha I-12	453
Abiebi	456
Abla	215, 217
Abraxas Abu Hilal	585
Abraxas Amir	585
Abraxas Habielaa	584
Abraxas Halamaa	584
Abraxas Halimaar	465
Abraxas Hilals Gemm	585
Abraxas Moon Glow	569
Abraxas Moonbeam	569
Abraxas Moonique	466
Abraxas Moonstruck	466
Abraxas Nejd Moon	466
Abraxas Prince	467
Abraxas-Maar-Hala	569
Adaweya	217
Adjbah Al Safinat	428
Adnan	542
Afifa	288
*Afifaa	305
Al Shahhaa Hamdan	556
Aischa	361
Aisha	541
Aissha	288, 292
AJ Ahlams Delight	450
Ajeeb Al Jazira	519
Ajmal Maghrebeia	425
Ajmal Obbeyah	543
AK Bint Dalia II	289, 296
AK Bint Maisa	247
*AK Dalia	289
AK Faressa	265
AK Malouma	303, 306
AK Princess Maisa	247
AK Rishafa	209
Akhtal	87
Akid Fa Mona	575
Akid Geshan	576
Al Aadeed Al Shaqab	358, 363, 523
Al Aangha Al Rayyan	409
Al Maymoon Al Aadeyat	434
Al Rayyan	432, 433
Al Rayyan mares	366, 368, 472, 536
Al Wajba Al Rayyan	406
Alaa El Din	43, 86
Alajneha Nahme	242
Alecsis S	470
Alenah	500
Aleseri	470
Alfabia Mameluka	455
Alfabia Salaa	597
Alfabia Sohar	455
Alfaya Kamar	550
Ali Jamila	236, 424
Ali Saroukh	395
Ali Zafir	528
Aliaa Al Shaqab	415
Alidaar	555, 559
Alidarra	391
Aliikat	428
Allah Ateyyah	275, 281
Almaas Al Sabah	398
Amaal	463
Amaar	529
Amal	303
Amal (*Bint El Bataa)	75, 257
Amani	328
Amar	361
Ameena/Amina	92
Ameer	89
Ameera	540
Amira Al Rayyan	488
Amira El Shaqab	414
Amrulla (Ziada)	489
Anas El Wogoud	484
Ansata Abbas Pasha	113, 261
Ansata Abu Halim	134
Ansata Abu Jamal	127
Ansata Abu Nazeer	277, 490
Ansata Abu Sudan	103, 292
Ansata Abu Tai	129
*Ansata Adeeba	262
Ansata Afifa	237
Ansata Aida	187
Ansata Al Bassam	146
Ansata Al Caliph	246
Ansata Al Halima	249
Ansata Al Murtajiz	150, 371, 381
Ansata Aladdin	297
Ansata Aldara	238
Ansata AlGhazzali	172
Ansata Ali Abbas	321
Ansata Ali Khan	272
Ansata Ali Pasha	277
Ansata Alida	242
Ansata Aliha	177
Ansata Alima	187
Ansata Allegra	163
Ansata Aly Jamil	187
Ansata Aly Sherif	209
Ansata Amir Zaman	175, 451
Ansata Amir Zarif	269
Ansata Amira	240
Ansata Amon Ra	137
Ansata Aniq	238
Ansata Anisah	293
Ansata Anna Maria	240
Ansata Araby Bey	272
Ansata Ariana	293
Ansata Ashraf	293
Ansata Asila Nile	199
Ansata Asmarra	189
Ansata Astra	247
Ansata Athena	178
Ansata Aya Halima	131, 162
Ansata Aya Maria	163, 452
Ansata Aya Nadira	163
Ansata Azali	333, 591
Ansata Azalia	280
Ansata Aziza	277, 491
Ansata Bint Aliha	179
Ansata Bint Asjah	234
*Ansata Bint Bukra (Hosnia)	122, 123, 124, 125
*Ansata Bint Elwya	244, 245, 246
Ansata Bint Faressa	265
Ansata Bint Halima	372
Ansata Bint Mabrouka	20, 21, 58, 76, 256, 258, 260
*Ansata Bint Misr	123, 126, 128
*Ansata Bint Nazeer	267
Ansata Bint Nefri	161
*Ansata Bint Sameh	217, 225
Ansata Bint Shahrezad	172
Ansata Bint Sokar	266
Ansata Bint Sudan	126, 164, 174
*Ansata Bint Zaafarana	20, 21, 73, 274, 276, 278, 490
Ansata Binthalima	138, 151, 152
*Ansata Bintmisuna (Bint Maysouna)	288, 290, 291
Ansata Blue Nile	182, 198
Ansata Cairo Bey	130
Ansata Chiron	294
Ansata Damietta	129, 132
Ansata Damitha	135
Ansata Deborah	139
Ansata Delilah	129, 136, 370
Ansata Desert Rose	179
Ansata Dia Halima	134
Ansata Diana	236
Ansata Divina	277
Ansata El Ajeeb	175
Ansata El Alim	277
Ansata El Arabi	320
Ansata El Aziz	279
Ansata El Emir	185
Ansata El Ghazal	169
Ansata El Hakim	246
Ansata El Halim	232
Ansata El Hassan	329
Ansata El Ibriz	167, 376
Ansata El Kaswa	248
Ansata El Khedive	277
Ansata El Mabrouk	134
Ansata El Mamluke	277
Ansata El Masri	292
Ansata El Muteyri	188
Ansata El Naseri	126
Ansata El Nasrany	130
Ansata El Nisr	277, 490
Ansata El Reyhan	184
Ansata El Salaam	141, 371
Ansata El Salim	319
Ansata El Shahkir	331
Ansata El Shahraf	148, 385
Ansata El Shahwan	165
Ansata El Sherif	126
Ansata El Tareef	175
*Ansata El Wazir	225
Ansata Emir Halim	138
Ansata Emir Sinan	161
Ansata Empress	235
Ansata Exemplar	234, 235
Ansata Exotica	170, 426
Ansata Fantastiq	263
Ansata Fantazi	263, 482
Ansata Farah	332
Ansata Fatima	277
Ansata Fay Jamila	321
Ansata Faye Maara	322
Ansata Fayrouz	333
Ansata Ghazala	170
Ansata Ghazia	171
Ansata Gloriana	171
Ansata Haalah	294
Ansata Haisam	242
Ansata Haji Halim	138
Ansata Haji Jamil	139
Ansata Halim Bay	130
Ansata Halim Bey	133
Ansata Halim Shah	51, 103, 113, 169, 336, 338, 340, 342, 344, 347, 348, 350, 352, 354, 355, 357
Ansata Halima Son	133
Ansata Halisha	187, 397
Ansata Haliwa	246
Ansata Hamama	294
Ansata Hejazi	105, 113, 155, 358, 373, 423
Ansata Helwa	171, 390
Ansata Ibn Aziza	279
Ansata Ibn Bukra	127
*Ansata Ibn Halima	19, 20, 22, 44, 53, 58, 71, 96, 101, 111, 338, 352
Ansata Ibn Jamila	232
Ansata Ibn Rashid	165
Ansata Ibn Shah	103, 112, 226
Ansata Ibn Sudan	24, 26, 52, 59, 101, 111, 259, 261
Ansata Ibn Zaman	226
Ansata Iemhotep	55, 107, 111, 114, 158
Ansata Imperial	137
Ansata Jacinda	226
Ansata Jahara	227
Ansata Jalala	230
Ansata Jamila	225, 228, 231
Ansata Jasmin	320
Ansata Jehan Shah	279
Ansata Jellabia	320
Ansata Jeylan	228
Ansata Jezebel	225, 228
Ansata Joy Amira	231
Ansata Joy Halima	226
Ansata Joy Jamila	230
Ansata Jubayl	229

Horse photo index continued

Ansata Judea	232
Ansata Julima	230
Ansata Julnar	235
Ansata Jumana	233, 234
Ansata Justina	233, 236
Ansata Kamal Bey	165
Ansata Kamriya	178
Ansata Ken Ranya	173, 378
Ansata Ken Rashik	173, 379
Ansata King Tut	175
Ansata Laila	297
Ansata Latifa	175
Ansata Magnifica	220, 425
Ansata Maharani	223
Ansata Mahasen	224
Ansata Mahroussa	322, 565, 572
Ansata Majdi	224
Ansata Majeeda	224
Ansata Majesta	210, 220, 354, 360, 437
Ansata Majestic	221
Ansata Majesty	218, 220, 360
Ansata Majid Shah	130
Ansata Majida	219
Ansata Malaha	220, 360, 418
Ansata Malaka	218, 220
Ansata Maliha	330
Ansata Malik Shah	52, 109, 117, 221, 359
Ansata Malika	219, 437
Ansata Malima	329
Ansata Manasseh	179, 362, 387
Ansata Mansoura	222
Ansata Marha	222, 437
Ansata Mari Isis	166
Ansata Mariam	223
Ansata Marietta	134
Ansata Mariha	224
Ansata Marjaneh	223
Ansata Maya	222
Ansata Meryta	219, 420
Ansata Millennia	221
Ansata Misriya	130
Ansata Misty Nile	204
Ansata Mital	221, 441
Ansata Monahalima	262
Ansata Mouna	265
Ansata Mourad Bey	177
Ansata Mumtaza	330
Ansata Munira	222
Ansata Nadra	153
Ansata Nafisa	158, 390
Ansata Nahema	175
Ansata Nahida	239, 241
Ansata Najdi	293, 526
Ansata Najiba	240
Ansata Nariya	153
Ansata Narjisa	264
Ansata Nashmia	157
Ansata Nasr Halim	329
Ansata Nasrina	241
Ansata Nawarra	152, 372
Ansata Nazira	268
Ansata Neamet	153, 384
Ansata Nefara	155, 159
Ansata Nefer	157
Ansata Nefer Isis	27, 157, 408
Ansata Nefertiti	155, 156, 360, 422
Ansata Nefri	155
Ansata Neoma	152
Ansata Nile Bay	200
Ansata Nile Charm	193, 196
Ansata Nile Comet	199
Ansata Nile Dawn	193, 200
Ansata Nile Diva	204
Ansata Nile Dove	200
Ansata Nile Dream	50, 193, 399
Ansata Nile Echo	203, 401
Ansata Nile Emir	196
Ansata Nile Fame	194
Ansata Nile Flame	194
Ansata Nile Gem	201
Ansata Nile Gift	202, 399
Ansata Nile Glory	194
Ansata Nile Gypsy	199
Ansata Nile Jade	198
Ansata Nile Jewel	192, 201
Ansata Nile Joy	203
Ansata Nile King	191
Ansata Nile Lace	205
Ansata Nile Magic	196
Ansata Nile Mist	192, 400, 401
Ansata Nile Moon	201
Ansata Nile Nadir	204
Ansata Nile Pasha	194, 401
Ansata Nile Pearl	203
Ansata Nile Pharo	192
Ansata Nile Queen	191
Ansata Nile Rose	195, 196
Ansata Nile Ruler	193
Ansata Nile Sand	196
Ansata Nile Satin	205
Ansata Nile Silk	197
Ansata Nile Spice	199
Ansata Nile Splendour	202
Ansata Nile Star	192
Ansata Nile Starr	198
Ansata Nile Storm	193
Ansata Nile Sun	192
Ansata Nile Wine	195
Ansata Omar Halim	169
Ansata Omari	323
Ansata Omnia	323
Ansata Orienta	272
Ansata Orion	143
Ansata Osiron	145, 383
Ansata Palmyra	227
Ansata Paloma	229
Ansata Petra	229
Ansata Prima Nile	197
Ansata Prima Rose	170, 172
Ansata Princessa	227
Ansata Qasim	266
Ansata Queen Nefr	156, 157
Ansata Ra Harakti	279
Ansata Rafi	173
Ansata Rahmah	173
Ansata Rahotep	173
Ansata Raja Halim	169
Ansata Ramazan	167
Ansata Ramesses	167
Ansata Ramona	166
Ansata Ranita	167
Ansata Raqessa	177
Ansata Rebecca	166
Ansata Regina	126, 176
Ansata Reza Shah	185
Ansata Rhodora	127, 164
Ansata Rhozira	165
Ansata Riva Nile	194
Ansata Riyadh	333
Ansata Riyal	332
Ansata Rose Queen	170
Ansata Rosetta	126, 168, 338
Ansata Saamir	143
Ansata Sabiha	184, 392
Ansata Sabika	146
Ansata Sabrina	185, 187
Ansata Safeer	149
Ansata Sahhara	270
Ansata Sahir	172
Ansata Sakkara	270
Ansata Samai	141
Ansata Samantha	138, 140, 371, 372, 383
Ansata Samari	143, 422
Ansata Samaria	139, 141
Ansata Samarra	141, 142, 147
Ansata Samiha	141
Ansata Samira	268
Ansata Samiri	150
Ansata Sammoura	143
Ansata Samsara	141, 371
Ansata Sarai	161
Ansata Saroya	269
Ansata Sekhmet	143, 147, 360
Ansata Selket	7, 144, 147, 360, 383, 408
Ansata Selma	149, 429
Ansata Selman	149, 407
Ansata Semiramis	238
Ansata Serena	177
Ansata Serqit	145, 382
Ansata Shaamis	265
Ansata Shah Abbas	179
Ansata Shah Zahir	232
Ansata Shah Zam	185
Ansata Shah Zaman	50, 101, 112, 261, 347
Ansata Shah Zeer	268
Ansata Shahlia	296
Ansata Shahnaz	146
Ansata Shahrazada	127
Ansata Shahrezade	172
Ansata Shahriyar	148
Ansata Shalim	160, 373
Ansata Shalimar	142
Ansata Shammar	160
Ansata Shams	297
Ansata Sharifa	34, 45, 141, 148, 406
Ansata Sherifa	127
Ansata Sherrara	150, 420
Ansata Sinan	55, 107, 158
Ansata Sinan Bey	333
Ansata Sirdar	160
Ansata Sirius	107, 111, 115, 146
Ansata Sokar	109, 116, 142, 371
Ansata Splendora	88, 393
Ansata Star O	189
Ansata Stari Nile	195
Ansata Starletta	135, 387
Ansata Sudarra	138, 154
Ansata Sudarra	373
Ansata Suhaya	264
Ansata Sulayman	272
Ansata Suleyma	150
Ansata Tabitha	331
Ansata Taya	264, 361
Ansata Tousson	157
Ansata Vali Nile	201
Ansata Wanisa	268
Ansata White Nile	202, 203, 421
Ansata Zaafina	280
Ansata Zaahira	280
Ansata Zahra	280, 281
Ansata Zaki Halim	264
Ansata Zareifa	277, 280
Antar (Anter)	74
Antara Shalima	403
Antigua Dance	531
Arabest Kalid	516
*Asadd	581
Asal Sirabba	183, 206
Asami Ruuh	550
Asfour	541
Ashhal Al Rayyan	409, 437
Asra Salaa	597
Assad	359
*Assiedeh	288
Aswan	516
Atfa	91
Awad	68
*Aziz	305
Aziza (x Galila)	289
*Aziza (x Negma)	317
Azure Moon	587
B	
B A Halamet	584
Badawia	557
Badia	245
Baheera	257
Bahiya Al Nour	398
Balance	71
Bikr NA	482
Bilal	91
*Bint Amal	303, 304, 543
Bint Amer	454

Horse photo index continued

Bint Arrieta	272
Bint Atallah	575
*Bint Baheera	257, 271
Bint Bint Hamamaa	589
Bint Bint Magidaa	559, 560, 561
Bint Bint Moniet	56
Bint Bint Riyala	328
*Bint Bint Sabbah	183
Bint Bukra	67
*Bint Dahma	216
Bint El Shaikh	248
Bint Faras Azali	328, 332
Bint Farazdac	217
Bint Farida	69, 217
Bint Gabriela CA	459
Bint Haseema	318
Bint Helwa	274, 317
Bint Kamla	288, 505
Bint Kateefa	85
Bint Maar Hala	565
Bint Magidaa	558, 559, 560, 562
*Bint Maisa El Saghira	72, 244, 245, 462
Bint Maysa	553
Bint Moftakhar	279
*Bint Mona (*Bint Mouna)	76, 110, 256, 475, 492
*Bint Moniet El Nefous	56, 76, 256
Bint Nashua	513
*Bint Nefisa	70
Bint Obeya	67, 123
Bint Radia	274, 489
Bint Rebecca	374
Bint Rissala	594
Bint Rustem	78, 289, 515
Bint Sabah	67, 123, 183
Bint Samiha	81, 288, 504
*Bint Serra	275
Bint Somaia	217
Bint Tahia	486
Bint Yamama	316
Bint Yosreia	80, 516
Bint Zaarina	530
Bint Zareefa	72
BKA Damilll	134
Black Halim	431
Black Watch	13, 14
Bukra	67, 123
C	
Cashai	467
Ceylan	15
Ciaffar	32
Classic Shadwan	497
CN Jericho	450
D	
Dahma II	70, 216
Dahmah Shahwania	464
Dal Macharia	216, 223
Dal Sheba	328
Dalal	256
Dalia Halim	296, 359, 533
Dalima Shah	296, 532
Dana Al Rayyan	408
*Dantilla	216, 237
*Deenaa	88
DeSha Java	443
Desire RB	478
DHS Halima Bint Hawaa	468
DHS Mabrouk	469
DHS Miraya El Nizr	469
DHS Shaakirah	552
E	
Egyptian Anjali Amanda	485
Egyptian Asinah	485
El Amira	492
El Arabi	91
El Bataa	75, 257
El Deree D.B.	77
El Halimaar	566, 567
El Hilal	53
El Maar	21, 317, 318, 319
El Sareei (El Sarie)	72
El Shaliah	524
El Thay Bint Kamla	439
El Thay Ibn Halim Shah	362, 438
El Thay Kamla	439
El Thay Mashour	439
218 Elf Layla Walayla	570
Elmiladi Zaynah	153
Emad	89
Emeraldd Moon	491
Enayah	217
Endow Blu	446
Entebbe CA	461
Enzahi	87
ET Crown Prince	566
F	
Fa Deene	275
Fa-Habba	191
Fa Halima	185
Fa Moniet	574
Fa Serr	317
Fa-Abba	183
Faarraah	482
Fada	574
Faddan	574
*Fadell	74
Fadila	73
*Fadl	317
Fa-Habba	21, 182, 183
*Fakher El Din	89
*Faleh	516
Falima	180, 183, 190, 191, 399
Famira KA	478
*Fantasia	256
Faras Azali	328, 582
Farasha (Farasa, Frasha, Frashah)	80, 289
Fardous	67
Farid Nile Moon	491, 495
Farida	216
Farida Al Shaqab	521
Farres	447
Farressa CA	458
Fasiha	289
Fatthaya (Fathia)	82
Fay El Dine	317
Fay Negma	317
Fay Roufa	318
Fayalia Nile Moon	470
Faye Roufa	315, 318, 321
Fay-Sabbah	21, 183, 184
Fayza II	70
*Fortun	398
Foze	82, 506
Frasera Futura	388
Frasera Ramses Shah	389
Futna	216
FW Zaariya	530
G	
G Dendera	134
G Fantazia	263
G Shafaria	148, 407
GA Moon Taj Mahal	279
GA Moon Tajhalima	495
Galal	90
Galila	79, 289
Gamil III	84
*Gamila	504
Gassir	73
Gawharah Al Shaqab	598
*Ghalii	81
Ghalion	89
Gharib	57
Ghazal	87
Ghazal I	69
Ghazal Sakr	391
Ghazala (x Hanan)	303, 541
*Ghazalah (El Beida)	274
*Ghazalahh (x Bint Farida)	71
Ghazallah	500
Ghazi	71
Ghowleh Al Rayyan	500
Gift of the Nile	202
Glenglade Lecsi	470
Glorieta Saafrana	424
Glorieta Sabdana	186
Glorieta Sermina	393
Glorietasayonaara	393, 394, 395
GR Amaretta	593
GR Amaretto	51, 593
GR Aya Farida	452
GR Marianah	496
GR Marietta	496
GR Marja	498
GR Shadiva	532
H	
Hadaya El Tareef	465
Haddiyah	288
Hadibah	288
Hafeed Anter	90
Hafiza	50, 303
Hagir	183
Halim El Mansour	565
Halim Pasha	206
Halima	69, 215, 216
Halima Al Shaqab	416
Halimas Rose	197
Halims Asmara	582, 593
Hamdan	74, 93
Hanan	303, 539
Hania	307
Hansan	581
Hawaa	468
*Hegrah	183
Heirogance	444
*Hekmat	581
Helala	171
Helwa	70, 217
Hemmat	288
HF Aida	493
HF Amneris	493
HF Tesa Reekh	586
Hi-Fashion Hitesa	586
*Higran	307
Hilal	316
Hilala Mystique	449
Hilalia Al Shaqab	520
Hind	78, 289
Hodhoda	307
Horoob	529
*HS Hero	583
*HS Tadeusz	405
I	
Ibn Dahman	464
Ibn El Nil	402
Ibn Galal I	540
*Ibn Galal I-7	288, 295, 505
*Ibn Hafiza	91
*Ibn Moniet El Nefous	53, 110
Ibn Rabdan	35, 78
Ibriiah	245, 249
Ibtisam	89, 256
Idaa	318
Ikhnatoon	92
Il Nadheena	242
Ilham	257
Immell	493
Imperial Al Kamar	595
Imperial Baarakah	448
Imperial Baarez	578
Imperial Im Tiarah	401
Imperial Impress	204
Imperial Kamaala	401
*Imperial Madheen	476, 477
Imperial Madori	53, 582
Imperial Mistilll	400
Imperial Mistry	401
Imperial Phanadah	412
Imperial Phanilah	413
Imperial Tali	401
Inas	88
Insignia DeSha	442
Izara Blue CA	459
J	
Jabarut	486

617

Horse photo index continued

Jaflah Al Rayyan	549
Jalila Al Rayyan	407
Jamal El Dine	378
Jamala Al Zamet	429
Jamil Ezzain	592
Jamila Al Rayyan	373
*Jamilll	105, 113, 305, 540
*JKB Bint Nehaya	264
*JKB Blue Belkies	257
*JKB Mamdouhah	262
*JKB Masouda	216, 219
Johara Al Naif	363, 602
Jubilllee	442

K

Kafrawi	430
Kamar (Kamr)	67, 183
Kamasayyah	522
Kamla	288, 504
Karima GAD	431
Kateefa	85
Kawsar	73
Kayed	92
Kazmeen/Kazmeyn	84
Ken Bint Bint Mahiba	361
Khafifa	75, 256
Khamsa	275
Khedena	275
Kheir	77
Kholeif	514
Kis Mahiba	476
Kismat/*Bint Dahma	70
Kodwa	535
Korima	257
Kyro KA	478

L

LA Selket	404
*Lancers Asmara	92
Latiefa	421
Latifa (Lateefa)	79
Layla	67, 183
Lohim	362
Looza	217
Lubna	76
Lutfia	

M

MA Belladonna	288
MA Victoria	590
Ma-Ajmala	590
Maar Bilahh	469
Maar Hala	517
*Maarauder MH	565
Maarena	567
Maarou	575
*Maaroufa	318
Maarqada	313, 316, 317, 318
Maar-Ree	318, 319
Mabrouk Manial	564
Mabrouka	83
Madallan-Madheen	76, 256, 474
Madinah	498
Madinah Bint Saariyah	476
Madkour I	434
Mahameh	438
*Magidaa	303, 558
Maha Al Jazira	501
Mahameh	438
Maharan	484
Maheeb	498
Mahiba	475
Mahomed	436
Mahroussa	42, 316, 317
Mahyubi	436
Maisa	72, 245, 462
Maisouna (Maisoona)	82
Makhnificent KA	479
Makhsous	531
Malacha	216
Malaka	84, 328, 361, 580
Malameh	484
*Malekat El Gamal	505
Malik	436
Malikah	216, 435

Mamlouka	328
Mangalee Ibn Shadwan	499
Mansour	83
Mansoura Al Rayyan	549
Mansoura El Halima	241
Marani El Malikah	436
Marei	71
Maria Halima	496
Mariyah	306
Marqueesa	451
Marquis I	531
Martessa	586
Mashhour	81
Massarrah Ezzain	572
*Mawaheb	329
Maydan Madheen	477
Mayhoub	529
Maymoon	436
Maysa	303, 553
Maysoun	359
Maysoun (x Shams) (Maisouna, Maisooona)	82, 288
Maysouna (x Kis Mahiba)	476
MB Adallah	205
MB Jacinta	486
MB Kariita	527
MB Moneena	266
MB Sataarka	457
MB Sateenha	266
Medallela	75, 256, 257
Mesaoud	317
Mesoudah-M	477
MFA Bint Monien	565, 571
Minstril's Gabriela	460
Moheba II	435
Mohga (Mohgat)	80
Mona (Inshass)	303
Mona (Mouna)	76, 256, 474
Mona II	435, 475
Moniet El Dine	377
Moniet El Nefous	56, 64, 75, 253, 256, 474
Montasar	451
*Morafic	18, 77, 110, 475
Morning Glorii	514
*Moshira	307
Mossa	216
Mossa RSI	236
Mouna Al Rayyan	488
Mudira	440

N

Naajah	521
Naama Al Rayyan	409
Nabajah El Chamsin	573
*Nabda	90
*Nabilahh	289, 516
Nabya	217
Nadeema	547
Nadia/Nadja	217
*Nadima	217, 24
Nadra El Saghira	216
Nafaa El Saghira	83
Nafairtiti	463
Nagda	557
*Nageia	217, 240
Nahma Bint Ibn Halim Shah	596
Naif Al Rayyan	393
Nashua	505
*Nasr	42, 66
Nawaf	92
Nayzak Ezzain	512
Nazeer	62, 82, 99
Nazeera	328, 580
Neamat	288, 506
Nefertiti	557
Nefisa	70, 217
Nefisa Al Safinat	428
Negma	316, 317
NF Naeema	153
NF Safiyyah	548
Nigmh	362
*Nihal	288
Nile Allure	197, 402

NK Aischa	543
NK Al Amirah	545
NK Hafid Jamil	374
NK Halla	547
NK Jurie	544
NK Nabeelah	509
NK Nada	510
NK Nadirah	507
NK Nakeebya	511
NK Nasrin	508
NK Yasmin	380
Noble Imdalia	550

O

Om El Fotooh Hamdan	518
Om El Saad	86
*Omnia	595
Orient Queen	209

P

Pasha Farid	452
Patinaa	467
Phaaros	577
*Pharrah	183, 411
Pretty Woman	430
Prince Fa Moniet	53, 105, 112, 575
Princess Shamira	460
PWA Asifa	293

Q

Queen Sheeba	459

R

RA Jehan	456
Raalima	206
Rabaab Al Shaqab	419
Rababa	483
Radames II	375
Radia	244, 274, 463
Rafica (Rafika)	86
Ragia	69, 216
Rahma	85
Rahmaa	245
*Ramses Fayek	88
Rana Al Safinat	427
RAS Amala	304, 548
RAS Najma	304
*Rashad Ibn Nazeer	86
Rasmoniet RSI	56
Rayhanah Bint Rebecca	375
Rayya	248
RDM Maar Hala	564, 565, 567
Richteous	559
Richter MH	516
Ridaa	328
Rihahna	245, 463
Riyala	328
RN Ajeeba	34, 538, 541
RN Ajlay	554
RN Dananeer	480
RN Farida	410
RN Hejaziah	198
RN Marwa	501
RN Rayana	427
RN Safia	393
RN Sultana	549
*Roda	274, 317
Rooda	257
Rose of Egypt	207
Rouda	79, 289, 515
Royal Jalliel	412
RSI Raya Del Sol	56
Ruminaja Ali	558
Ruminaja Bahjat	559
Ruminaja Majed	559
Rustem	84
RXR Lia Moniet	289, 297

S

Saab	87
Saada Al Rayyan	429
Sabah	67, 123, 183
Sabrah	183, 184
Safir	359, 538
Safir KA	553
Sahbine	394, 395
Sakani	257
Saklawia II	492

618

Horse photo index continued

Salaa El Dine	51, 358, 541
Salma	433
Salomy	328
Salwa	79
Samara SQR	374
Sameer	499
Sameh	77
Samha	288, 504
Samhan	77
Samia	85, 580
Samiha	288
Samira	274, 489
Samura	396
*Sanaaa	595
Sarjah SMF	396
Seef	88
Serene Bint Asra Salaa	597
Serene Carima	480
Serene Ciai Dii	404
Serene Dalila	480
Serene Desire	404
*Serenity Montaha	303
*Serenity Sonbolah	594
Serra (Sara)	275, 317
Serrada	330
Serrasab	183
SH Alleya	237
SH Say Anna	238, 239
Shaamisa Mystique	461
Shaarawi	90
Shaboura	394, 395
Shadwan	87
Shadwanah	517, 532
Shahbaa	91
Shaheen	588
Shahil	487
Shahin	541
Shahlika	322
Shahloul	74
Shaikh Al Badi	463
Shaker El Masri	90
Shamruk	464
Shams	81, 288, 506
Sheikh El Arab	69
Sheherezade (Shahrzada)	80
Sherifa Tamria	53, 209
Sid Abouhom	65, 68
Silima	206
Simeon Sadran	534
Simeon Sehavi	534
Simeon Sitri	434
Simeon Sukari	534
Simurgh	241
Sinan Al Rayyan	406
Sohret	17
Soja	216
Sotamm	47
Star Bint Hafiza	479
*Subhaya	257
Suleiman SMF	396
Sulifah	551
Sultana Al Shaqab	417
*Sultann	89
Sundar Alisha	186
Sundar Sabbahalim	186
Suror Ezzain	546
T	
Taghira B	455
Taher Sihr	456
Taira	519
*Talal	490, 492
Talmona	265
Tamria II	411
TB Qadifa	551
Tempest Moon	587
The Egyptian Prince	110
Thee Brigadier	445
Thee Cappuccino	443
Thee Desperado	525
Thee Infidel	561
Thee Revolution	530
Tranquility Moon	587
*Tuhotmos	52
Tuscan Keboush	527
V	
VA Ahlam	179
W	
W Sudans Nafisa	318, 323
Wafaa Elkuwait	421
Wahag	87
Wahag Al Shaqab	522
Wanisa	256
Ward Shah	552
Waseem	581
*Watfa	328
Y	
Yashmak	85
Yosreia	79, 289, 515
Z	
Zaafarana	73, 274, 489
Zahara Keela	481
Zandai Ibn Omar	386
Zandai Jadallah	494
Zandai Rubayana	568
Zareefa	245, 462
*Zarife	317
Zebeda	90

INDEX OF PHOTOS OF OTHER HORSES, PEOPLE AND PLACES

Abdelhamid, Amid	363
Al Sa'ud, H.R.H. Prince Abdullah Ibn Abdul Aziz	27
Al Sa'ud, Prince Bandar Ibn Sultan	27
Al Shaqab Stud	355
Al Thani, H.H. Sheikh Hamad Bin Khalifa	355
Al Thani, Sheikh Abdulaziz Bin Khaled Bin Hamad	28, 354
Al Thani, Sheikh Hamad Bin Ali	363
Aly, H.H. Prince Mohamed	66
Ansata at Chickasha, OK	23
Ansata at Lexington, KY	29
Ansata at Lufkin, TX	23
Ansata at Mena, AR	25
Black Watch	13, 14
Bradbury, David	144, 354
Bradbury, Nicole	336
Byatt, Michael	363
Clinton, Gov. Bill	26, 295
Cobb, James	342
Cranz, Dr. Wolfgang	344
E.A.O.	18
Entrance to Diyarbakir	17
Ferriss, Joe	9
Forbis, Donald	16, 24, 27, 29, 66, 122, 354, 363
Forbis, Judith	7, 12, 13, 14, 15, 16, 17, 24, 27, 28, 29, 66, 174, 354, 363
Hamza, Ahmed Pasha	93
Issa, Houssein	538
Jacobs, Glenn	363
Marsafi, Dr. Mohammed	19, 66
Marshall, Douglas	24
Marshall, Margaret	24
McNair, Tom	22
Nagel, Dr. Hans	344
Nagel, Katharina	344
Nagel, Martin	344
Qatar Desert	502
Sain, Linda	124
Sanders, Richard	158, 295, 342
Silver	12
Stodghill, Pat	342
Tribes in Iran	21
van Buggenhout, Bart	27
Van Natta, Yvette	342
Villa Akhnaton in Egypt	21
von Szandtner, General Tibor Pettko	64
Zaher, Dr. Ameen	65

INDEX OF ART

*Ansata Ibn Halima (DeClaviere painting)	44
1929 Painting of Ibn Rabdan	35
A chestnut Arab mare	325
Alfred DeDreux Painting	37
Ansata Halim Shah, pencil drawing (B. Lewis)	350
Ansata Halim Shah, sculpture (K. Kasper)	352
Aquarelle	119
Arab Warriors (Delacroix)	43
Arabe du desert, (A.J. Gros lithograph)	283
Arabian horse sent by Abbas Pasha to Duke of Leuchtenberg	31
Caravan on the March	11
Ciaffar (DeDreux)	32
Dahman (Victor Adam lithograph)	95
Fatima and groom (Victor Adam)	365
Koheil Aguse mare	309
Old prints & drawings (Vernet, Adam, DeDreux)	41
Omar (DeDreux)	251
Seti I from Temple of Karnak	61
The Nile at Luxor, Upper Egypt (Tyrone Lewis)	624
The Pilgrimage (Carl Göbel)	614
Tutankhamen sculpture (Edwin Bogucki)	39
Victor Adam lithograph	299

619

INDEX OF HORSE ENTRIES APPEARING IN CHAPTER 4 - THE ANSATA STUDBOOK

Aana	182, 207
*Afifaa	302, 305
Aissha	286, 292
AK Bint Dalia II	287, 296
AK Bint Maisa	247
AK Faressa	254, 265
AK Malouma	302, 306
AK Princess Maisa	244, 247
AK Rishafa	208, 209
Al Shahab	237
Alajneha Nahme	214, 242
Ali Asjah	247
Ali Jamila	236
Allah Ateyyah	273, 281
Anazeha	247
Andara Joy Amira	231
Ansara Saamir	142
Ansata Abbas Pasha	261
Ansata Abu Halim	133
Ansata Abu Jamal	122, 125
Ansata Abu Mansur	188
Ansata Abu Nazeer	276
Ansata Abu Simbel	135
Ansata Abu Sudan	102, 291
Ansata Abu Tai	129
*Ansata Adeeba	254, 262
Ansata Afifa	237
Ansata Ahmed Bey	263
Ansata Aida	186
Ansata Al Barak	172
Ansata Al Bassam	146
Ansata Al Caliph	246
Ansata Al Halima	249
Ansata Al Hamid	189
Ansata Al Haroun	271
Ansata Al Hasan	329
Ansata Al Kadir	161
Ansata Al Malik	331
Ansata Al Murtajiz	149
Ansata Al Nadim	242
Ansata Al Nitak	293
Ansata Al Rahim	173
Ansata Al Shahab	175
Ansata Al Zafar	280
Ansata Aladdin	297
Ansata Aldara	238
Ansata Alexandria	129
Ansata AlGhazzali	172
Ansata Ali Abbas	321
Ansata Ali Baba	189
Ansata Ali Halim	187
Ansata Ali Khan	271
Ansata Ali Pasha	276
Ansata Alida	214, 242
Ansata Aliha	177, 179
Ansata Alima	186
Ansata Allegra	163
Ansata Almira	175
Ansata Aly Halim	226
Ansata Aly Jamil	187
Ansata Aly Sherif	209
Ansata Alyssa	178
Ansata Amen Hotep	135
Ansata Amir Ali	227
Ansata Amir Fahim	152
Ansata Amir Sarin	145
Ansata Amir Zaman	174
Ansata Amir Zarif	269
Ansata Amira	238
Ansata Amon Ra	137
Ansata Angela	175
Ansata Aniq	238
Ansata Anisah	292
Ansata Anna Maria	238
Ansata Araby Bey	271
Ansata Ariana	293
Ansata Ashayet	292
Ansata Ashraf	293
Ansata Asila Nile	199
Ansata Asmarra	187, 189
Ansata Astarra	187, 189
Ansata Astoria	247
Ansata Astra	247
Ansata Atallah	179
Ansata Athena	178
Ansata Aya Adora	163
Ansata Aya Halima	129, 162
Ansata Aya Maria	162, 163
Ansata Aya Nabila	162
Ansata Aya Nadira	162
Ansata Azali	332
Ansata Azalia	280
Ansata Azeem	292
Ansata Aziza	273, 276, 279
Ansata Azza	264
Ansata Bint Aliha	179
Ansata Bint Asjah	234, 235
*Ansata Bint Bukra	121, 125
*Ansata Bint Elwya	245, 246
Ansata Bint Faressa	265
*Ansata Bint Mabrouka	253, 260
*Ansata Bint Misr	121, 125, 129
*Ansata Bint Nazeer	255, 267
Ansata Bint Nefri	160
Ansata Bint Rayya	248
*Ansata Bint Sameh	212, 225
Ansata Bint Shahrezad	172
Ansata Bint Sinan	292
Ansata Bint Sokar	266
Ansata Bint Sudan	121, 125, 174
*Ansata Bint Zaafarana	273, 276
Ansata Binthalima	137, 152
*Ansata Bintmisuna	285, 291
Ansata Blue Nile	197, 198
Ansata Cairo Bey	129
Ansata Caressa	270
Ansata Chandra	174, 175
Ansata Chiron	294
Ansata Cleopatra	237
Ansata Dafina	237
Ansata Dalala	269, 270
Ansata Damietta	129, 133
Ansata Damitha	135
Ansata Darius	129
Ansata Deborah	137
Ansata Delilah	129, 137
Ansata Desert Rose	179
Ansata Dia Halima	133, 135
Ansata Diana	236
Ansata Divina	273, 276
Ansata El Ajeeb	175
Ansata El Alim	276
Ansata El Arabi	320
Ansata El Aziz	279
Ansata El Azrak	171
Ansata El Baheer	271
Ansata El Emir	185
Ansata El Faheem	156
Ansata El Farhan	135
Ansata El Farid	237
Ansata El Ghazal	169
Ansata El Haddidi	330
Ansata El Hakim	246
Ansata El Halim	231
Ansata El Hamis	152
Ansata El Harik	270
Ansata El Ibriz	167
Ansata El Iman	172
Ansata El Jadib	233
Ansata El Jaffar	227
Ansata El Jalil	152, 231
Ansata El Jeddi	227
Ansata El Jedrani	266
Ansata El Karim	321
Ansata El Kaswa	248, 249
Ansata El Khalil	270
Ansata El Khedive	276
Ansata El Mabrouk	133
Ansata El Mamluke	276
Ansata El Masri	291
Ansata El Muteyri	188
Ansata El Naji	178
Ansata El Naseri	121, 125
Ansata El Nasrany	129
Ansata El Nisr	276
Ansata El Rahim	166
Ansata El Rakkad	268
Ansata El Reyhan	184
Ansata El Riyal	329
Ansata El Sadiqy	178
Ansata El Salaam	139
Ansata El Salim	319
Ansata El Sami	137
Ansata El Samir	149
Ansata El Sarin	178
Ansata El Shahir	166
Ansata El Shahkir	330
Ansata El Shahraf	146
Ansata El Shahwan	165
Ansata El Shams	135
Ansata El Shaqra	270
Ansata El Sherif	121, 125
Ansata El Simitar	189
Ansata El Sirhan	178
Ansata El Suhayli	155
Ansata El Tareef	174
Ansata El Waseem	237
*Ansata El Wazir	225
Ansata El Zareef	280
Ansata Emir Halim	137
Ansata Emir Sinan	160
Ansata Empress	235
Ansata Etherea	155
Ansata Exemplar	235
Ansata Exotica	170
Ansata Fakhir	189
Ansata Fantastiq	263
Ansata Fantazi	263
Ansata Farah	332
Ansata Fatima	273, 276, 279
Ansata Fay Jamila	321
Ansata Faye Maara	321, 322
Ansata Fayrouz	332
Ansata Fayza	263
Ansata Gala Nile	195
Ansata Galia	170
Ansata Ghazala	169, 170
Ansata Ghazi Bey	170
Ansata Ghazi Pasha	170
Ansata Ghazia	170, 171
Ansata Gloriana	171
Ansata Haalah	294
Ansata Haisam	242
Ansata Haji Halim	137
Ansata Haji Hamid	170
Ansata Haji Jamil	137
Ansata Haji Said	178
Ansata Haji Tahir	188
Ansata Hakima	263
Ansata Halim Bay	129

620

Chapter 4 - Ansata Studbook Index continued

Ansata Halim Bey	133
Ansata Halim Shah	102, 169
Ansata Halima Son	133
Ansata Halisha	186
Ansata Haliwa	246
Ansata Hamama	294
Ansata Hejazi	104, 155
Ansata Helwa	171
Ansata Ibn Asjah	248
Ansata Ibn Aziza	279
Ansata Ibn Bukra	122, 125
*Ansata Ibn Halima	100
Ansata Ibn Jamila	231
Ansata Ibn Meheyd	268
Ansata Ibn Rashid	165
Ansata Ibn Rayya	248
Ansata Ibn Shah	102, 226
Ansata Ibn Sudan	100, 261
Ansata Ibn Tulun	264
Ansata Ibn Zaman	226
Ansata Iemhotep	106, 158
Ansata Imperial	137
Ansata Jacinda	226, 227
Ansata Jahara	227
Ansata Jalala	230
Ansata Jalisa	233, 234
Ansata Jamila	213, 225, 231
Ansata Janina	234
Ansata Jasmin	320
Ansata Jasmina	226
Ansata Jasour	230
Ansata Jehan Shah	279
Ansata Jellabia	320
Ansata Jemlah	235
Ansata Jessamine	235
Ansata Jessica	233
Ansata Jeylan	229
Ansata Jezebel	212, 225, 226
Ansata Joy Amira	230, 231
Ansata Joy Halima	226, 230
Ansata Joy Jamila	230
Ansata Joy Moniet	230
Ansata Jubayl	229
Ansata Judea	231, 233
Ansata Julima	230
Ansata Julnar	235
Ansata Julnara	233
Ansata Jumana	233, 234
Ansata Justina	233, 236
Ansata Kamal Bey	165
Ansata Kamriya	178
Ansata Karim Shah	227
*Ansata Karima	255, 270
Ansata Ken Ranya	172, 173
Ansata Ken Rashik	172
Ansata Khaleed	249
Ansata Khanjar	249
Ansata King Tut	174
Ansata Laila	297
Ansata Latifa	175
Ansata Magid Shah	129
Ansata Magnifica	220
Ansata Maharani	223
Ansata Mahasen	223
Ansata Mahroussa	322
Ansata Majdi	224
Ansata Majeeda	223
Ansata Majesta	220
Ansata Majestic	221
Ansata Majesty	220, 221
Ansata Majida	219
Ansata Malaha	220
Ansata Malaka	220, 221
Ansata Maliha	330
Ansata Malik Shah	108, 221
Ansata Malika	212, 219, 220
Ansata Malima	329, 330
Ansata Mameluke Bey	221
Ansata Manasseh	179
Ansata Mansoura	222
Ansata Mara	189
Ansata Marduk	222
Ansata Marha	222
Ansata Mari Isis	166
Ansata Mariam	223
Ansata Marietta	133, 135
Ansata Mariha	224
Ansata Marjaneh	223, 224
Ansata Marvella	133
Ansata Maya	222
Ansata Mehemet Aly	271
Ansata Memnon	142
Ansata Menkaure	219
Ansata Meryta	212, 219, 222
Ansata Millennia	221
Ansata Misriya	129, 163
Ansata Misty Nile	204
Ansata Mital	221
Ansata Monahalima	262, 263
Ansata Monalima	135
Ansata Moniet Sudan	174
Ansata Montu	222
Ansata Mouna	265
Ansata Mourad Bey	177
Ansata Mumtaza	330
Ansata Munira	222
Ansata Nabiha	264
Ansata Nadir Shah	129
Ansata Nadra	153
Ansata Nafisa	158
Ansata Nahema	174, 175
Ansata Nahida	241
Ansata Najdi	293
Ansata Najiba	240, 241
Ansata Nariya	153
Ansata Narjisa	264
Ansata Nashmia	156
Ansata Nasr Halim	329
Ansata Nasrina	241
Ansata Nassri	306
Ansata Nawarra	152, 153
Ansata Nazeera	255, 268
Ansata Nazira	255, 268, 270
Ansata Neamet	153
Ansata Nefara	155, 158
Ansata Nefer	156
Ansata Nefer Isis	156
Ansata Nefertari	155, 160
Ansata Nefertiti	155, 156
Ansata Nefri	155, 160
Ansata Neoma	152, 153
Ansata Nibal	235
Ansata Nile Bay	200
Ansata Nile Beau	197
Ansata Nile Bey	201
Ansata Nile Blue	198
Ansata Nile Cameo	200
Ansata Nile Charm	193, 195
Ansata Nile Comet	199
Ansata Nile Dawn	193, 200
Ansata Nile Diva	204
Ansata Nile Dove	200
Ansata Nile Dream	193, 194
Ansata Nile Eagle	197
Ansata Nile Echo	203
Ansata Nile Emir	196
Ansata Nile Fame	194
Ansata Nile Flame	194
Ansata Nile Gem	201
Ansata Nile Genie	197
Ansata Nile Gift	201, 202
Ansata Nile Glory	194
Ansata Nile Gold	200
Ansata Nile Gypsy	198
Ansata Nile Hawk	200
Ansata Nile Jade	198
Ansata Nile Jewel	191, 201
Ansata Nile Joy	203
Ansata Nile King	191
Ansata Nile Lace	205
Ansata Nile Magic	196
Ansata Nile Mist	192
Ansata Nile Moon	201
Ansata Nile Nadir	204
Ansata Nile Pasha	194
Ansata Nile Pearl	203
Ansata Nile Pharo	192
Ansata Nile Pride	203
Ansata Nile Princess	205
Ansata Nile Queen	191, 193
Ansata Nile Rose	195, 196
Ansata Nile Ruler	193
Ansata Nile Sand	196
Ansata Nile Satin	205
Ansata Nile Sheik	195
Ansata Nile Silk	197
Ansata Nile Spark	194
Ansata Nile Spice	199
Ansata Nile Splendour	202
Ansata Nile Star	192, 203
Ansata Nile Starlight	199
Ansata Nile Starr	197, 198
Ansata Nile Storm	193
Ansata Nile Sun	192
Ansata Nile Wine	195
Ansata Omar Halim	169
Ansata Omar Kayam	135
Ansata Omar Zahir	280
Ansata Omari	323
Ansata Omnia	323
Ansata Orienta	271
Ansata Orion	142
Ansata Osiron	145
Ansata Palmyra	227, 229
Ansata Paloma	229, 230
Ansata Petra	229
Ansata Phaidra	229
Ansata Prima Nile	197, 200
Ansata Prima Rose	170, 172
Ansata Prince Hal	152
Ansata Princessa	227
Ansata Pristina	152
Ansata Qasim	266
Ansata Queen Nefr	156
Ansata Ra Harakti	279
Ansata Radia	135
Ansata Rafi	173
Ansata Rahmah	173
Ansata Rahotep	173
Ansata Raja Halim	169
Ansata Ramazan	167
Ansata Ramesses	167
Ansata Rami	234
Ansata Ramona	166, 167
Ansata Ramose	152
Ansata Ramzy	161
Ansata Ranita	167
Ansata Raqessa	177, 178
Ansata Rebecca	166
Ansata Regal Nile	195
Ansata Regina	121, 125, 177
Ansata Reza Shah	185

621

Chapter 4 - Ansata Studbook Index continued

Ansata Rhoda	166
Ansata Rhodora	121, 125, 165
Ansata Rhozira	165
Ansata Riva Nile	194
Ansata Riyadh	332
Ansata Riyal	332
Ansata Rose Queen	169
Ansata Rosetta	121, 125, 169
Ansata Sabbara	186
Ansata Sabiha	184, 185
Ansata Sabika	145
Ansata Sabrina	185, 187
Ansata Safari	142
Ansata Safeer	149
Ansata Safwan	145
Ansata Sahhara	269
Ansata Sahir	172
Ansata Sakkara	269
Ansata Saklawia	255, 268, 269
Ansata Salome	255, 268, 269
Ansata Samai	139
Ansata Samantha	137, 139
Ansata Samari	142
Ansata Samaria	139, 146
Ansata Samarra	139, 142
Ansata Samiha	142, 150
Ansata Samira	255, 268, 269
Ansata Samiri	149
Ansata Sammoura	142
Ansata Samsara	139, 149
Ansata Sarai	137, 161
Ansata Saroya	269
Ansata Sekhmet	142, 145
Ansata Selene	291
Ansata Selket	142, 145
Ansata Selma	149
Ansata Selman	149
Ansata Semiramis	238
Ansata Serena	177, 178
Ansata Serenada	178
Ansata Serqit	145
Ansata Shaamis	265
Ansata Shah Abbas	179
Ansata Shah Osman	177
Ansata Shah Zahir	231
Ansata Shah Zam	185
Ansata Shah Zaman	100, 261
Ansata Shah Zeer	268
Ansata Shahlia	296
Ansata Shahira	174
Ansata Shahnaz	145
Ansata Shahrazada	121, 125
Ansata Shahrezade	172
Ansata Shahriyar	148
Ansata Shakilah	241
Ansata Shalim	160
Ansata Shalimar	142, 150
Ansata Shammar	160
Ansata Shams	297
Ansata Sharifa	139, 148
Ansata Sherifa	121, 125
Ansata Sherrara	150
Ansata Shukri	235
Ansata Simoom	209
Ansata Sinan	106, 158
Ansata Sinan Bey	332
Ansata Sirdar	160
Ansata Sirius	106, 146
Ansata Sokar	108, 142
Ansata Splendora	188
Ansata Star O	189
Ansata Stari Nile	195, 197
Ansata Starletta	135
Ansata Sudarra	137, 155
Ansata Suhaya	264
Ansata Sulayman	271
Ansata Suleyma	149
Ansata Tabitha	331
Ansata Taj Malik	167
Ansata Taya	264
Ansata Tousson	156
Ansata Valeria	270
Ansata Vali Nile	200, 201
Ansata Vanessa	255, 268, 270
Ansata Vizier	234
Ansata Wanisa	255, 268, 269
Ansata White Nile	202, 203
Ansata Yildiz	230
Ansata Zaafara	281
Ansata Zaafina	280
Ansata Zaahira	280
Ansata Zaahy	281
Ansata Zahra	280, 281
Ansata Zaki Halim	264
Ansata Zareifa	273, 276, 280
Ansata Zia	281
Asal Sirabba	181, 206
Asjahs Black Jewl	234
Asjahs Dominion	234
Asmara Bintsamira	269
*Aziz	305
*Bint Amal	302, 304
Bint Arrieta	255, 272
*Bint Baheera	255, 271
Bint El Shaikh	248
Bint Faras Azali	327, 332
Bint Moftakhar	279
BKA Damilll	133
Dal Macharia	212, 223
Dalia Halim	296
Dalima Shah	296
*Dantilla	213, 237
DS Rawa	322
El Maar	314, 319
El Mais	247
El Malim	330
El Mousheer	171
El Nabil	167
El Wajeeb	231
El Yasmine	307
Elmiladi Zaynah	153
ESA Halim	163
Fa Halima	185
Fa-Habba	181, 191
Falima	191
Faye Roufa	314, 321
Fay-Sabbah	181, 184
G Ayyah	162
G Halisha	162
G Princess Halima	162
G. Dendera	133
G. Fantazia	254, 263
G. Princess Fadamiet	133
G. Shafaria	148, 149
GA Alexis	279
GA Fire Magic	270
GA Magic Melody	270
GA Moon Tajmahal	279
Gift of the Nile	202
Glorieta Sabdana	186
Glorieta Ali Sudan	186
Halim Pasha	206
Halimas Rose	197
Hasbah	286, 294
Helala	171
Ibn Galal I-7 ("Gala")	286, 294
Ibn Galal I-7 (unregistered)	287, 294
Ibriiah	244, 249
Il Nadheena	214, 242
Imperial Impress	204
*Jamilll	104, 305
*JKB Bint Nehaya	254, 264
*JKB Mamdouha	254, 262
*JKB Masouda	212, 219
Jubillee	236
LOM Jafar	234
LOM Karima	233
LOM Tajalli	234
Maarqada	319
Malika El Sheba	327, 331
Mansoura El Halima	214, 241
Mariyah	302, 306
*Mawaheb	327, 329
MB Adallah	205
MB Moneena	254, 266
MB Sateenha	254, 266
Mona El Ajzaa	246
Mossa RSI	213, 236
My Amir	270
*Nadima	214, 242
*Nageia	213, 240
Nazeer	99
NF Naeema	153
Nile Allure	196
Nile Splendor	269
Omar Bey	265
Orient Queen	209
Phalom	269
Prince Fa Amir	248
Prince Fa Moniet	104
Prince Hal	293
PWA Asherah	286, 293
PWA Asifa	286, 293
RA Rebecca	178
Raalima	206
RAS Amala	304
RAS Najma	302, 304
RAS Prince Fareed	304
Rayya	244, 248
RN Hejaziah	198
Rose of Egypt	182, 207
RXR Lia Moniet	287, 297
Sabrah	181, 184
Serrada	327, 330
SH Alleya	213, 237
SH Say Anna	213, 238
Shahliah	270
Shahlika	314, 322
Shams Al Sahara	174
Shams Al Shahira	174
Shams El Nejd	137
Shams El Nil	197
Sherifa Tamria	208, 209
Silima	206
Simurgh	214, 241
Sir Halim	206
Sundar Alisha	186
Sundar Sabbahalim	186
Talmona	254, 265
VA Ahlam	179
W Sudan's Nafisa	314, 323
Zandai Alexandria	264
Zandai Fa Sabbah	186
Zandai Tabitha	265
ZF Zalima	330

622

Bibliography

Al Khamsa, Inc; *Al Khamsa Arabians*, Al Khamsa, Inc. Capron, Ill. ,Vol. 1983 & Vol. II, 1993

Asil Club; *Asil Araber, Volumes 1-5,* Olms Press, Hildesheim, Germany.

Ashoub, Dr. Abdel Alim; *History of the Royal Agricultural Society's Stud of Authentic Arabian Horses,* The Royal Agricultural Society, Cairo, Egypt, 1948

Blunt, Lady Anne; *Journals and Correspondence 1878-1917,* Rosemary Archer and James Fleming, Ed. Alexander Heriot & Co. Ltd, Cheltenham, Glos. England, 1986

Blunt, Lady Anne; *Original Diaires - personal notes taken by J. Forbis* from the Blunt Diaries at the British Museum, London.

Brown, William Robinson; *The Horse of the Desert*, Derrydale Press, 1929

Dickinson, Col. H.R.P; *Catalog of Travelers Rest Arabian Horses,* Franklin, Tennessee, 1947

Forbis, Judith; *The Classic Arabian Horse*, Liveright, New York, 1976

Forbis, Judith; *Hoofbeats Along the Tigris*, J.A. Allen & Co., Ltd. London, 1970 and Ansata Publications, Mena, Arkansas, 1990

Forbis, Judith and Gülsun Sherif; *The Abbas Pasha Manuscript,* Ansata Publications, Mena, Arkansas, 1993

Forbis, Judith and Walter Schimanski; *The Royal Arabians of Egypt and the Stud of Henry B. Babson,* Thoth Publishers, Waco, Texas, 1976

Murdoch, Martha; *Manual of Straight Egyptian Horses*, The Pyramid Society, Lexington, KY 1997-2001

Nagel, Dr. Hans Joachim; *Hanan, the Story of An Arabian Mare and the History of the Arabian Breed,* Alexander Heriot & Co., Ltd, Northleach, Cheltenham, England, 1998

Nagel, Dr. Hans Joachim and Wolfgang Eberhardt; *Arabische Pferde aus dem Orient, Moniet el Nefous, Mona, Mahiba,* Püttlingen, Germany, 1988

Pearson, Colin with Kees Mol; *The Arabian Horse Families of Egypt*; Alexander Heriot & Co., Ltd. Cheltenham, Glos. England, 1988

Piduch, Dr. Erwin A; *Egypt's Arab Horses*, Kentauros-Verlag, Lienen/Germany, 1988

Raswan, Carl; *The Raswan Index*, original edition privately published in 7 volumes: also by Itex Pub. Co., Ames, Iowa, 1969 and Esperanza Raswan, California, 1990.

Tewfik, H.R.H. Prince Mohamed Aly; *Breeding of Purebred Arab Horses*, Cairo, Egypt, 1935 and 1936

The Pyramid Society, Inc; *The Reference Handbooks of Straight Egyptian Horses, Volumes 1-9,* The Pyramid Society, Inc., Lexington, Kentucky.

"Sailing Down the River of Life"
The Nile at Luxor, Upper Egypt. Painting by Tyrone Lewis. Forbis Collection.